# TECHNOLOGY AND CULTURE IN TWENTIETH-CENTURY MEXICO

EDITED BY
ARACELI TINAJERO
AND J. BRIAN FREEMAN

THE UNIVERSITY OF ALABAMA PRESS
Tuscaloosa

Copyright © 2013
The University of Alabama Press
Tuscaloosa, Alabama 35487-0380
All rights reserved
Manufactured in the United States of America

Typeface: AGaramond & Stone Sans

Cover illustration: Fotosearch Stock Photography
Cover design: Michele Myatt Quinn

∞

The paper on which this book is printed meets the minimum requirements of American National Standard for Information Sciences—Permanence of Paper for Printed Library Materials, ANSI Z39.48-1984.

Library of Congress Cataloging-in-Publication Data

Technology and culture in twentieth-century Mexico / edited by Araceli Tinajero and J. Brian Freeman.
    pages cm
Includes bibliographical references and index.
ISBN 978-0-8173-1796-6 (trade cloth : alkaline paper)
1. Technology—Social aspects—Mexico—History—20th century.
2. Technological innovations—Social aspects—Mexico—History—20th century. 3. Popular culture—Mexico—History—20th century.
4. Mass media—Social aspects—Mexico—History—20th century.
5. Transportation—Social aspects—Mexico—History—20th century.
6. Mexico—Social conditions—20th century. 7. Mexico—Intellectual life—20th century. I. Tinajero, Araceli, 1962– II. Freeman, J. Brian, 1982–
    T24.M6T425 2013
    303.48'309720904—dc23
    2012050730

# Contents

List of Illustrations  vii
Acknowledgments  ix
Introduction
*Araceli Tinajero and J. Brian Freeman*  1

## I. HEALTH, FOOD, AND THE HOME

1. Material Culture, Public Health, and the Technologies
of Hygiene in Modern Mexico, 1890s–1940s
*Claudia Agostoni*  25

2. Cooking Technologies and Electrical Appliances in 1940s and 1950s Mexico
*Sandra Aguilar-Rodríguez*  43

3. Domestic Technologies: Gender, Technology,
and Mexican Housewives, 1930–1950
*Joanne Hershfield*  55

## II. PHOTOGRAPHY, TELEVISION, AND THE INTERNET

4. Technologies of Seeing: Photography and Culture
*John Mraz*  73

5. The Early Years of *La Tele*
*Celeste González de Bustamante*  90

6. And Television Appeared among the Mexicans
*Carlos Monsiváis, translated by Lorna Scott Fox*  111

7. Revolt, Confusion, and the Cult of the Trivial in Mexican Cyberculture
*Naief Yehya*  124

## III. RADIO AND MUSIC

8. The Race for the Airwaves: Journalism and the
Radio Industry in Modern Mexico
*Viviane Mahieux* 143

9. Music Culture and Resistance in Mexico, 1968–1988:
Popular Music and Mass Media
*Ricardo Pérez Montfort* 160

10. Technology for Cultural Survival: Indigenous-Language Radio
at the End of the Twentieth Century
*Antoni Castells-Talens and José Manuel Ramos Rodríguez* 178

## IV. RAILROADS, AUTOMOBILES, AND THE METRO

11. Film, Time, and the Railway in Porfirian and Revolutionary Mexico
*David M. J. Wood* 197

12. "Los Hijos de Ford": Mexico in the Automobile Age, 1900–1930
*J. Brian Freeman* 214

13. Railroad Culture and Mobility in Twentieth-Century Mexico
*Guillermo Guajardo and Paolo Riguzzi, translated by Viviane Gomez* 233

14. From the Primordial Cave to Postmodern Velocity: The Mexico City Subway
*Juan Villoro, translated by Lorna Scott Fox* 249

## V. ART, LITERATURE, AND ARCHITECTURE

15. *Estridentismo*'s Technologies: Modernity's "Efficient Agents"
in Post-revolutionary Mexico
*Lynda Klich* 263

16. Technology, Labor, and Realism: Diego Rivera's Secretaría
de Educación Pública Murals
*Anna Indych-López* 283

17. Cyborg versus "Homo scribens": Mexican Literary
Expressions in the Era of "Technoculture"
*Erja Vettenranta* 302

18. Technology and the Architectural Culture of Mexico from
the 1968 Olympic Games to the Onset of the New Millennium
*Edward R. Burian* 316

Bibliography 333
List of Contributors 355
Index 359

# Illustrations

FIGURES

1.1. Juárez free public baths, 1922   31

1.2. A victim of influenza in Mexico City, 1918   35

1.3. Campaign against typhus in Mexico City, 1922   37

4.1. Guillermo Kahlo, *Train on the Metlac Bridge, 1903*   75

4.2. Eustasio Montoya, *Photographers and a Dead Man in Nuevo León after Being Attacked by Huertistas, 5 July 1913*   79

4.3. Hermanos Mayo, *Striking Teachers Battle with Police; Zócalo, Mexico City, 6 September 1958*   82

4.4. Pedro Meyer, *Migratory Mexican Farm Workers, California Highway, 1986/1990*   86

11.1. Frame enlargement from *Inauguración del tráfico internacional de Tehuantepec*, Salvador Toscano, Mexico, 1907   201

11.2. Frame enlargement from *La toma de Ciudad Juárez y el viaje del héroe de la Revolución D. Francisco I. Madero*, Salvador Toscano, Mexico, 1911   205

11.3. Frame enlargement from *México ante los ojos del mundo*, Miguel Chejade, Mexico, 1925   209

15.1. Fermín Revueltas, *Andamios exteriores* [Exterior scaffolding], 1923   268

15.2. Fermín Revueltas, *Paisaje con líneas de alta tensión* [Landscape with high tension wires], 1924   269

15.3. Ramón Alva de la Canal, cover for Salvador Gallardo's *El Pentagrama Eléctrico* (1925)   271

15.4. Jean Charlot, *Poet on Airplane*, woodcut for Manuel Maples Arce's *Urbe*, 1924   274

15.5. Jean Charlot, *Ocean Liner*, woodcut for Manuel Maples Arce's *Urbe*, 1924   274

15.6. Jean Charlot, cover woodcut for Germán List Arzubide's *Esquina*, 1923   277

16.1. Diego Rivera, *Electric Machine*, 1923, fresco   287

16.2. Diego Rivera, *The Investigation*, 1924, fresco   288

16.3. Diego Rivera, *X-Ray*, 1924, fresco   288

16.4. Diego Rivera, *Mechanization of the Country*, 1925, fresco   290

16.5. Diego Rivera, *End of the Corrido* (from the Corrido of the Agrarian Revolution), 1926, fresco   292

16.6. Diego Rivera, *The Cooperative* (from the Corrido of the Proletarian Revolution), 1928, fresco   297

18.1. Palacio Buenavista, Manuel Tolsá, Mexico City, 1803   318

18.2. Iglesia la Purísima, Enrique de la Mora y Palomar with Félix Candela, Monterrey, Nuevo León, 1940–1946   319

18.3. Aceros Planos Office Building, Rodolfo Barragán Schwarz, Monterrey, Nuevo León, 1973–1975   321

18.4. Episcopal Church, Carlos Mijares Bracho, Mexico City, 1992   323

18.5. Gimnasio Dojo, Agustín Landa, Monterrey, Nuevo León, 1997   324

TABLE

10.1. Indigenous radio stations created between 1979 and 1999   182

# Acknowledgments

The idea for this book grew out of a 2009 interdisciplinary seminar on civil society in Latin America, designed by Alfonso W. Quiroz and Araceli Tinajero, which led us to contemplate the lack of scholarship on the connections between technology and culture in modern Mexico. We would thus like to offer our particular thanks to Alfonso not only for his role in giving rise to this book, but for his support, his professionalism, and his friendship.

From early on we had the fortune to benefit from the sage advice and enthusiastic support of Gilbert Joseph, of Yale University, who offered us his generous counsel on numerous occasions. We thank him for his help and guidance. We also benefited from a variety of suggestions from Alan Knight, of Oxford University. Roberto González Echevarría, of Yale University, took the time to read the manuscript and give us his insights on the structure of the book. We thank Rubén Gallo, of Princeton University, for his advice as well as the inspiration we took from his work on technology, art, and literature.

During the course of editing this volume we had the good fortune to work with colleagues and friends through the Bildner Center for Western Hemisphere Studies, where we established the Mexico Study Group with the help of the center's director, Mauricio Font, and Eric Zolov, of Stony Brook University. We also thank the faculty and staff of the PhD program in Hispanic and Luso-Brazilian literatures and languages and the PhD program in history at the Graduate Center, CUNY, as well as the Department of Foreign Languages and Literatures at the City College of New York.

This book was partly published thanks to a grant from PSC-CUNY (Professional Staff Congress of the City University of New York), and throughout the course of the project we received judicious guidance from Ana Delgado (in the City College of New York's Office of Research Administration) and Paul Cole, Clark Lantz, and Richard Markgraf from the Research Foundation of CUNY.

We would like to thank the authors of this volume for their collaboration and patience, Lorna Scott Fox and Viviane Gomez for their fine translations, and the University of Alabama Press, our editors Joseph Powell and Joanna Jacobs, and the anonymous readers for their help and advice.

In Mexico we were fortunate to receive superb assistance in the acquisition of rights and permissions from Mayra Mendoza Avilés of the Fototeca Nacional, Alma Vázquez and Fabiola Hernández Díaz of the Archivo General de la Nación, Abigail Molleda of the Museo Nacional de Arte, and Carla Barri Rosendo of the Banco de México. We are indebted to Mrs. Beatriz Sánchez Monsiváis for permitting us to publish the chapter that her cousin, the late Carlos Monsiváis, sent to us in 2010. We are only sorry that Monsi is no longer with us to read these lines and this book. From the beginning he was most enthusiastic about the project and had, in fact, offered to write the prologue.

Brian would like to thank Carlos Marichal of the Colegio de México, Guillermo Guajardo of the Universidad Nacional Autónoma de México, Stephen Allen of Rutgers University, and Alberto Hernández Sánchez of the Universidad Iberoamericana for welcoming him in Mexico with open arms. Most particularly, he thanks his wife, Mariana, his parents, Terry and Deborah, and his sister, Laura, for their love and support.

Araceli would like to thank professors Mario Alberto Vega Centeno, Doris Sommer, William Luis, Lía Schwartz, Isaías Lerner, Oscar Montero, Floyd Merrell, Elzbieta Sklodowska, and literary critic Rafael Lemus for their support and advice. She also thanks her sisters, Josefina Tinajero and Dr. María Natividad López Tinajero, for their affection. But above all, she is indebted to Stephen Pollard for his endless encouragement and compassion.

# TECHNOLOGY AND CULTURE IN TWENTIETH-CENTURY MEXICO

# Introduction

*Araceli Tinajero and J. Brian Freeman*

In his introduction to *Renascent Mexico* (1935), American journalist Ernest Gruening recounted an imaginary trip across the US border into Mexico. Describing the experience of crossing the Rio Grande, he observed a "swift-moving, mechanized pattern of a modern society . . . give way to a simple, less regulated, earlier stage of development."[1] For Gruening, and for many of his contemporaries, to enter Mexico was to leave behind machine-civilization, to travel back in time and enter a land, as popular economist Stuart Chase famously claimed, of "machineless men."[2] Yet such impressions often failed to recognize that Mexico had, by the 1930s, experienced at least a half century of technology transfer from North Atlantic economies and a growing production of locally designed technologies.

By the beginning of the twentieth century, mechanization had already appeared in industries ranging from transportation, mining, and construction to textiles and food processing.[3] And as the twentieth century progressed, industry expanded, as did urbanization, transportation, and communications. Roads spread out from cities like tentacles. Some Mexicans purchased automobiles, while many others traveled by bus. They bought radios, televisions, home appliances, and many of the other accoutrements of life in the "machine-age." Meanwhile, the Mexican state engaged in modernizing projects that employed state-of-the-art technologies in an attempt to transform society and the economy, all the while putting a modern face on the nation. By the end of the century, an increasing number of Mexicans ventured into cyberspace, and countless more bought cell phones and DVD players. Far from negligible, throughout the course of the twentieth century and beyond, the use of these technological innovations has been intimately tied up with nation- and state-building economic activities and patterns of mobility, as well as transformations in identity and everyday life.[4]

Before moving on it is worth asking the question, what is "technology"? On some level it clearly involves artifacts or material objects. Indeed, the mere use of the word inspires an imagery of mechanical and electronic devices. Nevertheless, scholars have been quick to note that technology extends beyond the domain of the material and includes knowledge or know-how, ideology, and values, as well as the larger systems of which technological artifacts form a part.[5] Over the last few decades work on the history and sociology of technology has tended to pay particular attention to the processes by which technology is produced: that is, how artifacts and systems of artifacts come into being. Through studies of the success and failure of such diverse objects as the electric car, Tupperware, and the airplane, to name only a few, it has become clear that new technologies tend to emerge out of a complex matrix of cultural values, consumer preference, and political and economic context, as well as concerns over technical efficiency.

By situating technology in society and culture, much of this scholarship has aimed to demystify the perceived autonomous power of technology and challenge both popular and scholarly assumptions about its ability to unilaterally shape history, an idea embodied in the term "technological determinism." As Thomas Misa has observed, "empirical studies of technology by historians and sociologists," as opposed to "macro-level" philosophical treatises, tend to argue "that technology is best understood as a product of underlying social and cultural dynamics and is not—in itself—a compelling force for change." Yet Misa notes that "as citizens of a 'technological age' we know that technology in some way shapes our future, but we seem to lack robust insights into how it will do so and, crucially, whether and how we can exert significant influence over this future-shaping process."[6] Similarly, David Edgerton has suggested that many scholars tend to emphasize "the question of technology," or the nature of technology, at the expense of broader historical questions, a characteristic that has led Rosalind Williams to call for a history of technology "whose ultimate goal is understanding how history works."[7]

The vast majority of contributors to this volume are in some sense outsiders within the field of technology studies. Instead, most are scholars of Mexico who have found in technology a fresh perspective. Their concerns extend beyond technology in and of itself and are rooted in a concern over how technological artifacts and ideas have interacted with culture, politics, society, and, indeed, history. While the chapters in this volume do not adopt a uniform definition of technology, generally, they agree that it is a product of our own making and cannot be divorced from the larger sociocultural context from which it emerges.[8]

Alternatively, without accepting a crude determinism, they view technology as a critical element in the production and shaping of culture and social relations. Contributors draw on these insights as they address questions, including, but not limited to these: How do race, ethnicity, gender, and political cultures impact and find themselves impacted by the production and use of technology? How does power interact with and shape technological change? How does technology affect space and time? And, how does technology impact state formation, nationalism, and the idea of nation?[9]

Well over a century ago many Mexicans wrestled with the implications of an array of new technologies, from railroads, telegraphs, and electric trams to photographic cameras and motion pictures. Indeed, during the late nineteenth and early twentieth centuries, a good deal of the globe experienced a massive transfer of technology and technological know-how from the industrialized world to less industrialized areas, a process that some scholars have termed the first modern age of globalization.[10] The consequence of this flow of artifacts, ideas, and people was particularly significant for Mexico, and it left an indelible mark on economics, politics, society, and culture.

In 1876, during the early years of this era of global technological change, liberal general Porfirio Díaz came to power in Mexico. As president he helped usher in the first period of extended political and economic stability in the nation's history. For three and a half decades Díaz and his allies maintained control over the country through deft manipulation and management of rivals, suppression of rural rebellion, censorship of the press, and by significantly limiting labor organization. While a veneer of democracy remained in the form of campaigns and elections, the regime was characterized by a form of authoritarianism that was transparent to many if not all observers.

During his time in power, Díaz surrounded himself by a group of technocrats known as the *científicos*. These elite political and intellectual leaders called for the establishment of a strong state that would manage society through rational means.[11] Statistics, cartography, and criminology, among other state-of-the-art "sciences," were their tools.[12] Articulating a particular form of positivism, inspired by Auguste Comte, Herbert Spencer, and others, the *científicos* crafted a "scientific" politics—informed by their own class interests and biases—which encouraged capital accumulation, foreign direct investment, technology transfer, expansion of agribusiness and mineral extraction, and an early form of industrialization.

During these years, the railroad constituted in many ways the flagship project of the authoritarian president, and for many people it probably represented the most striking manifestation of technological change. Adopted, in part, to bolster exports and refashion the country's international reputation, from 1877 to

1910 construction, which was financed largely by foreign capital, mushroomed from 640 to over 19,000 kilometers of track. The new transportation system did much to help weave a patchwork of local and regional centers into a unified domestic market and allowed many businesses to market their goods on a national level for the first time. With the cost of shipping cargo reduced by an incredible 90 percent, Mexico experienced an unprecedented boom in commodity production and export.[13]

The speeding up of travel also gave the Porfirian government a new set of tools to expand the authority of the state. Able to quickly deploy troops and artillery, trains increased the state's capacity to crush insurgents and other antagonists, and the central government even developed an elaborate network of spies throughout the railway system. Cognizant of the political power of this new technology, Díaz predicted that "the steel of the rails would complete the task begun by the steel of the bayonets: national unity"; yet railway construction produced significant social tensions as well, as it encouraged the concentration of land ownership and often destabilized peasant economies.[14]

In addition to railroads, mechanization began to appear in mining, other forms of transportation, energy production, construction, manufacturing, food processing, and, in some cases, agriculture. Meanwhile items like sewing machines spread out across the country, and inventors even began to search for ways to mechanize tortilla production. As Edward Beatty has observed, many of these innovations had the effect of altering "not only the nature and scale of production but also the lives of many thousands of Mexicans who found work in new industries, who consumed the products of new factories, or whose lives were affected in many, less direct, and often involuntary and painful ways."[15]

The introduction of new technologies from abroad brought with them workers with specialized skills as well. During the late nineteenth and early twentieth centuries, engineers and technicians from the United States and other parts of the world poured into Mexico to help build and maintain new communications and transportation networks, as well as other newly mechanized industries. This migration of technical workers would soon be reflected in American popular culture with the proliferation of books about daring young engineers and explorers, including *The Motor Boys in Mexico; or, The Secret of the Buried City* (1908) and *The Young Engineers in Mexico; or, Fighting the Mine Swindlers* (1913). Although foreigners represented only about 5 percent of the industrial workforce by 1910, their presence was much greater in mining, on the railways, and in the nascent oil industry, and Mexican laborers often objected to being overlooked when it came to technical and other professional positions.[16]

New technologies were not only adopted in order to modernize the economy, but often embraced in an effort to transform human environments as well. Indeed, some of the more radical technological changes occurred in cities. By 1910 the nation's capital had installed telephone lines and electric lighting, built a system of electrified trams, constructed new steel-reinforced buildings, paved streets with macadam and asphalt, and installed sewage and plumbing systems. Inspired by the Haussmannian reconstruction of Paris, Mexican reformers sought to build cities that would be worthy of their contemporaries in Europe as well as the United States.[17]

A crucial facet of these reform efforts included the construction of sanitary infrastructure. During the last two decades of the Porfirian regime, city authorities engaged in a veritable "war against water" as they drained Lake Chalco and built a series of drainage canals and tunnels, efforts they hoped would finally solve the capital's perennial problems of flooding. Yet even these massive projects did not alter the fact that, due to the haphazard expansion of the city, many poor neighborhoods lacked the necessary sanitary facilities like running water and sewage systems. In such areas of the city, rather than improve infrastructure, governing elites turned to health education and "aimed to convey ideals of personal hygiene, disease avoidance, parenting, and conduct."[18]

The construction of tramways and railroad stations left a particular mark on the structure of Mexico City by allowing the middle and upper classes of the capital to begin moving away from the traditional downtown—inspired by real and imagined fears of overcrowding, poor hygiene, and crime—and resettle in the new western sections of the city. The system moved commuters and merchandise around the capital and even carried the dead to cemeteries. Yet the majority of the lines served the wealthy western side of the capital, and even the two lines on the eastern end could seldom be used legally by the masses, given the prohibitive cost of a ticket. The center of town, in turn, began to specialize in commercial and financial activities, thus forcing the working poor to the less desirable eastern and southern periphery, a low-lying area prone to flooding.[19]

Meanwhile, consumers did much to introduce their own technological novelties. Wealthy residents purchased bicycles, phonographs, modern stoves, the first automobiles, and a diversity of new mechanically produced goods like machine-rolled cigarettes and bottled beer.[20] This growing consumer culture was reflected in the proliferation of advertisements for the most recent technological wonders, including electric lights, cameras, heaters, and refrigerators, among many items, which appeared in newspapers and other publications. The ability to produce these images cheaply was itself a product of advances in printing technologies. Meanwhile, businesses like the tobacco company El Buen Tono attracted customers through spectacular demonstrations of the latest technologies, including dirigibles, airplanes, and motorcars; and department stores like Palacio de Hierro, Puerto de Liverpool, and Puerto de Veracruz impressed shoppers with elevators and escalators, pneumatic tube systems, electrical lighting, and indoor plumbing.[21]

This reformed, technologically embellished city was formally displayed to the world in September 1910 during celebrations to mark the centennial of Mexico's independence war. In the presence of hundreds of thousands of Mexicans and prominent visitors from around the world, the month-long spectacle featured historically themed parades and scientific congresses, as well as the inauguration of a modern mental hospital, a seismological station, various schools and government buildings, public parks, monuments (including one of Louis Pasteur), a new gunpowder factory, the National University, new hydraulic works, and an expanded modern penitentiary. Hidden from sight, however, were the numerous poor and indigenous citizens who failed to meet the Porfirian standard of modernity and were thus banned from the areas where festivities would take place.[22]

As the Porfirian regime celebrated the nation's alleged progress, the very foundation of its hegemony had already begun to falter. During the first decade of the new century, a series of crises converged and fueled a violent social revolution that would wrack the country for ten years. Since early on in the century, the country had suffered from periodic recession, a notable growth in the cost of living, and a series of devastating droughts. And as the central government continued to consolidate its authority, it increasingly encroached on longstanding traditions of local autonomy in towns and villages. Meanwhile, the economic boom created new regional elites and an emergent middle class who demanded a political voice, only to find themselves excluded from Porfirian political circles. All the more troubling, after three and a half decades in power, Díaz and his political allies had become old, and they desperately needed to introduce a new generation of leaders.

During the two years before the centennial celebrations, a multi-class movement gathered around wealthy northerner Francisco Madero, who sought an end to Díaz's rule and a return to the liberal political principles of the 1857 constitution. Having faced harassment and imprisonment, he called on Mexicans to take up arms on November 20, 1910, and oust the authoritarian president. During the first half of the next year a series of uprisings took place, which eventually forced Díaz to abandon the country and take refuge in France. The next ten years witnessed a violent armed revolution, in effect, a series of civil wars, which would definitively impact the course of twentieth-century Mexican history.

Even in the midst of the revolution, technological change continued apace. Francisco Madero became the first incumbent president in the world to fly in an airplane, rebel forces commandeered trains to move fighters and their families, airplanes and machine guns were deployed on battlefields, and the United States used trucks for the first time in combat during its "punitive expedition" into Mexico in search of revolutionary leader Pancho Villa. Meanwhile, as Hollywood emerged as a film power, themes of revolution and banditry proliferated in cinema, and Villa even starred in his own films produced by the Mutual Film Company.

By the early 1920s, the violence and disorder of armed revolution had led to the loss of well over a million people to death and emigration, and a massive amount of infrastructure had been destroyed. Entire sections of railway track had been ripped up, while numerous freight cars, passenger cars, and locomotives had been ruined. Additionally, over a thousand miles of telegraph lines had been destroyed, and the few roads built during the Porfiriato had fallen into disrepair.[23] It was in this context that the revolution entered what President Plutarco Elías Calles (1924–28) would term its "constructive phase."[24] Broadly, national reconstruction included the dual objectives of encouraging economic development and reestablishing political stability. The former called for building new infrastructure, nationalizing resources, and using the power of the state to stimulate economic growth. To achieve the latter, leaders sought to centralize power, pacify the provinces, restrain the military, encourage a common national identity, and thus solve the "crisis of order."[25]

Yet revolutionary leaders looked not only to modify political and economic circumstances, but to encourage a thoroughgoing cultural transformation of the Mexican people. The new, post-revolutionary Mexican was to be "sober, industrious, literate, and patriotic."[26] Through an elaborate project of mass education, old vices and superstitions—which in this view included Catholicism—were to be replaced with a new rational and national worldview.[27] Meanwhile, indigenous people were to learn Spanish and adopt "mestizo" ways.[28] However, in such a large country, with a notably challenging geography, poor communication and transportation worked against efforts to extend this cultural revolution beyond major cities. Faced with such obstacles, revolutionary leaders soon embraced both road building and the newly popularized technology of radio broadcasting.[29] Yet even with the help of new technologies, the molding of hearts and minds proved more difficult than imagined. Educational reforms and cultural missions to the countryside, as well as later socialist educational initiatives, all largely failed to produce the imagined, revolutionary "new man."[30]

Alongside such efforts, a transformation of a different sort was already beginning to take place. As roads were built, radios were installed, and cities expanded through rural-urban migration, a growing number of Mexicans found themselves drawn into a common, national market as producers and consumers. The agents of this alternative cultural revolution included Bayer tonic distributors, who attracted customers by playing Popeye movies, bus drivers who linked rural people to towns and cities, commercial radio stations that broadcast popular music, and many other unwitting revolutionaries. As the market spread, an incipient consumer culture of a much broader, inclusive sort than before the revolution began to take shape. Newly "Ford-conscious" peasants began to turn tires into *huaraches* (sandals) and old Ford motors into corn mills. They embraced new songs and musical forms, and Eyler Simpson even recalled hearing a rural brass band play "St. Louis Blues."

Back in Mexico City the constructive phase of the revolution took a particularly striking form as the once quaint capital found itself transformed into a budding metropolis. From 1921 to 1930 Mexico's population expanded from around 615,000 to over 1 million, and between 1910 and the early 1920s, the built environment more than tripled in size as new structures appeared along railway tracks, tramways, and roads.[31] Real estate speculators built new suburban developments, and an increasing number of industrial facilities were established, including Ford and Chevrolet assembly plants. In this growing city, parking lots and garages proliferated, auto dealerships and filling stations appeared, streets were extended and widened, cobblestones were replaced with asphalt, massive stadiums were built, and many of the old colonial structures were demolished. Billboards, movie theaters, tabloids, and taxis flourished, and in 1927 authorities entirely banned horse-drawn vehicles from the city center.[32] Meanwhile, new social types like taxi and bus drivers, mechanics, flappers, *chicas modernas* (modern women), and *fifís* (dandies) proliferated, and Mexico City experienced its own "roaring twenties" of sorts.[33]

Inspired by this dynamic urban world and the possibilities of technology, avant-garde artists and writers experimented with radically new notions of representation. Although many Mexican muralists during these years became famous for their depictions of folkloric scenes and indigenous people, other less well remembered individuals (and some of the very same muralists) experimented with themes ranging from outer space, electricity, aviation, motoring, transatlantic liners, dirigibles, skyscrapers, telephony, jazz, robots, and X-rays to Albert Einstein, Thomas Edison, and Guglielmo Marconi.[34]

As the 1920s came to an end, Plutarco Elías Calles aimed to channel the wide-ranging political, economic, and cultural forces that had emerged out of the revolution and established the government-endorsed National Revolutionary Party (PNR). Far from a traditional political party, the PNR acted as a forum for competing interests to be brokered and to thus avoid open violence. Until the middle of the next decade Calles maintained a significant presence in national politics, but upon the election of President Lázaro Cárdenas (1934–40) he would find himself sidelined and excluded. As president, Cárdenas made a name for himself as a proponent of land redistribution and a creator of large national trade unions. These actions did much to shift political loyalties from the local and regional to the national level and, after 1938, to the reorganized "official" party, the PRM (Party of the Mexican Revolution). Meanwhile, as a depression-era president, Cárdenas, like many of his contemporaries around the world, experimented with economic nationalism. While in office, he founded the National Polytechnic Institute, nationalized railways and the country's oil industry, and oversaw the growth in university-trained civil and agricultural engineers.[35] The end of the 1930s, however, witnessed a slow but sure shift to the right as Cárdenas moved away from the radicalism of his early years, a consequence of mounting domestic and international constraints.

With the election of his successor, Manuel Ávila Camacho (1940–46), the official party continued its rightward swing. It abandoned its earlier anti-clericalism, curbed independent unionism, and excluded the radical left from electoral participation. Over the following years, as Alan Knight has observed, "the Mexican state was consolidated on the basis of a new programme, which stressed private enterprise over collective ownership, capital accumulation over redistribution, social control over representation, marginalization over encouragement of the left, urbanization over rural development."[36]

The late 1940s and early 1950s witnessed the transformation of the PRM into the PRI (Institutional Revolutionary Party) and the onset of a postwar economic boom.[37] The state redoubled efforts to industrialize by protecting the domestic market and encouraging the importation of productive technologies. Under the presidency of Miguel Alemán (1946–52), it built a modernistic national university campus inspired by pre-Columbian aesthetics, constructed futuristic housing complexes, completed the Pan-American Highway (and inaugurated it with an international automobile race), and promoted Acapulco as a luxurious international tourist destination.[38] By 1956 Mexico City could also boast of having the tallest skyscraper in Latin America, the appropriately named Torre Latinoamericana.[39] Additionally, the regime sponsored massive irrigation projects, the drainage of wetlands, and the construction of hydroelectric dams. Meanwhile, in the northern city of Monterrey, local business interests established the Tecnológico de Monterrey (ITESM), an institution modeled on the famed Massachusetts and California institutes of technology.

If the middle decades of the twentieth century represented an era of industrialization, urbanization, and a generalized push to modernize the nation, these years also constituted what Gilbert Joseph, Anne Rubenstein, and Eric Zolov have termed a "golden age of consumerism."[40] Mexico City developed a vibrant nightlife, and media moguls like Emilio Azcárraga began to amass spectacular fortunes. Multinational companies increasingly entered the national market, where they created local sister companies like Sears Roebuck de México and Volkswagen Mexicana, joining older standbys like Ford and Palmolive. Advertisements for new domestic appliances and other devices like washing machines, blenders, and electric blankets proliferated on radio and in newspapers and magazines, and in the 1948 film *Una familia de tantas*, director Alejandro Galindo offered moviegoers the story of a conservative family transformed after the arrival of a modern vacuum cleaner. Indicative of a broader shift in culture, during the 1950s and '60s films increasingly dealt with urban themes, and Mexico even became one of the world's largest producers of science fiction movies.[41]

These very visible transformations, in both the economy and culture more generally, inspired many observers to begin speaking of a "Mexican miracle."[42] Of course, the appearance of development and progress continued to coexist with a long-standing image of the country as rural, indigenous, and "traditional," a notion that was rooted in a mix of reality, stereotype, and active promotion by the state. Since the 1920s foreigners had often visited the country in search of an alternative to the Western, industrial societies of the United States and Europe. No doubt cognizant of this desire, both government and private business appealed to tourists through ad campaigns that featured idyllic settings populated by peasants and indigenous people. During the 1950s the Beat poets and many other tourists continued to associate Mexico with escapism, and during the 1960s and '70s hippies would similarly turn to the country in search of alternative lifestyles, often to the dismay of authorities.

A new round of modernization took place in the 1960s and culminated in the 1968 Mexico City Olympics, the first to be held in the Third World.[43] The years leading up to this event witnessed a laborious effort to demonstrate the nation's technological and organizational capacity. In 1964, President Adolfo López Mateos (1958–64) inaugurated the state-of-the-art National Museum of Anthropology, and during the administration of Gustavo Díaz Ordaz (1964–70) construction was begun on the capital's subway system and an impressive Olympic village was built, as were a variety of new sports facilities. Noting the changes in Mexico City since the early 1950s, Salvador Novo observed that by the late 1960s, the capital had increased its footprint by 79 percent. New freeways, cloverleaves, under- and overpasses, and bridges had been built, and in some cases rivers had been "tubed" to make way for highways like the Viaducto Miguel Alemán and Avenida Río Churubusco.[44] However, as the state attempted to replace the imagery of Mexico as a sleepy land of *mañana* with that of a forward-looking "land of tomorrow," a number of young people rejected the very idea of progress as they "looked for mushrooms in Huautla, walked around in peasant sandals, and changed the very image of Mexican youth."[45] These middle- and upper-class young people, dubbed *jipitecas*, offered a critique of the system that had, in a sense, given birth to them.[46]

As the 1968 Olympic Games approached, some of these young people—many of them students at the national university (UNAM)—took part in a social movement that grew out of a series of confrontations with the government over its heavy-handed treatment of dissidents. As the inaugural ceremonies drew near, authorities crushed the protesters during a gathering at Mexico City's Plaza de las Tres Culturas in Tlatelolco, where hundreds of people were killed and many more injured. Although national news organizations largely censored reporting on the brutal incident, for many people the massacre at Tlatelolco marked the definitive end to the "miraculous" decades of the mid-twentieth century.[47]

With the erosion of political legitimacy in the wake of the events of 1968, President Díaz Ordaz's successor, Luis Echeverría (1970–76), pursued a populist agenda that aimed to co-opt radical elements in society and, where necessary, remove them through violent tactics of "dirty" war.[48] He moved to release imprisoned students and channeled former student movement participants into government and university positions. Indeed, it was Echeverría who in 1970 founded the National Council of Science and Technology (CONACYT), an organization that provided an unprecedented number of scholarships for students to study abroad. Meanwhile, he supported the expansion of the state's role in the economy and called for a "shared" form of development, while on the international front he pursued an activist foreign policy and professed his support of both Chile's Salvador Allende and Cuba's Fidel Castro.

The growth in public spending under Echeverría and especially under his successor, José López Portillo (1976–82), had been supported through oil revenues and low-interest loans, which gave the appearance of a robust economy, but when global oil prices plummeted in the early 1980s, things soon fell apart. Borrowing had skyrocketed during these years with the assumption that oil profits would continue to fill state coffers, and as interest rates on these loans rose, massive and unsustainable deficits began to appear. By the early 1980s Mexico faced declining revenues, rising inflation, and capital flight, which depleted foreign reserves and forced the country to declare a moratorium on debt payments during the summer of 1982. When the new president, Miguel de la Madrid (1982–88), came into office, he slashed public spending in order to encourage exports and stabilize the country's finances. Instead of a quick recovery, Mexico experienced a decade of economic stagnation, a lost decade in some estimates. As public employment declined, a growing number of people entered the informal economy; and, with few options, rural people continued to head to cities and, increasingly, to the United States.[49]

The crisis was only made worse by a devastating earthquake that hit Mexico

City in 1985. On the morning of September 19, the earthquake, which registered 8.1 on the Richter scale, destroyed over 370 buildings, killed an estimated 5,000 to 20,000 people, and left at least 250,000 homeless. The government response was negligible, and the homeless, or *damnificados*, mobilized to put together an informal relief effort. It quickly became clear that the event had represented as much a social and technological disaster as it had a natural one. Many of the collapsed buildings had been built by the government, and it turned out that contractors, due to corruption and wanton negligence, had failed to employ earthquake-proof building methods. Underneath these structures—many of which featured impressive modernistic façades—lay brittle foundations, a fact that in turn became a metaphor for the regime itself.

By the 1980s, when José Emilio Pacheco looked back on the euphoric years of the mid-twentieth century in his story *Battles in the Desert*, the acclaimed writer saw a developmentalist fantasy that was never realized, possibly never real.[50] Since then Mexico City had grown into a chaotic megalopolis. In just ten years, from 1970 to 1980, the number of cars in the city had tripled.[51] Rural-urban migration reached seemingly unsustainable proportions, and shantytowns dotted the outskirts of the city. During the last decades of the century, writers and artists experimented increasingly with themes of dystopia, inspired all the more by the apocalyptic feel of the 1985 earthquake.[52] In the novel *Christopher Unborn*, for example, Carlos Fuentes writes of an ecological disaster that hits Mexico in 1990.[53] And by the closing decade of the twentieth century, Carlos Monsiváis would come to refer to Mexico City as "post-apocalyptic"; it had survived the apocalypse and lived on.[54]

During the final dozen years of the twentieth century, under the presidencies of Carlos Salinas de Gortari and his successor Ernesto Zedillo, the state aimed to increase economic growth by restructuring the national debt, engaging in widespread privatizations of public firms, deregulating the economy, and signing trade deals like the North American Free Trade Agreement (NAFTA). During these years words like *technology*, *efficiency*, and *global competitiveness* became mainstream, and from 1990 to 2000 public expenditure on science and technology doubled.[55]

Not everyone accepted this new model of government and economic development. On January 1, 1994, the day NAFTA came into effect, a group of masked indigenous rebels in the southeastern state of Chiapas rejected the trade agreement and declared war on the Mexican state. They took over five municipalities and engaged the military over the course of two weeks, after which a ceasefire was declared. The Zapatistas employed unorthodox methods of war, cognizant of their inability to match the physical power of the Mexican state. Rather than fight the government on the battlefield alone, the rebels turned to both old and new communications technologies—from radio to the Internet—as they waged a simultaneous media war.

The closing of the century coincided with the first peaceful transition of power from the ruling party to the opposition since the revolution itself had been unleashed. In 2000, former Coca-Cola executive and Partido Acción Nacional (PAN) candidate Vicente Fox (2000–2006) defeated his challengers from both the PRI and the center-left Party of the Democratic Revolution (PRD). Although the Fox presidency largely failed to live up to expectations, his party retained executive control after the 2006 elections with the election of Felipe Calderón.

Over the last two decades, cell phones and Internet cafés have penetrated many of the most remote corners of the nation. Migrant laborers in the United States are now able to maintain a presence in their home communities through long-distance phone services and email and by shipping home videos back and forth across borders. New technologies have even begun to mediate violence associated with drug trafficking as Mexican citizens increasingly turn to Twitter and other social networking forums to share information about clashes between rival drug organizations, even as traffickers themselves employ Twitter, Facebook, and YouTube to exchange information and gather intelligence, all the while instilling fear in the public.[56] Meanwhile, filmmakers have ever more embraced technological themes in their work, as in Alex Rivera's *Sleep Dealer* (2009), which offers the story of a young man who migrates not to the United States, due to militarization and technological fortification of the border, but to a futuristic labor center in Mexico where he and hundreds of others remotely operate robots in the United States.

This volume, written by historians, literary scholars, social scientists, and cultural critics, attempts to begin telling the long-neglected story of modern Mexico's wide-ranging technological transformations. Contributors examine topics ranging from the introduction of new forms of travel, innovations in media, and transformations within the home and the body to the relationship between technology, literature, art, and architecture. Coverage of manufacturing, agriculture, and mining—topics undoubtedly in need of further study—have been left aside due to limited space and in order to give fuller attention to the above themes.

The first part addresses issues related to health, food, and the home. In the first chapter, Claudia Agostoni, historian of medicine and public health, examines the implementation of a "culture of hygiene" in Mexico through the use of novel tools, technologies, scientific theories, and methods from the late nineteenth and through the early twentieth century. In her consideration of how daily practices within the home were shaped by the consumption of domestic technologies, Sandra Aguilar-Rodríguez studies the adoption and use of the modern refrigerator and blender. The section concludes with Joanne Hershfield's provocative examination of household technologies and the ways in which they provided Mexican housewives with "new ways of being women."

The second part, comprised of chapters dealing with photography, television, and the Internet, opens with visual culture historian John Mraz's panoramic study of photographic technologies during Mexico's twentieth century. In the following chapter, Celeste González de Bustamante examines the development of television from 1950 through the early 1970s, paying particular attention to how it was shaped by policy makers, producers, and media magnets, as well as viewers. The third chapter, by the late cultural critic Carlos Monsiváis, contemplates Mexico's changing television culture from its birth through the onset of the twenty-first century. Finally, writer, essayist, and trained engineer Naief Yehya examines the emergence and spread of a Mexican cyberculture in his study of the Internet.

The third part offers three chapters on radio and music. The first, by literary scholar Viviane Mahieux, reconstructs a 1923 "race for the airwaves" in order to understand how two intellectual circles envisioned the role of radio as well as how a newly industrializing press embraced the potential of this new means of communication. In the following chapter, renowned cultural historian Ricardo Pérez Montfort contemplates the relationship between music, mass media, and cultural resistance during and after the 1968 student movement. Bringing the section to a close, Antoni Castells-Talens and José Manuel Ramos Rodríguez examine the introduction of indigenous radio broadcasting by the Mexican state and its eventual appropriation by indigenous groups themselves.

Railroads, automobiles, and the metro are considered in the fourth part, which opens with David M. J. Wood's reflection on the dialogue established between film, mechanized transport, time, and space in Mexican silent film. The chapter by J. Brian Freeman examines the changing uses and meanings of the automobile in early twentieth-century Mexico. Paying particular attention to the relations between business, the state, and labor, Guillermo Guajardo and Paolo Riguzzi examine the country's railway system during the twentieth century, the challenges it faced as well as its impact on the "culture of mobility." Finally, acclaimed writer and critic Juan Villoro probes the subsoil of the city—its "final frontier," as he dubs it—in his study of the Mexico City subway.

The fifth and final part examines the use of technology in literature, art, and architecture. Lynda Klich studies the depiction and meaning of technology in the work of the post-revolutionary avant-garde writers and artists associated with the Estridentista group. Anna Indych-López examines representations of modern machinery in Estridentista collaborator Diego Rivera's Secretaría de Educación Pública murals, as well as the broader relationship between technology, workers, politics, and realism. In the next chapter, Erja Vettenranta studies narrative responses to technology in Mexican literature, paying particular attention to works by contemporary writers Cristina Rivera Garza, Naief Yehya, and Carmen Boullosa. Finally, Edward R. Burian examines the tensions between Mexican building practices and changing architectural paradigms in his study of architecture from the 1968 Olympic Games to the beginning of the twenty-first century.

Covering the twentieth century and beyond, the chapters in this volume illustrate the invention, use, adaptation, and representation of technology, as well as the diverse ways in which technology has interacted with culture, politics, and society. It is our hope that together they will encourage scholars to continue expanding their understanding of technology and further reveal its relationship to and role in the formation of modern Mexico.

## Notes

1. Hubert Clinton Herring and Herbert Weinstock, eds., *Renascent Mexico* (New York: Covici, Friede, 1935), 1. Gruening was an American journalist and politician.

2. Stuart Chase, *Mexico: A Study of Two Americas* (New York: Macmillan, 1931). Chase was a popular American economist.

3. Edward Beatty, "Approaches to Technology Transfer in History and the Case of Nineteenth-Century Mexico," *Comparative Technology Transfer and Society* 1, no. 2 (August 2003): 171.

4. Anglophone scholars have largely neglected the history of technology in Mexico. Edward Beatty stands out as one of the few exceptions to this rule. In addition to "Approaches to Technology," see, for example, his "Patents and Technological Change in Late Industrialization: Nineteenth-Century Mexico in Comparative Perspective," *History of Technology* 24 (2002): 121–50.

Researchers in Mexico, by contrast, have published a variety of innovative studies on technology. See, for example, Ramón Sánchez Flores's classic and encyclopedic *Historia de la tecnología y la invención en México* (México: Fomento Cultural BANAMEX, 1980). See also María del Carmen Aguirre Anaya, *El horizonte tecnológico de México bajo la mirada de Jesús Rivero Quijano* (Puebla: Benemérita Univ. Autónoma, Inst. de Ciencias Sociales y Humanidades, 1999); and Alejandro Tortolero, *De la coa a la máquina de vapor: Actividad agrícola e innovación tecnológica en las haciendas mexicanas, 1880–1914* (Mexico City: Siglo XXI, 1995). Guillermo Guajardo has published a number of essays on technology and engineering in nineteenth- and twentieth-century Mexico. See his "Tecnología y campesinos en la Revolución

Mexicana," *Mexican Studies/Estudios Mexicanos* 15, no. 2 (1999). For a fascinating study of technology sharing between indigenous people and colonizers during the conquest and colonial period, see Enrique Florescano, Virginia García Acosta, and Magdalena A. Garcia Sánchez, *Mestizajes tecnólogicos y cambios culturales en México* (Mexico City: CIESAS, 2004).

5. Ann Johnson, "Revisiting Technology as Knowledge," *Perspectives on Science* 13, no. 4 (2005).

6. Thomas J. Misa, "Findings Follow Framings: Navigating the Empirical Turn," *Synthese* 168, no.3 (2009): 358.

7. David Edgerton, "Creole Technologies and Global Histories: Rethinking How Things Travel in Space and Time," HoST 1 (Summer 2007): 82–83. Rosalind Williams, "Opening the Big Box," *Technology and Culture* 48 (2007): 104.

8. Studies that pay particular attention to the links between technology and culture include David Nye, *Technology Matters: Questions to Live With* (Cambridge: MIT Press, 2006); Thomas Parke Hughes, *Human-Built World: How to Think about Technology and Culture* (Chicago: University of Chicago Press, 2004); Wiebe E. Bijker, Thomas P. Hughes, and Trevor J. Pinch, eds., *The Social Construction of Technological Systems: New Directions in the Sociology and History of Technology* (Cambridge: MIT Press, 1987). For a thoughtful essay on how American studies scholars might approach the theme of technology, see Carolyn de la Peña, "Slow and Low Progress: Why American Studies Should Do Technology," *American Quarterly* 58, no. 3 (September 2006). For a classic critique of the idea of "autonomous technology," see Langdon Winner, *Autonomous Technology: Technics-out-of-Control as a Theme in Political Thought* (Cambridge: MIT Press, 1977).

9. For discussion on the cultural implications of technology as well as the technological implications of culture, see Carolyn Thomas de la Peña and Siva Vaidhyanathan, *Rewiring the "Nation": The Place of Technology in American Studies* (Johns Hopkins University Press, 2007).

10. Kevin H. O'Rourke and Jeffrey G. Williamson, *Globalization and History: The Evolution of a Nineteenth-Century Atlantic Economy* (Cambridge: MIT Press, 1999); Beatty, "Approaches to Technology," 171.

11. For an intellectual history of Porfirian-era liberalism and the development of a "scientific politics," see Charles Hale, *The Transformation of Liberalism in Late Nineteenth-Century Mexico* (Princeton: Princeton University Press, 1989).

12. On statistics and cartography, see Raymond B. Craib, *Cartographic Mexico: A History of State Fixations and Fugitive Landscapes* (Durham, NC: Duke University Press, 2004); on criminology see Pablo Piccato, *City of Suspects: Crime in Mexico City, 1900–1931* (Durham, NC: Duke University Press, 2001).

13. See John H. Coatsworth, *Growth against Development: The Economic Impact of Railroads in Porfirian Mexico* (Dekalb: Northern Illinois University Press,

1976). See also Stephen Haber, *Industry and Underdevelopment: The Industrialization of Mexico, 1890–1940* (Stanford: Stanford University Press, 1989).

14. Díaz, quoted in Alan Knight, "The Weight of the State in Modern Mexico," in *Studies in the Formation of the Nation-State in Latin America*, ed. James Dunkerley (London: Institute of Latin American Studies, 2002), 227; John H. Coatsworth, "Railroads, Landholding, and Agrarian Protest in the Early Porfiriato," *Hispanic American Historical Review* 54, no. 1 (February 1974): 48–71. For a revisionist approach, see Teresa Van Hoy, *A Social History of Mexico's Railroads: Peons, Prisoners, and Priests* (New York: Rowman & Littlefield Publishers, 2008).

15. Beatty, "Approaches to Technology," 171.

16. Jonathan C. Brown, "Foreign and Native-Born Workers in Porfirian Mexico," *American Historical Review* 98, no. 3 (June 1993): 790; David E. Lorey, *The University System and Economic Development in Mexico since 1929* (Stanford: Stanford University Press, 1993), 21–22.

17. Information on infrastructural changes in Mexico City may be found in Claudia Agostoni, *Monuments of Progress: Modernization and Public Health in Mexico City, 1876–1910* (Calgary: University of Calgary Press, 2003).

18. Gustavo G. Garza Merodio, "Technological Innovation and the Expansion of Mexico City, 1870–1920," *Journal of Latin American Geography* 5, no. 2 (2006): 118. On the history of the Mexico City "desagüe," or drainage system, during the Porfirian era, see Agostoni, *Monuments of Progress*, 115–53; Claudia Agostoni, "Popular Health Education and Propaganda in Times of Peace and War in Mexico City, 1890s–1920s," *American Journal of Public Health* 96 (2006): 54–55.

19. On changes brought about by tram construction, see John Lear, "Mexico City: Space and Class in the Porfirian Capital, 1884–1910," *Journal of Urban History* 22, no. 4 (May 1996): 454–92; and John Lear, *Workers, Neighbors, and Citizens: The Revolution in Mexico City* (Lincoln: University of Nebraska Press, 2001). On the impact of the tram, see also Pablo Piccato, "Urbanistas, Ambulantes, and Mendigos: The Dispute for Urban Space in Mexico City, 1890–1930," in *Reconstructing Criminality in Latin America*, ed. Carlos Aguirre and Robert Buffington (Wilmington, DE: Scholarly Resources, 2000), 113–48.

20. On bicycles, see William H. Beezley, *Judas at the Jockey Club and Other Episodes of Porfirian Mexico*, 2nd ed. (Lincoln: University of Nebraska Press, 2004), 41–52. On bottled beer, see Edward Beatty, "Bottles for Beer: Business Strategy and the Challenge of Technology Transfer in Mexico," *Business History Review* 83 (Summer 2009): 317–48.

21. On advertisements, see Julieta Ortiz Gaitán, *Imágenes del deseo: Arte y publicidad en la prensa ilustrada mexicana (1894–1939)* (Mexico City: UNAM, 2003), 68, as well as her "Arte, publicidad y consumo en la prensa: Del porfirismo a la posrevolución," *Historia Mexicana* 48, no. 2 (1998): 411–35. On department stores,

see Steven B. Bunker, "Creating Mexican Consumer Culture in the Age of Porfirio Diaz, 1876–1911" (PhD diss., Texas Christian University, 2006), 49–56, 233–34.

22. Mauricio Tenorio-Trillo, "1910 Mexico City: Space and Nation in the City of the Centenario," *Journal of Latin American Studies* 28, no. 1 (Feb., 1996): 75–104. In addition to Tenorio-Trillo's treatment of the Centenario, see Michael J. Gonzales, "Imagining Mexico in 1910: Visions of the Patria in the Centennial Celebration in Mexico City," *Journal of Latin American Studies* 39, no. 3 (2007): 495–533.

23. Héctor Aguilar Camín and Lorenzo Meyer, *In the Shadow of the Mexican Revolution*, trans. Luis Alberto Fierro (Austin: University of Texas Press, 1989), 71–72.

24. Plutarco Elías Calles, quoted in Patrice Elizabeth Olsen, *Artifacts of Revolution: Architecture, Society, and Politics in Mexico City, 1920–1940* (Lanham, MD: Rowman & Littlefield, 2008), 35.

25. Alan Knight, "Popular Culture and the Revolutionary State in Mexico, 1910–1940," *Hispanic American Historical Review* 74, no. 3 (August 1994): 399. On the arrival of the Sonorans in Mexico City and the decade of reconstruction, see Jean Meyer, "Revolution and Reconstruction in the 1920s," in *Mexico since Independence*, ed. Leslie Bethell (Cambridge: Cambridge University Press, 1991), 201–40. For a statistical assessment of death and exile during the Mexican Revolution, see Robert McCaa, "Missing Millions: The Demographic Costs of the Mexican Revolution," *Mexican Studies/Estudios Mexicanos* 19, no. 2 (Summer 2003): 367–400. Recently, scholars have begun to question the notion that the 1910s was a lost decade in terms of economic growth. See, for example, Sandra Kuntz Flicker, "The Export Boom of the Mexican Revolution: Characteristics and Contributing Factors," *Journal of Latin American Studies* 36 (2004): 267–96.

26. Alan Knight, "Revolutionary Project, Recalcitrant People: Mexico, 1910–40," in *The Revolutionary Process in Mexico: Essays on Political and Social Change, 1880–1940*, ed. Jaime E Rodríguez O (Los Angeles: University of California Press, 1990), 243.

27. Ibid., 244.

28. A central component of the post-revolutionary project was the integration and assimilation of indigenous people. Although the 1920s witnessed a growing interest in indigenous cultures, the goal of the state continued to emphasize modernization and assimilation, rather than multiculturalism. See Alan Knight, "Racism, Revolution, and Indigenismo: Mexico, 1910–1940," in *The Idea of Race in Latin America, 1870–1940*, ed. Richard Graham (Austin: University of Texas Press, 1990).

29. Knight, "The Weight of the State," 233. On radio, see Joy Elizabeth Hayes, *Radio Nation: Communication, Popular Culture, and Nationalism in Mexico, 1920–1950* (Tucson: University of Arizona Press, 2000). On roads, see Wendy Waters, "Re-

mapping Identities: Road Construction and Nation Building in Post-revolutionary Mexico," in *Nation and Cultural Revolution in Mexico, 1920–1940*, ed. Mary Kay Vaughan and Stephen E. Lewis (Durham: Duke University Press, 2006), 221–43.

30. Knight, "Revolutionary Project," 253–55. For a more positive take on the impact of this cultural revolution, see Mary Kay Vaughan, *Cultural Politics in Revolution: Teachers, Peasants, and Schools in Mexico, 1930–1940* (Tucson: University of Arizona Press, 1997).

31. Garza Merodio, "Technological Innovation," 119.

32. Olsen, *Artifacts of Revolution*, 35, 109. On the use of concrete, see Matthew Fry, "Mexico's Concrete Block Landscape: A Modern Legacy in the Vernacular," *Journal of Latin American Geography* 7, no. 2 (2008): 35–58; and Rubén Gallo, *Mexican Modernity: The Avant-Garde and the Technological Revolution* (Cambridge, MA: MIT Press, 2005).

33. See Joanne Hershfield, *Imagining la Chica Moderna: Women, Nation, and Visual Culture in Mexico, 1917–1936* (Durham: Duke University Press, 2008).

34. See Gallo, *Mexican Modernity*; Tatiana Flores, "Clamoring for Attention in Mexico City: Manuel Maples Arce's Avant-Garde Manifesto Actual No. 1," Review: *Literature and Arts of the Americas*, no. 69 (Fall 2004): 208–20; Elissa J. Rashkin, *The Stridentist Movement in Mexico: The Avant-Garde and Cultural Change in the 1920s* (Lanham, MD: Rowman & Littlefield, 2009). On the use of technology in the work of Diego Rivera, see Anna Indych-López in this volume. For a study of technology and the avant-garde in Argentina, see Beatriz Sarlo, *The Technical Imagination: Argentine Culture's Modern Dreams*, trans. Xavier Callahan (Stanford, CA: Stanford University Press, 2008).

35. Lorey, *The University System*, 64. On the history of agronomy, see Joseph Cotter, *Troubled Harvest: Agronomy and Revolution in Mexico, 1880–2002* (Westport, CT: Praeger, 2003).

36. Alan Knight, "State Power and Political Stability in Mexico," in *Dilemmas of Transition*, ed. Neil Harvey (London: Institute of Latin American Studies, University of London and British Academic Press, 1993), 52–53.

37. In recent years the post-1940 period has received growing attention from scholars. See Gilbert M. Joseph, Anne Rubenstein, and Eric Zolov, eds., *Fragments of a Golden Age: The Politics of Culture in Mexico since 1940* (Durham: Duke University Press, 2001).

38. Claudio Lomnitz-Adler, *Deep Mexico, Silent Mexico: An Anthropology of Nationalism* (Minneapolis: University of Minnesota Press, 2001), xviii.

39. For an examination of how physical changes in the city were represented in visual culture, see Erica Segre, "Reframing the City: Images of Displacement in Mexican Urban Films of the 1940s and 1950s," *Journal of Latin American Cultural Studies* 10, no. 2 (2001): 205–22.

40. Joseph, Rubenstein, and Zolov, *Fragments of a Golden Age*, 10.

41. Andrew M. Butler, Adam Roberts, and Sherryl Vint, eds., *The Routledge Companion to Science Fiction* (New York: Routledge, 2009), 99. The study of science fiction in Mexico is a growing field. See, for example, Gabriel Trujillo Muñoz's *Biografías del futuro: La ciencia ficción mexicana y sus autores* (Mexico City: UABC, 2000) and *El futuro en llamas: Cuentos clásicos de la ciencia ficción mexicana* (Mexico City: Grupo Editorial Vid, 1997). For broader studies of Latin America, see Jerry Hoeg, *Science, Technology, and Latin American Narrative in the Twentieth Century and Beyond* (Bethlehem, PA: Lehigh University Press, 2000); and Jane Robinett, *This Rough Magic: Technology in Latin American Fiction* (New York: P. Lang, 1994).

42. The term "Mexican miracle" appears to have emerged in the late 1960s and early 1970s, though its origins remain unclear.

43. In recent years the Mexico City Olympics has attracted growing attention from scholars. See, for example, Elaine Carey, *Plaza of Sacrifices: Gender, Power, and Terror in 1968 Mexico* (Albuquerque: University of New Mexico Press, 2005); Eric Zolov, "Showcasing the 'Mexico of Tomorrow': Mexico and the 1968 Olympics," *The Americas* 61, no. 2 (October 2004): 159–88.

44. Salvador Novo, *New Mexican Grandeur*, 5th ed. (Mexico City: Petróleos Mexicanos, 1967), 135–36. On the transformation of Mexico City during the 1950s and 1960s, see Diane E. Davis, "The Social Construction of Mexico City: Political Conflict and Urban Development, 1950–1966," *Journal of Urban History* 24, no. 3 (March 1998): 364–415.

45. Lomnitz-Adler, *Deep Mexico*, 134–35.

46. On *jipitecas*, see Eric Zolov, *Refried Elvis: The Rise of the Mexican Counterculture* (Berkeley: University of California Press, 1999), 132–50.

47. On Tlatelolco see Carey, *Plaza of Sacrifices*. The number of dead continues to be debated.

48. Work on Mexico's "dirty" war has seen growing attention with the declassification of archival sources in both the United States and Mexico. See, for example, Kate Doyle, "Official Report Released on Mexico's 'Dirty War,'" National Security Archive Project, http://www.gwu.edu/~nsarchiv/NSAEBB/NSAEBB209/index.htm (accessed August 6, 2011).

49. The 1980s has been studied largely by social scientists. For an anthropologist's perspective, see Claudio Lomnitz-Adler, "Times of Crisis: Historicity, Sacrifice, and the Spectacle of Debacle in Mexico City," *Public Culture* 15, no. 1 (2003): 127–47.

50. José Emilio Pacheco, *Battles in the Desert and Other Stories*, trans. Katherine Silver (New York: New Directions, 1987).

51. Diane E. Davis, *Urban Leviathan: Mexico City in the Twentieth Century* (Philadelphia: Temple University Press, 1994), 231–32.

52. On dystopia in Mexican literature, see Miguel López Lozano, *Utopian Dreams, Apocalyptic Nightmares: Globalization in Recent Mexican and Chicano Narrative* (West Lafayette: Purdue University Press, 2008).

53. Carlos Fuentes, *Christopher Unborn* (New York: Farrar Straus Giroux, 1989).

54. Carlos Monsiváis, *Mexican Postcards* (London: Verso, 1997), 35.

55. Jordy Micheli Thirion and Ruben Oliver Espinoza, "Changing Patterns in Mexican Science and Technology Policy (1990–2003): Still Far from Economic Development," in *Changing Structure of Mexico: Political, Social, and Economic Prospects*, ed. Laura Randall (Armonk, NY: M. E. Sharpe, 2006), 198.

56. Michael E. Miller, "'Twitteros' Are Mexico's Latest Outlaws," GlobalPost, February 2, 2010, http://www.globalpost.com/dispatch/mexico/100128/twitter-crackdown (accessed August 1, 2010).

# I
# HEALTH, FOOD, AND THE HOME

# 1
# Material Culture, Public Health, and the Technologies of Hygiene in Modern Mexico, 1890s–1940s

*Claudia Agostoni*

During the late nineteenth and early twentieth centuries, the improvement of health and the prevention of disease led to profound transformations in the most populated towns and cities of the country. Through public health legislation and sanitation and engineering projects, as well as via carefully crafted hygienic surveys conducted with precise instructions and techniques to identify and isolate the threats of disease, both the material environment and the codes of behavior expected of men, women, and children went through a process of radical transformations. According to some Mexican physicians, hygienists, and sanitary engineers, only through cleanliness and hygiene would it be possible to improve the health of the people and contribute toward the order and progress of the nation.

During the long Porfirio Díaz regime (1877–1910), a gradual centralization, professionalization, and expansion of medical services took place. The enactment of the first comprehensive public health legislation comprised in the Sanitary Code of the United States of Mexico in 1891 (reformed in 1894 and 1903) led to detailed sanitary surveys by public health officials and to numerous projects, plans, and activities to improve and transform the material and sanitary conditions of homes, factories, schools, hospitals, markets, and any other sites where people gathered. Also, particular attention was placed on examining the customs and habits of the urban populations and on stressing the measures, procedures, and technologies required to maintain good order, cleanliness, and health.

Public health was of particular concern after the armed phase of the Mexican Revolution (1910–1920), when the improvement of health became a pivotal obligation of the revolutionary state, as established in the 1917 constitution.

Numerous public health and health education campaigns were organized, such as the first massive and compulsory vaccination campaigns against smallpox, the control and eradication of yellow fever, and the strict imposition of diverse measures to combat the propagation of typhus in the main urban and rural settings. Those public health programs were followed with interest and promoted with vigor in newspapers, magazines, radio, and cinema, media that sought to disseminate the messages and symbols of the value of health throughout urban and rural Mexico. In this chapter I offer an overview of how novel tools and scientific theories, technologies, and regimes of hygiene led to implementing a wide array of strategies aimed at forging a veritable culture of hygiene in Mexico during the course of the late nineteenth and early twentieth centuries.

## Modern Cities and Hygienic Bodies

The *Guía Indispensable del Forastero en la Ciudad de México y Calendario para 1905* (The essential traveler's guide to Mexico City and calendar for 1905) established that June 24, the Feast Day of Saint John the Baptist, was none other than cleanliness day, due to the large outpouring of people to the bathing houses of the capital city.[1] According to diverse physicians, writers, and observers of Mexican customs and habits, the celebration of Saint John the Baptist since Colonial times was a lively event, when men, women, and children would splash or submerge their bodies in water. Water would pour into the life of Mexico City's inhabitants at diverse bathhouses, such as El Harem, San Felipe de Jesús, or Amor de Dios, where bathers would forget their inhibitions and fears about taking a dip, transforming the contact with water into an exciting and joyous occasion. Owners decorated their establishments with flags, garlands, and colorful crowns; musicians played string and wind instruments to animate the event; and visitors returned home with small gifts, such as bars of soap and small bottles with perfumed water.[2]

The Saint John the Baptist celebration had throughout centuries served a ritualistic function that was both enjoyable and entertaining, associated with water's ancient purity and numerous pagan and Catholic purification rites. However, by 1905 that festivity was associated with a series of habits and practices regarded as essential to the preservation of health, and numerous social observers considered that the immersion or contact with water should be permanent and not limited to one day of the year. During the course of the nineteenth century, as Jean-Pierre Goubert has established, the old relationship with water, water imbued with powers and a symbol of transition, was radically transformed. Water gradually became an everyday consumer, sanitary, and industrial product and was defined as an essential element in the pursuit of health.[3]

The preoccupation with water in Mexico City during the late nineteenth and early twentieth centuries led diverse physicians and sanitary engineers to emphasize that urban sanitation was key to the materialization of a modern and hygienic city. The prominence that urban sanitation acquired also led some members of the Superior Council of Public Health, the premier health authority of the country between 1841 and 1918, to stress that it was imperative to regulate the flow or stagnation of water in the main urban settings. Unprecedented investments in new urban technologies, such as drainage, sewer systems, and running water, were destined for the capital city and the main ports and provincial cities of the country, with the goal of sanitizing urban modern living.[4]

Public health and the scientific management of society and its ills were further enhanced in Mexico, as in other countries of Latin America, due to the bacteriological discoveries of the final decades of the nineteenth and early twentieth centuries. The germ theory of disease causation impinged not only upon the perceptions of the origins and prevention of disease, but also on the emergence of new devices and technologies. A veritable war was fought against the invisible enemies—germs and bacteria—a struggle that was sustained on new diagnostic hypotheses and an increasing array of novel therapeutic and prophylactic agents that emerged from the laboratory. For instance, tuberculin (for tuberculosis) and diphtheria antitoxin were new tools available to the Mexican health authorities from the late 1890s onward. Those technologies transformed health policies and programs; impinged on the daily lives, beliefs, and expectations of men, women, and children; and led to an obsessive insistence on cleanliness, regarded as a synonym for hygiene.[5]

Luis E. Ruiz, one of the most prominent physicians during the final decades of the nineteenth century, established in 1904 that hygiene was none other than the "scientific art of maintaining health and increasing well-being." In his view, each individual had to assume responsibility for personal hygiene; household hygiene was the responsibility of the family, and the sanitation, order, and cleanliness of the city had to be guaranteed by the local authorities.[6] The importance of hygiene and of health education was underlined by physician Eduardo Liceaga, president of Mexico's Superior Council of Public Health during the Díaz regime, and by engineer Alberto J. Pani during the armed phase of the Mexican Revolution.[7] This importance was also endorsed by numerous physicians, sanitary engineers, and health officials throughout the course of the first half of the twentieth century, notably by physician Alfonso Pruneda.[8]

The Porfirian health authorities aimed to transform certain aspects of everyday life so that they would conform to the guidelines set down by the science of hygiene. Houses or living quarters were required to meet certain criteria so that they became healthy environments; the body was required to be cleansed in a certain manner; and men, women, and children were expected to wear hygienic clothing and to consume adequate food. These and other themes became an essential part of a wide-ranging catalog of hygienic behaviors and practices that required the urban and rural dwellers to possess a "hygienic instinct."[9] But also and equally essential, people began to have access to new products, services, and technologies, such as sanitary installations, water closets, water filters, and bathing and washing facilities. The importance attributed to cleanliness and the care of the body led to the production and promotion of numerous products, services, and furniture designed to carry out a profound hygienic revolution during the course of the twentieth century.[10]

## Vigorous Bodies and Hygienic Housing

According to some members of the Porfirian medical community, homes were the most favorable places for the development and propagation of endemic and epidemic diseases, among which tuberculosis, typhus, and smallpox stood out. In order to eliminate those threats within a home or room, physicians and hygienists established estimates derived from studies carried out by scientists and engineers regarding the amount of cubic meters that a house or room required having in order to guarantee the health of each family member. Thus, the number of windows per room, the proposed times of the day that it was convenient to, or not to, open them, and the minimum ceiling height, as well as the materials that should be used to build houses, schools, hospitals, and jails, were carefully addressed.

Ventilation, or unrestricted air circulation, was presented as an essential requirement for the hygiene of dwellings. In the specific case of Mexico City, physicians established that rooms had to have a minimum of 30 square meters per person, and that ceiling height should not be less than 3.75 meters.[11] Likewise, residences were expected to have good solar or electrical illumination "so that everything is bright and we can see the dirt in order to maintain a clean place, which is the initial step toward hygiene."[12] Solar light was said to be particularly beneficial because it destroyed the "bacilli from cholera and typhoid fever," as well as those from tuberculosis.[13] Other illnesses, such as laryngitis, croup, and bronchitis, frequently afflicted people living in humid dwellings.[14] Therefore, some physicians recommended that residential owners or tenants should resort to waterproofing ceilings, walls, and floors via the application of substances such as oil varnish, as was increasingly done in most hospitals.[15] Another option was to request the services offered by the Alfredo Perez-Gil Company, established in 1892 to waterproof homes and counteract the effects of humidity.[16] However, it is important to underline that the recommendations and suggestions previously mentioned were practically impossible to carry out. According to the 1910 national census, more than 50 percent of homes throughout the country were listed as "shanties": living spaces with dirt floors that lacked internal subdivisions.[17] The predominant living conditions in both urban and rural Mexico involved people crowded into small and poorly ventilated rooms, which led to the propagation of numerous diseases.[18]

Physicians and engineers also scrutinized the sanitary habits and practices within the households through detailed surveys conducted in accordance with precise instructions and techniques. Doctors recommended that homemakers sweep the floors at least three times per day and that furniture, clothing, and kitchen utensils should be thoroughly and carefully cleansed. Physician Rafael Domínguez y Pastor asserted that the walls of a forty-five-square-meter room could contain nearly one million microorganisms and recommended that people should refrain from hanging pictures and decorations on walls, using very heavy curtains, or installing carpet or wallpaper because such additions would become receptacles for dust and microbes.[19]

Another stipulation, perhaps one of the most important, required healthy persons to avoid any contact with sick individuals. When a family member had typhus, for instance, his or her belongings had to be thoroughly disinfected and all utensils had to be cleansed. Furthermore, the sick person had to remain in isolation in a clean and ventilated room.[20] When disinfection was not sufficient and the threat of the spread of disease remained, physicians ordered that the personal belongings of the sick be destroyed and that the ill be taken to a hospital to stop the chain of contagion.

The impurity of water for food preparation and drinking was a crucial topic of concern, being the principal cause of gastrointestinal ailments and one of the main causes of mortality. The urban population was advised to refrain from consuming water from artesian wells, springs, and reservoirs and to avoid buying water from the water carriers who sold it in the main plazas throughout the country. In the specific case of Mexico City, particular anxiety arose if the water consumed had its origin in La Viga Canal or the lakes Xochimilco, Tlahuac, and Texcoco.[21] And in an attempt to assess the quality of water consumed, the city's inhabitants were invited to resort to various techniques. One of them was the simple household method of "leaving water for a few days in a tightly closed bottle, and then check[ing] for any horrible smells that would appear if there were actually organic substances in it."[22] Another option was to filter water using the Pasteur or Delfín filters, devices that were advertised as having the capacity of eliminating germs and that were manufactured for homes, schools, hospitals, and jails.[23] In addition, another simple and inexpensive technique was to boil the water.[24]

Some members of the medical community considered that lacking proper bodily hygiene would lead to the closing of pores by dust and grease, turning a person into a breeding ground for microbes.[25] Keeping a clean body was also presented as a feature of a person's dignity, decorum, honor, and self-respect.[26] However, the material infrastructure required for bodily cleanliness was not available to the majority of the Mexican population. To install a bathroom in a house required a constant water supply or a service that would bring water to one's home from the public water sources in diverse cities; both systems were inaccessible to most of the urban and rural populations well into the 1950s. Thus, a full-body bath in a private room was reserved for a minority of the urban population.

During the initial years of the twentieth century, the novel technologies of plumbing, drainage, and the provision of water made possible the inclusion of bathrooms in some houses located in the most modern neighborhoods of the capital city, those occupied by the well-to-do members of society.[27] In the Santa María la Ribera neighborhood of Mexico City, houses were equipped with enamel cast-iron bathtubs, toilets, and iron and porcelain sinks imported from Belgium, France, and England that worked with bronze faucets or porcelain spigots.[28] In 1907, the magazine *Albúm de Damas* established that bathrooms should have tubs and special bathtubs for children, as well as sponges, soaps, and towels within easy reach in order to guarantee the private ritual of personal hygiene.[29] Such was the importance that the personal and private bathroom acquired that the Gerber-Carlisle Company launched a new product on the Mexican marketplace: folding bathtubs that offered people a bathroom of their own for the reasonable price of twenty-five pesos. According to the advertisement, people of all sizes, with only a little bit of water, could use these tubs and take a "splendid bath."[30]

During the initial years of the twentieth century, the capital city also had more than forty-eight public baths.[31] Those establishments offered men, women, and children a great variety of services designed to cater to specific sectors of society.[32] The baths of El Harem or those at the Iturbide Hotel offered "decent people" the services of warm, hydrotherapeutic, Russian or Turkish-Roman baths.[33] Some first-class public baths also had large and beautiful gardens, billiard rooms, and separate areas for men and women and offered complimentary haircut and shoeshine services and had rooms for gymnastic exercises, while second- and third-rate baths were located in the most marginal areas of the urban centers.[34] There were also free public baths in the capital city, such as the Baños del Dormitorio Público Gratuito de la Primera Demarcación (Free Public Baths of the Dormitory of the First District), established in 1889 by the Pri-

1.1. Baños Juárez gratuitos, 1922 (Juárez free public baths, 1922). Inv. #141381, Sistema Nacional de Fototecas (SINAFO)–Fototeca Nacional del Instituto Nacional de Antropología e Historia (INAH), Mexico City.

vate Patriotic Board and sponsored by the Charity Board.[35] Other third-class establishments, such as the Baños y Lavaderos de Vapor de la Plazuela de la Lagunilla (Vapor Baths and Laundry Areas of La Lagunilla Square), offered reasonably priced services to the urban poor, where it was hoped they would learn "the immense benefits, both for individuals and for the species, of immersing the body in clean water, and of the truly wonderful caresses the skin feels when covered in bright white clothes purified by soap and water."[36]

Health and hygiene, key priorities of the health authorities during the initial years of the twentieth century, led to forging a link between health and beauty: "True and complete beauty always includes perfect health," as expressed by physician Máximo Silva.[37] The latter association was materialized in 1907 during the inauguration of the Hygeia Institute in Mexico City. Located on the prestigious Juárez Avenue, the Hygeia Institute attempted to reconcile female beauty with hygiene; or, as its advertisement noted, its aim was "the harmonious improvement of the female figure that leads to good health."[38] The establishment offered both traditional and novel services and technologies: bathtub and steam, dry air, hot air, sponge baths. Sunbaths were one of its major attractions and were said to have notable effects on skin illnesses.[39] The advertisements of the Hygeia Institute in the popular press underlined that hygiene was no longer reduced to the word cleanliness: "In our culturally advanced times, hygiene demands . . . many more requirements to achieve good health. Contemporary life is not as tranquil as that of earlier times; it is agitated and causes the appearance of diseases with names that our ancestors would have found amusing, such as neurosis and other terrible ailments that can be radically cured in this establishment."[40]

A clean and healthy body also had to be properly and neatly covered, not solely in the name of health, but also to maintain a good appearance, as the Venezuelan Manuel Antonio Carreño affirmed in his book *Manual de Urbanidad y Buenas Maneras* (The urbanity and good manners handbook). In his opinion, "our clothing may be more or less luxurious . . . but it will never be right to avoid any expense or care to ensure cleanliness."[41] Given that the cleanliness of the body and clothes was interconnected, physicians and hygienists also focused on establishing the proper materials for clothes and on the ways in which men, women, and children should dress, denouncing that nudity and filth predominated throughout both urban and rural Mexico. According to Julio Guerrero, "miserable men and women that lacked the necessary means to survive," whose nudity and filth was contrary to the most basic notions of hygiene, plagued Mexico City. Furthermore, within the city's streets, numerous indigenous people could be seen wearing "underwear, a shirt, and a *sábana de manta* (traditional coarse cotton sheet and clothing)," while their women covered their bodies with *huipiles* (the traditional embroidered dress) as they wandered barefoot through the capital's neighborhoods.[42] Guerrero also noted that most blue-collar workers did wear shoes and pants, as did many artisans.[43]

Whether it was shoes, pants, skirts, or dresses, from a hygienic standpoint, they were required to protect the person from the changes in the weather and should be kept absolutely clean in order to avoid the transmission of germs.[44] A proper item of clothing required being impermeable to air and water and should allow freedom of movement.[45] To satisfy this condition, some physicians recommended the use of woolen clothing to maintain heat. But if the goal was to favor the equilibrium of body heat, then it was necessary to wear linen clothing.[46] Two or more fabrics were better protection against the weather than only one. Therefore, wearing two items of clothing made of different materials was the best way to avoid changes in body temperature that could result in diverse respiratory ailments. It was also important to consider the clothing's colors. Underwear was an essential item for keeping the body clean and protected, and hygienists believed that it should always be white because that color prevented heat absorption in hot weather while permitting the emission of heat in cold weather. Furthermore, white denoted cleanliness and neatness.[47]

## Health for Sale

The manufacturing and sale of national and international medicinal products, from France and the United States in particular, as well as services and furnishings for a hygienic upkeep of the body, multiplied at a rapid pace from the late nineteenth century up to the end of the Second World War. In 1891, during the month of January alone, the newspaper *Diario del Hogar* included more than fifty-eight advertisements that promoted the virtues and benefits of a great diversity of products. Powders, syrups, pills, oils, and liqueurs that promised to reestablish and strengthen the person and ointments and creams that guaranteed the beautification of the skin expanded. For lack of appetite, headaches, blackouts, and congestions, the urban population could find a cure in the *Verdaderos Granos de Salud del Dr. Franck* (True grains of health by Dr. Franck).[48] The Vino de Bugeaud (Bugeaud wine), mixed with quinine and cacao, was recommended for anemic or convalescent women.[49] The marvelous Crème de Bismuth Quesneville was presented as the correct antidote for diarrhea occurring with cholera, dysentery, and abdominal ailments.[50] For people suffering from insomnia, Jarabe de Follet was recommended, and its advertisement underlined that the product lacked the drawback of opium or morphine: addiction.[51] Those and many other products and services were advertised on a daily basis and could be purchased in stores and pharmacies.[52] Although by law it was determined that any establishment that sold medicines should have the presence of a qualified pharmacist on its premises, that stipulation was seldom observed, causing alarm and concern among medical practitioners. Two additional causes of anxiety among physicians were the unrestricted sale of medicinal herbs on street corners and in markets and the fact that many individuals openly lied about having a medical degree and publicized their services in the press. Various physicians also deplored the promotion of the so-called infallible and miracle products in the market, affirming that they were harmful both to the health of the population and to the reputation of the medical community at large. From the 1920s onward, newspapers, magazines, almanacs, and calendars, as well as radio and cinema, promoted and advertised diverse remedies and medications to restore health, as well as products directed toward men, women, and children and intended for embellishing the body.

## Health in Revolution, 1910–1940

During the Mexican Revolution—a series of civil wars between 1910 and 1920—many rural dwellers fled to the capital in search of a safe haven, only to encounter unemployment, hunger, chaos, violence, and death. Until 1913 Mexico City had been spared the worst devastation of the revolution, but in the month of February large areas of the capital were transformed into a battlefield. Thousands of people became homeless, food and water were scarce, and the dead and wounded lay on the plazas and streets. To aggravate the situation even further, a typhus epidemic spread throughout the city during the second half of 1915, and the Spanish influenza pandemic reached Mexico in 1918.[53] The military and health authorities, the press, and the public asked if it was the bullets or the microbes that were causing the largest number of casualties.[54] According to the newspaper *El Demócrata*, it was imperative to save the lives of those who were surviving in the battlefields, a task that only hygiene could possibly accomplish.[55]

In October 1915, when the city was in the midst of the typhus epidemic, the health authorities required physicians to deliver brief informal talks and lectures in schools, plazas, and public gardens so that the urban population could value the importance that cleanliness, temperance, rest, fresh air, and exercise had as methods of prevention.[56] Doctor Alfonso Pruneda, responsible for the antityphus campaign, appointed forty physicians to traverse the capital and search for possible carriers. Two hairdressers accompanied each physician, and when a suspect was found, the person in question had to have his or her hair cut.[57] The health authorities also determined that any soldier in the capital with typhus had to be taken to San Joaquín Lazaretto and remain in isolation. Cinemas and theaters were ordered to close, and a campaign against charlatanism was launched due to the proliferation of "miraculous and infallible" remedies on the market.[58] Furthermore, churches were only allowed to open for one hour on weekdays and two on Sundays because churches, the military authorities stressed, were a threat to public health. By December 1915, due to the fact that the number of people with typhus was increasing, the Ministry of the Interior established a special sanitary police; the retail sale of alcoholic drinks was forbidden; all public meetings were ordered to end by 11:00 p.m., and the city's inhabitants were ordered not to have pigeons, hens, dogs, or any other animals inside their homes. People of "any social class who by their notorious dirtiness could carry on their body or clothes parasitic animals that are transmissible" were prohibited from having access to public places as well.[59] As can be seen, health education, hygienic propaganda, and advice on personal hygiene were not disrupted during the violent years of the Mexican Revolution. Health advice was particularly prominent in the newspaper *El Demócrata*, which, among other things, published two hundred pamphlets with instructions on how to prevent typhus; these were distributed free of charge among the city's inhabitants.[60]

The Superior Council of Public Health acknowledged in January 1916 that

1.2. Víctima de influenza, ciudad de México, 1918 (A victim of influenza in Mexico City, 1918). Inv. #75735, Sistema Nacional de Fototecas (SINAFO)–Fototeca Nacional del Instituto Nacional de Antropología e Historia (INAH), Mexico City.

the typhus epidemic had caused 2,001 deaths during that month alone, a situation that was of particular concern to both civilian and military authorities. The health authorities, led by deputy, general, and physician José Maria Rodríguez from 1914 to 1917 and who became president of the Department of Public Health, reorganized the sanitary personnel and established a program of home visits as well as a home-to-home disinfection campaign. According to Rodríguez, what the country required in the face of the epidemic and the death and destruction caused by the civil war was nothing less than a "sanitary dictatorship."[61] Rodríguez considered that the executive power had to intervene in all health issues and impose the adequate rules of health that should be followed by the urban and rural populations.[62] Rodríguez believed that the struggle for health would only be won through legislation, health education, hygienic propaganda, and the use of force.[63]

The 1917 constitution did not include the notion of a "sanitary dictatorship," but it did establish that all individuals had the right to enjoy physical and mental health and that individuals could not endanger the health of the community.[64] The Superior Council of Public Health was transformed into the General Sanitary Council, with jurisdiction over the entire country, and was placed under the direct orders of the executive power. Also the Department of Public Health, with federal jurisdiction, was created, and it was precisely this department that launched some of the most important public health programs and health education campaigns between 1920 and 1943. Those programs were destined to reach not only the urban populations but also to rural ones, targeting children and the indigenous populations in particular.

It should be noted that rural health education became essential for the state and for the educational authorities from 1921 onward, when rural schools were created with the goal to transform the peasantry into patriotic and scientifically informed individuals. Among the topics included in that educational endeavor, emphasis was placed on personal hygiene and on the importance that avoiding unlicensed healers had for the general health of the rural populations.[65] However, the transformation of personal habits of hygiene required not only the appropriate information, but also new technologies such as plumbing, bathrooms, and drinking water, which most rural dwellers did not possess. If what was sought by the post-revolutionary governments was to modernize, educate, and sanitize both the urban and rural populations, then health propaganda and health education were regarded, again, as the strategies that would allow them to fulfill that aim.

Another important feature of the post-revolutionary public health policies and programs was children's health. During the month of September 1921, Doctor Gabriel Malda, head of the Department of Public Health from 1920 to 1924, organized the first National Week of the Child. A children's exhibition was organized; and through graphs, charts, posters, and diverse objects, the place of health in the moral education of children was stressed. Pamphlets and calendars for 1923 with hygiene maxims and sayings were also distributed among the public, and an automobile parade with "healthy" children traversed the most important avenues of Mexico City on September 13. The celebration of children's health also included hygiene festivals in schools and the distribution of toothbrushes and other personal grooming objects to mothers and children, as well as an ingenious play, "Health Fairy," that aimed to teach the principles of public health to school-aged children.

Health education and hygienic propaganda were further enhanced when the Section of Hygienic Education and Propaganda was created at the Department of Public Health in 1922 under the leadership of Dr. Alfonso Pruneda, with the goal of disseminating hygienic and health advice in schools, homes, and factories through the press, public lectures, radio, and film. Posters with advice on personal hygiene, healthy food, and sexually transmitted diseases were displayed in streetcars and markets; and circulars and pamphlets that underlined the importance of cleanliness, hand-washing, vaccination, pasteurized milk, and exercise were distributed among schoolchildren and mothers. The Section of Hy-

1.3. Campaña contra el tifo en la ciudad de México, 1922 (Campaign against typhus in Mexico City, 1922). Inv.#652295, Sistema Nacional de Fototecas (SINAFO)–Fototeca Nacional del Instituto Nacional de Antropología e Historia (INAH), Mexico City.

gienic Education and Propaganda also published the journal *El Mensajero de la Salud* (The health messenger) and, through short stories, poems, and recipes, advised the urban and rural populations how to preserve their health.

The struggle for health became clearly established in the 1926 Sanitary Code, which stressed that popular health education was one of the main objectives of the health authorities. Throughout the following decades, intense health education campaigns were carried out in rural and urban settings. Through pamphlets, magazines, newspapers, cinema, radio, and, later, television, the scientifically directed efforts of public health experts reached numerous urban and rural localities throughout the country.

The importance that fostering rural health had for the future of the country began to be thoroughly addressed during the 1920s. In 1927 the first cooperative sanitary units were created in the state of Veracruz, financed by the Department of Public Health, the state health authorities, and the International Health Board of the Rockefeller Foundation.[66] The aim of those sanitary units was to provide health services, in particular, public health education, infant hygiene and child health, and hookworm diagnosis and treatment, as well as the training of local sanitary personnel, in order to achieve the proposed objectives.

In addition, "dental hygiene, prenatal care and routine smallpox vaccinations were incorporated," and in 1931 the Rural Hygiene Service was established.[67]

Health educational campaigns continued to resort to the press, but even more so to radio and cinema. In addition, one of the main concerns of the health authorities was to train a large and well-qualified health personnel, which included nurses, visiting nurses, bacteriologists, sanitary engineers, and vaccination agents. Thus, as of the 1920s, a new figure began to enter the homes of numerous urban dwellers for the first time: the visiting nurse, who, in addition to supervising the health of mothers, mothers-to-be, and children, provided guidelines for good hygienic housekeeping practices and for the proper and hygienic way to store food (in particular, dairy and meat products) and underlined the importance of avoiding the presence of fowl, pigs, or sheep on the patios or inside the living quarters.[68] Health education activities increased throughout the course of the 1930s and 1940s, when educational films began to be utilized in specific public health programs, such as in the compulsory smallpox vaccination campaign undertaken throughout the country. Printed materials were also produced and distributed, covering topics such as the prevention of smallpox, malaria, tuberculosis, sexually transmitted diseases, diphtheria, and typhoid fever, among others.[69]

The insistence on the virtues and benefits of maintaining good hygienic practices was overwhelming in Mexico from the late nineteenth century onward. The triumph of the ideal of good health, hygiene, and sanitation emphasized that the highest threat to individual and collective health was found in unsanitary households and in the absence of personal hygienic habits and practices. The discovery of the microbe led to the establishment of prolonged and systematic hygienic procedures and supervision tactics and to the adoption of prophylactic measures that resulted in multiple recommendations and prescriptions set down by physicians and by persons outside the medical world. A consequence of such a transformation was the production of numerous products, services, and novel therapeutics directed toward promoting good health and the cleanliness of the individual and the collective body.[70] Indeed, a profound revolution regarding the value of health in modern Mexico took place, a process that is far from complete but that began in earnest during the late nineteenth and early twentieth centuries.

## Notes

An earlier version of portions of this chapter appeared in "Las delicias de la limpieza: La higiene en la ciudad de México," in *Historia de la vida cotidiana en México*, vol. 4, *Bienes y vivencias: El siglo XIX*, ed. Anne Staples (Mexico City: Fondo de Cultura Económica–El Colegio de México, 2005) 563–97. I want to thank the anonymous readers for their comments and recommendations and the editors of this volume for their timely suggestions. Finally, I acknowledge the financial support of the DGAPA-UNAM Research Project PAPIIT IN403010.

1. Francisco José Zamora, *Guía indispensable del forastero en la ciudad de México y calendario para 1905* (Mexico City: J. Ricardo Garrido y Hermanos Editores, 1905), 36.

2. Artemio Valle Arizpe, *Calle vieja y calle nueva* (Mexico City: Editorial Jus, 1949), 436.

3. Jean-Pierre Goubert, *The Conquest of Water: The Advent of Health in the Industrial Age* (Princeton: Princeton University Press, 1988), 259.

4. Claudia Agostoni, *Monuments of Progress: Modernization and Public Health in Mexico City, 1876–1910* (Calgary: University of Calgary Press, 2003).

5. Nancy Tomes, *The Gospel of Germs: Men, Women, and the Microbe in American Life* (Cambridge: Cambridge University Press, 1988).

6. Luis E. Ruiz, *Tratado elemental de higiene* (Mexico City: Oficina Tipográfica de la Secretaría de Fomento, 1904), 166.

7. Alberto J. Pani, *La higiene en México* (Mexico City: Impr. de J. Ballescá, 1916).

8. Alfonso Pruneda, *Higiene de los trabajadores, por el Dr. Alfonso Pruneda* (Mexico City: Ediciones de la Universidad Nacional de México, 1937).

9. Physician Porfirio Parra employed the term "hygienic instinct." Porfirio Parra, "Pecados mortales contra la higiene," *Revista positiva* 12, no. 12 (December 1901): 500.

10. Georges Vigarello, *Lo limpio y lo sucio: La higiene del cuerpo desde la Edad Media* (Madrid: Alianza Editorial, 1991), 253.

11. Ruiz, *Tratado elemental*, 126–27.

12. Ibid., 134.

13. Ibid., 134.

14. Antonio Velasco, *Medicina doméstica o tratado elemental y práctico del arte de curar: Obra muy importante, útil y provechosa para las familias* (Mexico City: Oficina Tipográfica de la Secretaría de Fomento, 1886), 143, 157, 166, 176.

15. Ruiz, *Tratado elemental*, 128.

16. "Empresa contra la humedad y el salitre." *La Nación*, July 17, 1892, 4.

17. *Estadísticas sociales del porfiriato, 1877–1910* (Mexico City: Talleres Gráficos de la Nación, 1956), 134–36.

18. Vicente Martín Hernández, *Arquitectura doméstica de la ciudad de México (1890–1925)* (Mexico City: UNAM, 1981); Ledislao de Belina, "Influencia del clima en México sobre la tuberculosis pulmonary," *Gaceta Médica de México* 8, no. 13 (May 1878): 267.

19. Rafael Domínguez y Pastor, *Breves apuntes acerca de la higiene del enfermo* (Mexico City: Imprenta de Federico Gayosso, 1896).

20. Domínguez y Pastor, *Breves apuntes*, 1896, 19.

21. Antonio Peñafiel, *Memoria sobre las aguas potables de la ciudad de México* (Mexico City: Oficina Tipográfica de la Secretaría de Fomento, 1884), 125.

22. Máximo Silva, *Higiene popular: Colección de conocimientos y consejos indispensables para evitar las enfermedades y prolongar la vida, arreglada para uso de las familias* (Mexico City: Departamento de Talleres Gráficos, 1917), 573.

23. Ibid., 583–84.

24. Peñafiel, *Memoria sobre las aguas*, 158.

25. Ruiz, *Tratado elemental*, 300.

26. Silva, *Higiene popular*, 463.

27. Berta Tello Peón, "Intención decorativa en los objetos de uso cotidiano de los interiores domésticos del porfiriato," in *El arte y la vida cotidiana: XVI Coloquio Internacional de Historia del Arte*, ed. Elena Estrada de Gerlero (Mexico City: Instituto de Investigaciones Estéticas, UNAM, 1995), 143.

28. Ibid., 91.

29. "La Sala de Baño," *Album de Damas*, no. 19 (October 1, 1907): 14.

30. See the advertisement "El Placer de Tomar un Buen Baño—Tinas Plegadas" (The pleasure of taking a good bath—folding bathtubs), *Album de Damas*, no. 31 (February 16, 1908): 24.

31. Adolfo Prantl and José Groso, *La ciudad de México: Novísima guía universal de la capital de la República Mexicana; Directorio clasificado de vecinos y prontuario de la organización y funciones del gobierno federal y oficinas de su dependencia* (Mexico City: Juan Buxó y Compañía Editores–Librería Madrileña, 1901), 39; Valle Arizpe, *Calle vieja*, 411; Zamora, *Guía indispensable*, 108.

32. Valle Arizpe, *Calle vieja*, 411.

33. Prantl and Groso, *La ciudad de México*, 39.

34. José María Marroquí, *La ciudad de México* (Mexico City: Ediciones La Europea, 1900), 2:261–62, 286; Antonio García Cubas, *El libro de mis recuerdos: Narraciones históricas, anecdóticas y de costumbres mexicanas anteriores al actual estado social* (1904; repr., Mexico City: Editorial Porrúa, 1986), 372; Valle Arizpe, *Calle vieja*, 414–17.

35. Manuel Marroquín y Rivera, *Memoria descriptiva de las obras de provisión de aguas potables para la ciudad de México* (Mexico City: Imprenta y Litografía de Müller Hnos., 1914), 481.

36. Silva, *Higiene popular*, 818.

37. Ibid., 463.

38. "Instituto Hygeia," *Album de Damas* 1, no. 12 (June 1907): 34.

39. Ibid.

40. Ibid., 35.

41. Manuel Antonio Carreño, *Manual de urbanidad y buenas maneras* (1854; repr., Mexico City: Editora Nacional, 1979), 51.

42. Julio Guerrero, *La génesis del crimen en México* (1901; repr., Mexico City: Consejo Nacional para la Cultura y las Artes, 1996), 132–33.

43. Ibid., 136.

44. Ruiz, *Tratado elemental*, 186.

45. Silva, *Higiene popular*, 855.

46. Guerrero, *La génesis*; Ruiz, *Tratado elemental*, 187.

47. Ruiz, *Tratado elemental*, 187.

48. *Diario del Hogar*, January 9, 1891, 4, published in Mexico City.

49. Ibid., January 6, 1891, 4.

50. Ibid., January 13, 1891, 4.

51. Ibid., January 3, 1891, 4.

52. Ricardo Pérez Montfort, "Fragmento de historia de las drogas en México, 1870–1920," in *Hábitos, normas y escándalo: Prensa y criminalidad durante el Porfiriato tardío*, ed. Ricardo Pérez Montfort (Mexico City: CIESAS, 1997), 155.

53. John Womack, "The Mexican Revolution, 1910–1920," in *Mexico since Independence*, ed. Leslie Bethell (Cambridge: Cambridge University Press, 1991), 185.

54. "Bacilos o balas," *El Demócrata*, September 15, 1915, 1. *El Demócrata* was a newspaper published in Mexico City.

55. "La campaña contra el tifo: Urge higienizarnos," *El Demócrata*, December 29, 1915, 3.

56. "La forma segura de evitar la propagación de tifo será la de que se observe la higiene en cada hogar," *El Demócrata*, October 22, 1915, 1.

57. "Cuarenta médicos y numerosos peluqueros combaten al tifo," *El Demócrata*, December 24, 1915, 3.

58. "Iniciativas para coadyuvar a la campaña emprendida contra el tifo," *El Demócrata*, December 25, 1915, 5.

59. José Álvarez Amézquita et al., *Historia de la salubridad y de la asistencia en México* (Mexico City: Secretaría de Salubridad y Asistencia, 1960), 2:44.

60. "El Consejo Superior de Salubridad acepta nuestro ofrecimiento," *El Demócrata*, December 28, 1915, 1.

61. José Maria Rodríguez, quoted in Álvarez Amézquita et al., *Historia de la salubridad*, 2:104.

62. José María Rodríguez, "Informe que rinde el Jefe del Departamento de Salubridad de los trabajos efectuados por el Departamento a su cargo en 1917 al C. Presidente de la República," in *Memoria de los trabajos efectuados por el Departamento de Salubridad Pública en el año de 1917*, ed. Departamento de Salubridad Pública (Mexico City: Imprenta Victoria, 1918), vi–vii.

63. Ernesto Aréchiga Córdoba, "Dictadura sanitaria, educación y propaganda higiénica en el México Revolucionario, 1917–1934," *DYNAMIS* 25 (2005): 117–43.

64. Ibid., 121–22.

65. María Rosa Gudiño, "Educación higiénica y consejos de salud para campesinos en El Sembrador y El Maestro Rural, 1929–1934," in *Curar, sanar y educar: Enfermedad y sociedad en México, siglos XIX y XX*, ed. Claudia Agostoni (Mexico City: IIH–UNAM/BUAP, 2008), 71–97.

66. Anne-Emanuelle Birn, *Marriage of Convenience: Rockefeller International Health and Revolutionary Mexico* (Rochester, NY: University of Rochester Press, 2006).

67. Ibid., 131.

68. Claudia Agostoni, "Las mensajeras de la salud: Enfermeras visitadoras en la ciudad de México durante la década de 1920," *Estudios de Historia Moderna y Contemporánea de México* 33 (2007): 89–120.

69. Manuel González Rivera, "Health Education in Mexico," *American Journal of Public Health* 37 (1947): 850.

70. Álvaro Matute, "De la tecnología al orden doméstico en el México de la posguerra," in *Siglo XX: La imagen, ¿espejo de la vida?*, ed. Aurelio de los Reyes, vol. 5 of *Historia de la vida cotidiana en México* (Mexico City: El Colegio de México–FCE, 2006), 157–76.

# 2
# Cooking Technologies and Electrical Appliances in 1940s and 1950s Mexico

*Sandra Aguilar-Rodríguez*

In mid-twentieth-century Mexico, reformers called for a modernization of daily practices while the domestic technology industry launched publicity campaigns extolling the virtues of its products. Home economics and the science of nutrition provided the knowledge and impetus for adapting industrial technologies, such as refrigeration, to daily life. These technologies were considered the epitome of modernization due to their association with material well-being and national development. Diligent and well-informed mothers were behind the creation of healthy and hard-working citizens, and this process was facilitated with the help of domestic appliances. Therefore, the process of modernization was intimately tied up with the development of consumerism.

As Rita Felski points out, consumer culture "helped to break down fixed and seemingly natural hierarchies which assigned those groups a fixed place in the social order."[1] Consequently, as this chapter demonstrates, electrical appliances not only facilitated daily chores, but also worked as symbols of social improvement. Women learned about new lifestyles and fashion through the mass media, particularly publicity. Class and race played key roles in representations of modernity in Mexico as advertisements portrayed light-skinned women using gas stoves while dark-skinned or indigenous women appeared using traditional cooking methods. As historian Julio Moreno shows, 1940s advertisements defined middle-class status through consumption and lifestyle. Moreover, publicity portrayed US products and brands as Mexican and deployed a nationalist discourse claiming that the progress of the nation depended on people who bought them.[2]

This chapter analyzes women's memories across class to understand how they appropriated or resisted modernizing discourses in their daily use of technology. It explores two of the most life changing food-preserving and food-processing technologies: the refrigerator and the blender. Life stories allow us to penetrate into the past in ways which no other source can do, thus this chapter relies to a great extent on thirty-five interviews carried out by the author in 2005. In recognition of all women who worked hard to raise their families inside and outside their home, I do keep their names, which are given in full in the notes. Along with life stories, this study draws from advertising images published in cookbooks and women's magazines of the period to look at ideal practices, class, and gender perceptions. This work focuses on the capital city, or Federal District, in both its urban downtown and rural outskirts. It also draws comparisons with Guanajuato, the capital city of the eponymous state, located 220 miles northwest of Mexico City. Although there were differences in the experience of women from both Guanajuato and Mexico City, social class was the main factor that influenced women's access to consumption goods and their use of technology.

After the Mexican Revolution, industrialization accelerated, fostering urbanization and migration, particularly to Mexico City. The population of the nation's capital jumped from 1.7 million in 1940 to more than 3 million in 1950.[3] While it was largely urbanized by 1950, rural inhabitants living on the outskirts of the city represented 5.4 percent of the total population.[4] Although internal migration to Mexico City continued apace, by the middle of the century most Mexicans still lived in towns or villages of less than 2,500 inhabitants.[5] As a result, most migrants to the capital city were peasants who took up residence in crowded tenements in the downtown or around the surrounding industrial zone. Upon arrival, most newcomers were lucky to find menial factory jobs or to work as street peddlers or domestic servants.

The transition from subsistence agriculture to a consumerist society was a complex process. Most people changed their practices once they had money or migrated to urban areas, but some preserved their rural lifestyles. Transforming peasants into workers was a challenge that reformers sought to address, in part, through education and welfare. State institutions aimed to "domesticate" rural migrants by teaching them how to live and organize their household. Drawing from the home economics movement, welfare advocates and marketers promoted women's access to science and technology in their efforts at social transformation.

Homemaking became a body of knowledge that required training, which, in turn, discredited female traditional knowledge. The rhetoric of home economics entailed a paradoxical reaffirmation of women's role as housewives. Home economists implied that the ability to do housework was not a "natural" female characteristic. Thus, a healthy and productive family was, in this view, the result of an informed housewife who knew how to manage her household, while an ignorant mother sowed vice and crime as a result of unhygienic conditions and poor diet. Moreover, women were encouraged to use technology at home, just as men were operating machines in factories.[6] Home economics sought to professionalize housework to mirror the world of men, but without transgressing gender roles or social hierarchies.

The adoption of domestic technology in Mexico was a complex phenomenon in which class, gender, and perceptions about tradition and modernity came into play. Time became something that women needed to be aware of, so housewives could maximize their productivity by doing things fast. Space also changed as kitchens had to be adapted to accommodate new appliances. The use of technology transformed cooking and provisioning practices, but this change was not immediate. The aspiring working class bought blenders to prepare tomato sauce, the basis of daily soups and salsas, yet they kept their *molcajetes* to make salsa as a weekend treat.[7] Upper-class women bought refrigerators but did not store cooked food longer than overnight. They used gas stoves but preferred stews that were slow-cooked on charcoal and hand-made tortillas produced by their domestic servants.

Using electrical appliances, however, did not transform gender roles. In fact, pressure on women increased as they were expected to become professional mothers who were knowledgeable about nutrition, health, and home economics, while those in poverty also had to do paid work. The analysis of women's use of domestic technology provides a new window into the politics of daily life, gender, class, and race, as well as the connection between private and public. So let us now enter into the kitchen amid the vapors of boiling beans and the smell of chilies roasting over the *comal*, to find out what women thought.[8]

## The Refrigerator

It was a chilly February morning in 1958. Luisa woke up to prepare breakfast for her ten children, some of whom were about to return to school after the winter vacation. Luisa went to the local dairy to buy a liter of milk, which she added to herbal-infused drinks or coffee so that all her children could drink some. She lifted the pot of beans from a hole she had dug in the kitchen's dirt floor, a practice that served to preserve them, and put the pot on the griddle. She patted tortillas while yesterday's beans were reheating. Preserving food in a cool spot inside the kitchen, in a sink or a bucket with cold water or in a basket hung from the ceiling, was a common practice among women. Domestic appliances remained expensive and beyond the means of most workers and peasants. Moreover, many poor neighborhoods and rural areas did not have a power supply; this applied to 30 percent of households in Mexico City in 1947.[9]

Before having a refrigerator, Dolores cooked every day; and, as food was not abundant, they did not have leftovers. Esther shared that experience: "We were used to going to the market and cooking every day; even nowadays I only use the fridge to keep beans, ham, and cheese."[10] In the 1930s, better-off women had an icebox, but they did not use it to keep food for more than one day. Carolina said, "We used the icebox to keep beans, as usually my mother did the shopping every day."[11] When Esperanza bought a fridge in the late 1950s, her mother still cooked fresh food on a daily basis: "We used the fridge to preserve cheese, cream, milk, or some chorizo, but we never put vegetables inside the fridge as we bought them daily; people did not think about buying five kilos of tomatoes to store."[12] Some upper-middle-class women reported that they stored a week's worth of fruit, vegetables, and meat in the refrigerator, but most of them never used this appliance to store weekly groceries because domestic servants went to the market and cooked every day for them. Ángela said, "We used the refrigerator to preserve milk, cream, and perhaps cheese, but that was it. There was nothing else that could become spoiled as domestic servants cooked fresh food every day."[13]

Since the late 1940s, favored by the Import Substitution Industrialization policy, domestic technology companies established plants in Mexico.[14] In 1948, Industria Eléctrica Mexicana (IEM; Mexican Electric Industry) and Friem manufactured the first Mexican refrigerators. Although some of these companies were subsidiaries of US brands, like IEM of Westinghouse, and most parts and technical knowledge came from the United States, these appliances were advertised as being "made in Mexico." Producers and advertisers deployed a nationalist discourse, suggesting that the purchase of Mexican goods would facilitate the progress of the nation.[15]

Articles and advertisements in women's magazines, newspapers, and cookbooks praised the advantages of having a refrigerator and even a freezer. A 1953 advertisement published in *Madame* magazine claimed that women could save time and money by storing food in an IEM freezer: "You can buy meat, fish, vegetables, and fruit wholesale and keep it for months and months without losing the food's consistency, color, vitamins and flavor. You can also freeze cooked food and serve it as fresh and tasty as if it had just been served out of the pot."[16] In 1952, an article published in *Mujer* magazine claimed: "Ice boxes and refrigerators are the most hygienic place to preserve meat. For this reason, housewives should consider the expense of buying either of them as it would guarantee a healthy diet for their family."[17]

IEM refrigerator's handbook stressed that the investment families made in buying a refrigerator benefited their health, time, and budget: "Enjoy a better life. IEM refrigerator is big enough to accommodate a week's worth of groceries. It is modern and large. It contains all the accessories to give you a better life for years. You will have everything at hand straightaway. From now on your dream of serving fresh, nutritious, and healthy food to your family will come true. At any time and within just minutes you could arrange a plentiful and delicious meal. You will save money as you will not throw away spoiled food rotted by bacteria."[18] A refrigerator was, however, too expensive for most people and, therefore, inaccessible to most housewives. In 1951, a Philco refrigerator cost 3,995 pesos, while the average minimum wage per day was 3.35 pesos in urban areas and 2.66 pesos in rural areas.[19]

The introduction of refrigerators varied according to the economic situation of each family; the better-off families bought them in the late 1930s, while working-class households did not acquire them until the 1970s. Although Jeffrey Pilcher and Lilia López point out that "Mexicans used refrigerators to store soft drinks and beer instead of a week's worth of groceries," most interviewees recalled that they stored dairy products, beans, and leftovers.[20] This difference could be gender-biased as men recalled using the refrigerator to store beer while women mentioned food first. Although changing provisioning or cooking practices took decades, having a refrigerator facilitated the modernization of the kitchen and its practices. Refrigerators allowed women to prepare jelly or cold milkshakes, they could store food for longer periods, and they had the option of shopping only once a week.

Chepina (b. 1930), a middle-class woman, recalled, "In the 1950s I went to the Juárez market once a week to buy groceries. I had a refrigerator, so I could keep everything there. My weekly shopping infuriated my mother-in-law, as she was used to buying fresh foodstuffs every day. But my husband did not have another option."[21] In the 1940s, Aurora, a middle-upper-class housewife from Guanajuato, also did her shopping every weekend and stored groceries in the refrigerator. Aurora recalled that her husband was not particular about food, so her weekly shopping did not represent a problem to him. Nevertheless, storing cooked food or groceries was an unusual practice because most women shopped and cooked on a daily basis. But still, by 1960 the majority of women in Mexico could not afford this technology. Refrigerators continued to be a class marker that symbolized an improvement in a family's living standard.

Blenders were more prevalent than refrigerators because they were cheaper and did not require constant power, which diminished the cost of electricity and the chances of breaking down due to a change in voltage. But the main reason for becoming the most popular domestic appliance was that blenders facilitated

Mexican cooking as they were used to prepare tomato and chili sauce and the ubiquitous salsas. Therefore, blenders simplified traditional cooking while favoring the creation of new dishes and drinks such as *licuados* (fruit shakes).

## The Blender

Children ran out of their classroom after another school day came to an end. Esperanza, a schoolteacher, picked up her books and walked home. She passed by Salinas y Rocha, a furniture and electrical appliances store, and stopped to look at the shiny refrigerators, blenders, mixers, and pressure cookers showcased in the store window.[22] After buying a gas cooker, she wanted a blender for her mother, so in 1956, Esperanza bought an Osterizer. She did not know how it worked, but a sales clerk showed her and gave her a booklet. Esperanza found the blender very easy to operate because it only had one button.

Carolina recalled that in the 1940s her brother told her, "Soon you will stop using the stone [to grind], a machine in which you can grind everything will come."[23] Carolina did not believe him, but soon after she heard about blenders. According to Anahi Ballent, the blender was the most important appliance of the late 1940s because its low price made it accessible to most budgets.[24] But the majority of my interviewees did not have their first blender until the 1950s or even later. Nevertheless, since the 1940s advertisements extolled the virtues of blenders in women's magazines, newspapers, cookbooks, and on the radio.

In 1950, Josefina Velázquez de León, renowned chef, teacher, and author of several cookbooks, published a two-volume guide entitled *Cómo cocinar en los aparatos modernos* (How to cook using modern appliances).[25] In the first volume she claimed that her work was a response to questions from several women about how to use kitchen appliances. She therefore described the use of pressure cookers, ovens, mixers, and blenders. She informed readers about how to use a Birtman blender, the brand that probably sponsored her book along with other companies, and gave recipes such as mayonnaise, butter, baby food, sauces, shakes, fruit juices, and cocktails. Women could also use blenders to grind coffee, rice, wheat, maize, and nuts. In her book, Velázquez de León included several illustrations drawn by herself. In the blender section she portrays an experienced cook teaching her students the advantages of a Birtman. Pupils were young and elegant women who listened attentively to their teacher, who was also a young women wearing an apron and a chef's cap. The teacher is behind a kitchen table that looks like a desk, where the blender is located, just at the center of the image.[26]

Velázquez de León argued that elaborate dishes could be cooked easily and in less time with the help of a blender. She claimed that Birtman blenders made it possible to cook complex desserts such as those prepared by our grandmothers and even by nuns in Colonial convents. Another illustration shows how Velázquez de León integrated traditional cooking and modern technology by juxtaposing the image of a professional cook dressed in an apron and chef's cap using a blender with the image of a nun wearing her religious outfit while preparing food by hand. She insisted that electrical and modern appliances allowed housewives to preserve traditional cooking, bringing it to everyday meals. According to Velázquez de León, modern women had to use electrical appliances to keep cooking what their ancestors cooked. Modern Mexico had to preserve its traditional culture but reproduce it with the aid of technology. Velázquez de León's discourse shows, in Ana María Alonso's words, that tradition is "a form of understanding and legitimating the present in relation to a past that is never passively reproduced but always actively produced in relation to the 'modern.'"[27] Although modern women were trained and informed about hygiene and nutrition, this knowledge was meant to be used at home to benefit the family.

Another image in this cookbook portrays an old, fat, and untidy woman using her hands to mix ingredients while a well-groomed cook pushes a button without touching the mixture. Hence, with the help of blenders, women cooked in less time and with less effort. Consequently, they were not so tired and had time to look after their appearance and be rested and refreshed to welcome their husband or guests. This image also reveals social and racial differences. Grinding or mixing with the hands was portrayed as an unhygienic and uncivilized practice identified with the poor, who usually had indigenous features. In contrast, those who used modern appliances had fair skin and possessed knowledge about hygiene and money to buy domestic technology. Cookbooks and advertisements reproduced and took advantage of racial and class stereotypes, promoting the idea that upward mobility depended on the acquisition of electrical appliances. At the same time, the mass media reinforced gender roles as women remained responsible for food preparation.

Blenders were described as a blessing for modern housewives, either for those who had to cook by themselves or for those who had domestic servants.[28] A Waring advertisement portrayed a light-skinned and stylish woman posting a sign stating, "Looking for a domestic servant." A servant, featured as a short, dark-skinned woman with indigenous features, black hair, and squinted eyes, replied, "'Everything is very nice, the room, the radio, the holidays . . . but you do not have a Waring. I also need some help and a Waring makes sauces, purées, baby food, and all sorts of dishes.' So they bought her a Waring blender."[29] The media portrayed having a blender as a way to attract or keep domestic servants at work. Servants, however, were not always eager to change their metate for a blender. According to Chepina, in the 1940s, her "mother's cook made a drama out of using the blender. Doña Chole liked to make *pacholas* (minced meat steaks) in her metate, so she never used the blender. She stuck to her metate and molcajete."[30] Esperanza recalled that her mother kept using her molcajete to prepare salsa because tomatoes, onions, and chilies ground with a mortar stone had a different flavor and texture from those prepared in a blender.

Catalina, from Guanajuato, recalled that her husband did not like to have his salsa ground in a blender: "He said that it did not have the same flavor, so I made his salsa always in the molcajete."[31] Men's hostility toward blenders and, particularly, their refusal to eat salsa that was not prepared in a molcajete expressed not only their yearning for traditional flavors, but also gender anxieties. The use of electrical appliances simplified women's tasks, which apparently reduced the time they spent in the kitchen. Men feared that without the discipline of the metate and molcajete women would go astray; so in demanding certain dishes and textures, men tried to control women's activities.[32] For those who grew up eating freshly cooked food or food prepared in metates and molcajetes, there was a clear difference. Most women, however, valued blenders and refrigerators because this technology simplified their daily tasks.

## Conclusions

In the 1940s, the Import Substitution Industrialization program was launched using a nationalist discourse that encouraged consumers to buy products made in Mexico. Domestic appliances also had a symbolic value as they were identified with material progress, social mobility, and modernization. Women became instrumental in modernizing family life due to their role as consumers. Although not all women acquired appliances for themselves, as on many occasions young men and women bought refrigerators or blenders as presents for their mothers, women were nevertheless the target of publicity campaigns. The domestic technology industry and its advertisements associated modern appliances with rationalization, organization, and control. Furthermore, using electrical appliances, as the media suggested, resulted in a hygienic and sanitized kitchen and a better life.

Using technology at home symbolized a middle-class lifestyle. Advertisers portrayed owners of electrical appliances as light-skinned, well-groomed, urban women, in opposition to dark-skinned, rural, and untidy women. Marketers implied that to move up the social ladder women had to own a blender and a refrigerator. Advertisements thus linked modernity and material progress with Western (European and United States) practices, while indigenous and peasant lifestyles were identified with poverty and backwardness. Most advertising agencies producing campaigns for the Mexican market were based in the United States, so they tended to produce translated but otherwise unchanged versions of advertisements already published in the US.[33] Portraying light-skinned and even blonde women, however, worked as a marketing strategy in itself in a nation like Mexico, where race and class are almost interchangeable.

Therefore, the modernization process did not imply a challenge to social hierarchy or the refusal of tradition, which had different meanings across class. Working-class women felt relieved once they could grind chilies in a blender, while upper-class women continued eating salsa ground in a molcajete by their domestic servants. New appliances generated new traditions, such as preparing *licuados* or milkshakes. In making different use of electrical appliances and giving them a particular meaning, women generated their own view of modernity. The rhetoric of modernization, however, underscored that women's place was at home and that their main responsibility was bringing up healthy and hardworking citizens. The media depicted kitchen and housework as "naturally" rewarding to women, both emotionally and aesthetically. Good mothers who were concerned about catering to their family deserved the help of technology. The time women saved had to be employed in doing more housework, cooking elaborate dishes, looking after children, and making homes comfortable for men.[34] Women had to participate in modernizing the nation but without transgressing established gender roles. They had to collaborate in the construction of a modern nation from home.

The modernization of the household entailed a change in terms of domestic technologies and cooking practices; however, this process did not seek to alter gender roles. Patriarchy was reinforced by the media and by state rhetoric by underscoring the importance of motherhood as women became responsible for improving the living standards of their families and, eventually, of the country. Housewives learned through formal education, cookbooks, and the print media how to imitate a living style that was portrayed to be superior to their own way of life. But women's memories revealed that using domestic technology and adopting new cooking practices took longer than what reformers, marketers, and the domestic technology industry expected. After the 1960s, the number of working women grew, generating deep changes in cooking and eating practices. Although by then appliances such as refrigerators and blenders were owned by most working-class households in urban areas, they did not always bring a better living standard or a more healthy diet. Nevertheless, domestic technology did facilitate women's daily chores, particularly for those who could not afford domestic servants. One of my interviewees remembered the day she received a blender as "the day I reached heaven."

## Notes

I would like to thank Patience A. Schell and Penny Tinkler for their guidance and support throughout my years at the University of Manchester. Paulo Drinot and Rebecca Earle provided excellent feedback and advice on this piece of work. Jeffrey M. Pilcher, whose work inspired my scholarship, has been a great mentor. The University of Lehigh granted me a fellowship which allowed me to start working on this chapter.

I am grateful to Moravian College, particularly my colleagues at the History Department, for being very supportive of my work. I am also grateful to Araceli Tinajero and J. Brian Freeman for inviting me to collaborate in this book and for their comments and suggestions.

1. Rita Felski, *The Gender of Modernity* (Cambridge: Harvard University Press, 1995), 62–88.

2. Julio Moreno, *Yankee Don't Go Home! Mexican Nationalism, American Business Culture, and the Shaping of Modern Mexico, 1920–1950* (Chapel Hill: University of North Carolina Press, 2003), 135, 212.

3. The population went from 1,757,530 inhabitants in 1940 to 3,050,442 in 1950. Dirección General de Estadística, *Séptimo censo general de población* (Mexico City: Secretaría de Economía—Dirección General de Estadística, 1950); INEGI, *Anuarios estadísticos de las entidades federativas* (Mexico City: INEGI, 1998).

4. Dirección General de Estadística, *Séptimo censo general de población*; INEGI, *Anuarios estadísticos de las entidades federativas*.

5. Ibid. Fifty-seven percent of the population lived in rural areas.

6. Harmke Kamminga and Andrew Cunningham, *The Science and Culture of Nutrition, 1840–1940, Clio medica* (Amsterdam: Rodopi, 1995), 32.

7. *Molcajete* is a three-legged basalt mortar with a pestle, called a *tejolote*, used for grinding.

8. *Comal* is an earthenware griddle.

9. Banco Nacional Hipotecario Urbano y de Obras Públicas, "El problema de la habitación," *Espacios*, no. 2 (1949): n.p.

10. Esther Bonilla (b. 1925), interview by the author, Mexico City, August 24, 2005. All translations are mine.

11. Carolina Basave (b. 1930), interview by the author, Mexico City, August 31, 2005.

12. Esperanza Sortibrán (b. 1934), interview by the author, Mexico City, October 14, 2005.

13. Ángela Malo (b. 1925), interview by the author, Guanajuato, October 27, 2005.

14. Lorenzo Meyer, "La encrucijada," in *Historia General de México*, ed. Daniel Cosío Villegas (Mexico City: COLMEX, 1998), 1280.

15. In 1949, an advertisement by the National Economic Movement claimed: "There was a time in which Mexicans preferred to buy goods with foreign names. That is why we find several brands that seem to be foreign such as Walter, Metropolitan Mills, Cadillac, Helen Harper, Jarman, and Boston, but in fact they are Mexican. From now onwards, when buying such goods Mexicans will know they are buying goods 'made in Mexico.' To the confidence people already had in these brands, it will be added the satisfaction of buying Mexican products which contribute to the progress of our nation." See the magazine *La Familia*, Mexico City, no. 350, year 17, December 30, 1949, n.p. On the Mexicanization of US brands and companies, see Julio Moreno, "J. Walter Thompson, the Good Neighbor Policy, and Lessons in Mexican Business Culture, 1920–1950," *Enterprise & Society* 5, no. 2 (2004): 254–80; Moreno, *Yankee Don't Go Home!*

16. *Madame* (Mexico City) 3, no. 24 (July 1953): n.p.

17. *Mujer: Belleza, cultura, honestidad*, Mexico City, no. 15, January 16, 1952, 40.

18. IEM, *Guía y recetario del refrigerador IEM* (Mexico City: IEM subsidiary of Westinghouse, [1950?]).

19. *La Familia*, Mexico City, no. 397, year 20, December 15, 1951. Philco deluxe refrigerator was sold at Fábricas Universales, a department store. It had a five-year warranty. Presidencia de la República, *50 años de Revolución Mexicana en cifras* (Mexico City: Presidencia de la República–NAFINSA, 1963), 112.

20. Jeffrey M. Pilcher, "Industrial Tortillas and Folkloric Pepsi," in *Food Nations: Selling Taste in Consumer Societies*, ed. Warren Belasco and Philip Scranton (New York: Routledge, 2001), 222–26. López Ferman also mentioned how men filled fridges with beer and soft drinks. Lilia Isabel López Ferman, "Las cosas que nos transformaron: Usos, apropiaciones y significados de la tecnología doméstica en la ciudad de México (1940–1970)" (MA diss., Centro de Investigaciones y Estudios Superiores en Antropología Social, 2005), 102.

21. Chepina Peralta (b. 1930), interview by the author, Mexico City, October 13, 2005.

22. Salinas y Rocha was founded in 1906 as a furniture store. It was famous for selling domestic appliances.

23. Carolina Basave (b.1930), Mexico City, August 31, 2005.

24. Anahí Ballent, "La publicidad de los ámbitos de la vida privada: Representaciones de la modernización del hogar en la prensa de los años cuarenta y cincuenta en México," *Alteridades* 6, no. 2 (1996): 57.

25. On Velázquez de León, see Jeffrey M. Pilcher, "Josefina Velázquez de León: Apostle of the Enchilada," in *The Human Tradition in Mexico*, ed. Jeffrey M. Pilcher (Wilmington, DE: SR Books, 2003).

26. Josefina Velázquez de León, *Como cocinar en los aparatos modernos*, vol. 1 (Mexico City: Academia de Cocina Velázquez de León, 1950), 254. In 1950 a Birt-

man blender cost $250. José Luis Juárez López, "Innovaciones en la cocina," *Cuadernos de Nutrición*, no. 2 (2001): 56.

27. Ana María Alonso, *Thread of Blood: Colonialism, Revolution, and Gender on Mexico's Northern Frontier* (Tucson: University of Arizona Press, 1995), 10.

28. For more advertisements, see *La Familia*, Mexico City, no. 357, year 17, April 15, 1950.

29. *Madame* (Mexico City) 1, no. 1 (September 1950).

30. Chepina Peralta (b. 1930), interview by the author, Mexico City, October 13, 2005.

31. Catalina Amador Espinoza (b. 1937), interview by the author, Guanajuato, October 26, 2005.

32. Jeffrey M. Pilcher, *¡Que vivan los tamales! Food and the Making of Mexican Identity* (Albuquerque: University of New Mexico Press, 1998).

33. On the history of J. Walter Thompson, see Moreno, "J. Walter Thompson, the Good Neighbor Policy, and Lessons in Mexican Business Culture."

34. Ruth Schwartz Cowan, *More Work for Mother: The Ironies of Household Technology from the Open Hearth to the Microwave* (New York: Basic, 1983).

# 3
# Domestic Technologies
## Gender, Technology, and Mexican Housewives, 1930–1950

*Joanne Hershfield*

The city center of Mexico City in the early 1930s was a bustling metropolis, an eclectic mix of Colonial cathedrals and government palaces, art deco bank facades, modern department stores, traffic jams of cars, trucks, and buses, all fighting to cross cobblestone streets designed for horse and buggies and pedestrians, *vendedores* selling fruits and vegetables, and signs advertising US-made products such as cars and home appliances.[1] Old buildings were constantly being razed, and contemporary architects were hired to design new hotels and modern buildings to house local and international businesses. Modern suburban residential development projects for middle- and upper-class families, which drew on designs borrowed from the housing boom in the southwestern United States, sprung up in the 1920s and expanded throughout the '30s and '40s. Patrice Elizabeth Olsen describes a number of these developments, including the *colonias* of Chapultepec Heights and Colonia Reforma. Lot sizes ranged from 800 to 1,000 square meters, and homes included running water and sophisticated sewage systems. Architects of these projects built family homes with European-influenced art deco facades or in the "California Colonial" and "Mission" styles favored in California.[2] These houses were decorated according to the latest international fashion and filled with all the trappings of the modern home, including the new technology of mechanical and electric domestic appliances such as refrigerators, gas and electric stoves, vacuum cleaners, and radio consoles.

The availability of running water and electricity and widespread and affordable access to oil aided the diffusion of new home technologies in the United States, Great Britain, and other western European nations after World War I. Although electric appliances were available for home use in the early 1900s in

Mexico, most homes were still not wired for electricity by the 1930s, and in 1940 "in small towns and in the countryside, the dark of night enveloped everything."[3] In the shadow of the metropolitan milieu described above, the majority of urban dwellers lived in crowded and unsanitary *vecindades* (tenements) that ringed the city center, lacking access to electricity, running water, sewer systems, and other forms of modern technology that reduced the incidence of disease and chronic death. Only a very small percentage of the urban population had access to or could afford to purchase energy in the form of electricity, oil, and running water.[4] However, as early as the 1920s, middle- and upper-class Mexican women who could afford them would have been able to purchase household appliances such as *planchas eléctricas* (electric irons), *hornos eléctricos* (electric ovens), and Frigidaires (electric refrigerators) that were guaranteed to be safe, economical, and time-saving.

As Sandra Aguilar-Rodríguez argues in her chapter in this volume, the introduction of domestic technologies for the home symbolized "social improvement" for many women, a democratization of housework. The introduction of new stoves, refrigerators, and sewing machines was not limited to urban households. Even though the reality of the experience of the majority of women in Mexico during these decades was that they worked either for wages in factories or in subsistence agriculture, some families in villages and small towns were able to purchase *molinos*, hand-operated corn-grinding mills for the making of tortillas.[5] In a description of Tepoztlán material culture, for example, Robert Redfield writes that "a conspicuous minority" of households "enjoy elements of modern industrial civilization"; but he also notes that the sewing machine "has become a part of the general Tepoztecan material culture; it is found in all parts of the village and in houses otherwise Indian in character."[6] The ability to purchase a home sewing machine offered many women in villages and urban centers the opportunity to partake in the "piecework" industry, providing outside income to support their families.

When President Lázaro Cárdenas took office in 1934, Mexico was beginning to recover from a period of economic downturn and the majority of the economically active population still worked in agriculture. The following two decades celebrated recovery and growth and rising consumer incomes so that by 1950 almost as many people were employed in industry, mining, services, and the public sector.[7] Economic and industrial expansion resulted in the growth of consumer culture and the promotion of new technologies and practices of domesticity that were shaped by global discourses of the modern woman that sometimes conflicted with conventional ideologies of woman as the keeper of hearth and home and a civilizing influence in the face of modern ideas that threatened the sanctity of the family. Although the state had a stake in promoting consumer culture, many of the policies it enacted in the labor sector tacitly reinforced traditional notions of gender.[8]

While Cárdenas promoted industrial and agrarian reform under the mantle of "revolution," his modernizing projects aimed at the rural population concurrently reinforced the conservative, Catholic ideology of the family asserted during the Porfiriato that insisted that in order to preserve the integrity of the family, the woman's place was in the home. An editorial, "La Madre: Alma de las Sociedades," in *Todo*, one of the prominent Mexico City weeklies, articulates the dominant discourse that reinforced the importance of the mother's role to the success of the nation. Defining the home as "both a temple and a school" and the mother as "the priest and the preceptor" of the family, the article proclaims her as "the moral force in the development and education of the family . . . the grand motivator of the community, the crux of its metaphysical rotation and the neural center of its activities."[9]

The domestic sphere was articulated as central to the welfare of the nation, and housewives were given a central role in the reproduction of the family's health, safety, and emotional well-being. State policy, public discourse, and private industry participated in the advancement of a nationalist ideology that characterized a woman's primary role as manager and caretaker of home and family. In the early 1920s, the federal government sponsored and promoted vocational education for women, opening schools across the nation, ostensibly to provide them with a skilled trade. Patience A. Schell writes that these schools "sought to train women for real life while using the model of the revolutionary family" in which the woman "was a selfless, prudent, dedicated housewife and mother who did not have to engage in paid work because her sober, industrious husband supported her on this family wage." In Schell's analysis, this meant that, "for women, a skilled trade meant an activity shoehorned between domestic chores to generate supplemental income."[10] Classes were offered in cleaning, cooking, and sewing; and although these skills could be translated into wage-earning work as a domestic servant, the primary objective was to promote the model of the middle-class citizen housewife.[11]

Despite the official rhetoric that reinforced traditional, conservative ideologies of gender, my analysis of the promotion of domestic technologies shows that discourses of modernity and the modern woman did reshape public conceptions of gender and gender roles in significant ways. Here, I take issue with Aguilar-Rodríguez's conclusion, articulated in her contribution to this volume, that the modernization of the household "did not seek to alter gender roles." Although "concern with the role of women as mother, nurturer, and protector of moral values was central to the formation of the new nation" and to the imperatives of the marketplace, it is evident that the discourse of modernity provided women with opportunities to refashion, challenge, or reject those traditional roles.[12] An advertisement for "the electric living room," for example, features a modern woman dressed for going out, wearing a knee-length daytime dress, high heels, a fur stole, and a cloche hat over her bobbed hair. She is surrounded by smaller tableaux of family life that show her cooking on an electric stove, vacuuming, ironing, reading by electric light, and serving coffee to her husband from an electric coffee pot. The contrast between the dutiful housewife and the modern woman of leisure reminds women that "electricity is the most valuable aid for the woman of the house" because it allows her to complete all her tasks and still have free time to spend on other activities such as shopping or lunching with her friends.[13]

Socioeconomic progress, industrialization and the emergence of industrial systems that favored mass production and the division of labor, and the burgeoning growth of urban space describe the major processes of modernization in Mexico during the decades following the Mexican Revolution. The introduction of new technologies and their dissemination in the workplace, public space, and the home prompted vigorous public debates in relation to commerce and public policy, but also with respect to cultural conventions and ideologies. Technological change confronted people's values and beliefs about the world, challenging deeply felt ideas about what was possible and what was "right." This debate circulated information and advanced particular interpretations about the purpose and meaning of technology in everyday life. The state and private industry sanctioned technology for its progressive possibilities in the realm of nation building and the expansion of national trade, but it figured also in public proclamations about the importance of the home in the progress and health of the nation. As one journalist put it in a column entitled "El Hogar y la Economía," which was addressed to housewives and appeared regularly in the popular weekly magazine *Hoy*, "Do you know that the future of our country is in our hands? Because we manage and use more than half the national wealth for domestic needs, exactly 64%. There is in the Mexican Republic no group of workers so big, so numerous, as the group that we housewives make up, and our action is continuous and is updated daily."[14]

Thus, one area in which public debates about technology reached a fevered pitch in Mexico was that of the restructuring of the meanings and material practices related to domesticity. Domesticity is a system that is about homemaking and housework. It is a function of historically contingent aspects of culture, technology, commerce, and, most significantly, gender. The "job" of the housewife was redefined through this process, producing new kinds of identities related to femininity. The new housewife was imagined not only as a wife, mother, and homemaker, but also as the major household consumer. In order to create the ideal home, she was expected to educate herself on subjects related to the family's well-being, including health, hygiene, child-rearing, cooking, decorating, and shopping. Each of these subjects required her to learn a new set of skills, one of which was the operation of mechanical and electric household appliances or "domestic technologies." While the dominant political and social discourse may have privileged conservative ideals of domesticity, the commercial sector of public life challenged those ideals to a certain extent. Advertising campaigns addressed women as mothers and wives, promising that electric appliances would bring cleanliness and orderliness to the home; but they also appealed to women's sense of aesthetics and pleasure and identified them as knowledgeable, skillful, and technically competent.

In her 1991 overview of the literature on the history and sociology of technology and modernity, Judy Wajcman argued that what was missing was the dimension of gender. In much of this literature "the traditional conception of technology too readily defines technology in terms of male activities," ignoring the contributions women have made to "the invention of such crucial machines as the cotton gin, the sewing machine, the small electric motor," and other important technological advancements.[15] Feminist revisions brought a feminist perspective to technology studies, arguing, for example, that "Western technology itself embodies patriarchal values."[16] However, Wajcman finds some of this work to be essentialist in its calling for a technology based on women's values, as well as overly deterministic in its interpretation of technology as an agent of modern social change.

In her later book, *TechnoFeminism*, Wajcman expands her critique of feminist essentialism and technological determinism. She identifies "gender and technology [as] relational categories" that are mutually constitutive and co-configured in social relations.[17] In order to understand technology and gender in their constitutive relationship, she argues, we need to expand our understanding of technologies as merely a set of "objects" or things and consider technology as implicated within a broader framework of interconnecting social systems. According to Wajcman, "a technological system is never merely technical: its real-world functioning has technical, economic, organizational, political and even cultural elements. . . . Gender roles and sexual divisions are part of the socio-technical system or network."[18] The relationship between gender and technology in the context of modernity is co-constructive, and their relationship and mutual influence are fluid and historically specific. Technological innovation does not necessarily transform gender relations. As Barbara Marshall puts it, "the philosophical and institutional transformations of modernity, including its technological dimensions, were not just mapped onto already existing gender differences, but actively constructed and invoked difference."[19]

The co-constructive nature of technology and gender are central to understanding the transformation of domesticity in the early decades of the twentieth century in the Western world. Domesticity is most productively understood not as a set of practices, but as "*gynotechnics*: a technical system that produces ideas about women, and therefore about a gender system and about hierarchical relations in general."[20] Francesca Bray, who coined the term "gynotechnics," describes this system as one in which gender ideologies and social relations of power shape technology and its uses. In turn, technology participates in the redefinition of gender and gender roles and relations. According to Bray, technology is a form of cultural expression, and gynotechnics communicate and transmit gendered ideologies of everyday domestic technologies as they are articulated within local contexts.

In the following discussion, I explore the gendering of modern domestic technologies in Mexico from the 1930s through the early '50s. I consider the mechanisms through which marketing strategies worked alongside and sometimes in disagreement with official and unofficial policies of political nationalism to shape ideas about a woman's role as a housewife and purchasing agent for her household in the context of the emergence of consumer culture. Because the intention of advertisements is to sell products, promotional campaigns also sought to change women's behaviors in relation to consumption practices as well as housekeeping habits. In the following discussion, I look at advertisements promoting household appliances and articles addressed to women in the popular press that commented on issues related to domesticity. I identify a distinctive model of the modern homemaker that was produced through discursive strategies: an independent, intelligent, middle-class woman, devoted to home and family, who possessed the resources and talents to modernize the family's household and adequate leisure time to undertake this transformation.

Recent studies of domesticity challenge conventional ideologies of separate spheres, wherein the "home as a bounded and rigidly ordered interior space is opposed to the boundless and undifferentiated space of an infinitely expanding nation." Women had a role to play in the reconstruction of the nation through their civilizing influence, their position as the family's primary consumer, and as a "home worker," in charge of industrial and specialized housework.[21] Public discourse conceptualized the modern Mexican home as the center of a woman's life: clean and well ordered, tastefully decorated, and professionally managed by the housewife. The promotion of household technology promised women that they would be able to transform their lives through a reduction of housework in ways that were more efficient and hygienic.

Discourses of domesticity enabled certain domestic practices while constraining others. They emphasized novelty, luxury, and individual expressiveness by appealing to individual desire and fantasy and drew from scientific and business discourses of rationality and efficiency. Scientific discourse made the home into a workplace and highlighted professionalism, efficiency, organization, hygiene, technology, and progress. Ads configured women as conversant with the operation of technology, a role that challenged conventional notions of femininity. Promotional campaigns and weekly advice columns conceived of the modern homemaker as a model wife and mother, but also as a businessperson and an expert in the modern, economic systems of domesticity.

A 1938 advertisement for El Nuevo Compuesto Vegetal de Lydia E. Pinkham features "the most talked about woman in the world," the modern housewife who does everything. She is around thirty years old, she is pictured holding a broom and wearing an apron, and she is reminded that "no one is a wife half-time, no one is a mother half-time."[22] The ad articulates the dominant national discourse while also serving a purpose for the company that manufactured Lydia E. Pinkham products. This job required a full-time commitment on the part of women, and in order to fulfill the job's demands, housewives were required to purchase the accoutrements of homemaking. The family home was envisioned as the primary socioeconomic unit and organized as a place where women could realize individual creativity and initiative, and housework was given more status in the public eye. It was both a space of production—of children and social values, for example—and of consumption of goods and services. And the housewife was elevated to the position of manager of both practices.

Another advertisement in a 1928 issue of *Mexican Folkways*, "La electricidad es el auxiliar más valioso de las señoras de casa," touted the importance of electricity for the modern home. The ad features a woman dressed in a knee-length daytime dress, high heels, a fur stole, and a cloche hat over her bobbed hair, ready for a day of shopping or lunching with friends. She is surrounded by a series of images of family life that picture her cooking on an electric stove, vacuuming, ironing, reading by electric light, and serving coffee to her husband from an electric coffee pot. This advertisement suggests that thanks to the wonders of electricity, the housewife is free to expand her identity in the public sphere. One might argue that the *Folkways* ad demonstrates two dominant yet seemingly contradictory narratives of the modern Mexican housewife: one bound by the traditional discourse that confines her in the domain of private domestic space and the other enabling her newfound freedom to participate in a public sphere as a consumer.[23] Interpreting the ad through a gynotechnics lens, we can see the complexity of a system's operation in which the various elements work in constant tension as part of an ongoing process in which gender, as well as technology and the nexus of private/public, is continually reconceptualized.

Advertisements and articles written in popular magazines are, of course, not indicators of actual material practices that took place in Mexican households during the 1930s and '40s. They do, however, reveal how the private sector worked to "constrain or facilitate the development of particular gendered identities, agency, and contexts" in the realm of the marketplace as well as in that of ideologies.[24] While there is no way to know what choices women actually made, we can analyze the various levels of mediation, paying particular attention to the visual and narrative strategies that advertisers employed in order to understand how domestic technologies came to be interpreted in gendered ways.

The promotion of household technology played a major role in the redefinition of domesticity and in the production of a modern domestic feminine subject, providing information and advice that informed housewives about the knowledge and skills they would need to become competent professionals. An article addressed to women, "Charlas Femeninas: Orden y Economía," which appeared in a 1938 issue of the popular weekly magazine *Hoy*, emphasizes the need for new wives to learn "domestic economy," described as the "prudent management of the resources that are necessary for the upkeep of the home." The article defines domestic economy as a "scientific practice" that requires a wide-ranging knowledge of "the kitchen, the pantry, and laundry." It advises housewives that "knowing what to buy constitutes a true science that involves tact, discretion, patience, and careful observation." The housewife is also counseled to develop industriousness and good taste and to learn how to conserve resources and repair domestic equipment "in anticipation of bad times."[25] At the same time, the article imagines the modern housewife as a business person, an expert in the modern, economic systems of domesticity, and as conversant with the operation of technology, roles that challenged conventional notions of femininity by educating women on the use of electrical appliances that facilitated modern methods of accomplishing household tasks.

This new gynotechnics of domesticity did not necessarily mean that Mexican women's roles in the home changed as a direct effect of the introduction of new home appliances. As discussed above, conservative ideologies of family life remained stubbornly intransient. And as Ruth Schwartz Cowan and others have demonstrated, although promotional campaigns often promised women that technology would translate into less work for housewives, in fact, studies revealed "rising expectations of the housewife's role which generated more domestic work for women."[26] However, the ideal of the modern home as a *technological* home did contribute to transforming what it meant to be a good housewife and a good citizen in the context of the modernizing, revolutionary state and the growing influence of a culture of consumption. As I argue above, conservative ideals of gender roles were challenged by alternative discourses that required women to expand their knowledge and their skills in order to manage the modern household. Homemaking was conceived of as a science that was "useful and beautiful." It required women to learn the "laws of boiling and thermal conductivity, elementary electricity, chemistry . . . and the phenomena of combustion, the natural sciences the reveal the secrets of hygiene and cooking."[27]

Domestic spaces came to be understood as private, family spaces as opposed to public spaces occupied by crowds of unrelated people. The home was a feminized space of consumption—of food and other commodity products. Within the space of the home, practices informed by social discourses of gender took place. One of the inventions of the twentieth century, for example, was the idea of the "family kitchen" as a distinct space, modernized through the introduction of technological inventions such as electricity and running water. This space became the center of women's lives, despite the fact that most Mexican homes, especially rural homes, did not have separate rooms for cooking. But the modern kitchen was not only a room for cooking, it was a feminized location in which women were pictured acting out the socially prescribed role of housewife. And, above all, it was a productive space, the primary location of "women's work"—women were always pictured in the kitchen cooking or cleaning.

Earlier advertisements released during the first two decades of the century generally pictured only the technological artifacts, an electric stove or an electric flat iron, for example. By the late 1920s, the majority of visual ads featured images of women using this technology in a way that imagined these artifacts as an extension of the female body, a kind of cybernetic organism whose function was to cook, clean, and sew—a woman's body aligned with technological artifacts. While the female body was often pictured as the user of home technology, the vacuum cleaner or the electric or gas stove, it frequently served as a symbol of technology itself, a "mythic embodiment of the modern electric age."[28] In an advertisement for Solar light bulbs, the face of a beautiful woman is superimposed onto the bulb itself, envisioning a modern, gendered, technological subjectivity, the "electric woman."[29] The woman/light bulb serves as an allegorical figure: radiant, luminescent, and glowing, the "light" of the household.

Arguably, one of the most widely used metaphors was that of beauty and elegance. A full-page advertisement for RCA Victor radios in *Hoy* features photographs that promote three of the forty-six "wonderful models" available for 1939: the "Radio Consola," the "Super-Consola," and the combination "Radiofonógrafo." Each model is promoted for its distinctive capabilities and is shown next to a different female figure. A seated woman in an elegant halter-topped evening gown leans against the Radio Consola that is described as "captivatingly beautiful" but also is an example of electronic sophistication: the front of the console features six bulbs and a "three-band tuner" that provides improved electrical tuning, a button that allows for listening to records, and "many other features of high value." Another woman in evening wear gazes at the Super-Consola-97 KG. This more-advanced radio console is a "work of perfection and impressive in its presentation," just like the female model. "With seven bulbs, this magnificent instrument captures foreign stations as easily as local ones."[30] Interestingly, the RCA ad assumes that women consumers are both feminine and technologically knowledgeable. The reader is constituted as a complex subject, in charge of home purchases, intelligent, sophisticated, and elegant. She is interested in the mechanical dependability of the radio, its inner workings, and its design, as well as its functionality.

This example helps us understand that technology is not just a set of mechanical or electronic objects but also a "form of knowledge [and] a set of activities."[31] The discursive and iconic strategies of promotional campaigns for home appliances feminized housework and domestic labor and produced a Mexican female subjectivity that was understood to be a housewife. At the same time, these campaigns also constituted housewives as professionals with a special set of knowledge, vocabularies, and skills exclusive to housekeeping and family maintenance. Advertisements pictured women competently interacting with technology and assumed they possessed the intelligence and aptitude to actively participate in the modernization of their homes.

Many of these ads call attention to images of women using domestic appliances in a way that suggests that technological artifacts are functionally an extension of the female body. A full-page ad promoting the new 1938 model of the Norge refrigerator features a woman in an elegant evening gown admiring the refrigerator that is set upon a pedestal. The visual conflation of the woman and the Norge as elegant objects provides a compelling metaphor of gendered technology and the technologization of gendered bodies. The article "Charlas Femeninas: Orden y Economía," discussed above, is illustrated by a photograph of a housewife cooking something in a frying pan on a stovetop. The photograph projects a vision of what I identify as the "techno-woman:" the woman, the frying pan, and the stove are literally enjoined as one organism, suspended in the space of the text and engaged in the gendered work of cooking.[32] One might say that the stove is an extension of the woman's body and that the woman's body is an extension of the stove. And an advertisement for the department store El Centro Mercantil presents another example of this techno-woman or feminized technology. The ad "offers housewives the five most useful items for the home that will make every woman a queen." A housewife wearing an apron is positioned in the center of a five-pointed star. Five electric household appliances surround her and the star: a washing machine, a radio, an electric clothing press, a gas stove, and an electric refrigerator, structuring a woman-machine relationship that envisions woman as just another household appliance, a labor-saving device.³³ The visual conflation of woman and technology in each of these ads provides a compelling metaphor of gendered technology and the technologization of gendered bodies.

Although women's relationship to technology was primarily defined through the gynotechnics of domesticity, other discourses of gendered technologies circulated in popular culture, introducing new notions of how to be a modern woman. One example can be found in automobile advertising. Automobile advertising in the 1910s and 1920s was addressed to upper- or middle-class men, emphasizing "appeals to dependability, economy, simplicity, and ease of operation."³⁴ Within a few years, the industry had found another group of customers, women. And by the late 1920s, "photographs of women and automobiles had also become a central cultural emblem of women's modernity, independence and mobility," and female celebrities were regularly featured in the driver's seat.³⁵ A 1937 Chevrolet advertisement in the Mexican weekly magazine *Todo* features four photographs: in the first, a smiling, well-dressed woman stands next to her new car and waves to a friend outside the frame; a second photo pictures her in the driver's seat; in the final two, she is holding her young child. The ad's copy calls attention to a number of the car's features, including its beauty, its comfort, its power, its safety features, and its economical price, and tells the reader that it "offers a steel body and an armored roof . . . shatterproof glass that is clear and perfectly transparent." This advertisement assumes that its female reader is interested in not only the car's aesthetics but also its technological features; it is marketing not merely a beautiful new automobile but also the idea that the modern woman is interested in and knowledgeable about modern technology.³⁶ Moreover, the automobile is feminized through its pictorial and textual association with a woman, while the figure of the woman is structured as a technological subject. This mediation works to produce the "techno-woman," a subjectivity formed through the production of a gynotechnics that situates women and automobiles within a new transportation system that includes not only cars and drivers but also ideologies about the relationship between gender, transportation, and technology.

An illustrated article in *Todo*, entitled "Cuando Eva se Vuelve Adán," talks about "Rosie the Riveter," women in the United States who took over male jobs as their men went off to war. Mexican readers are informed that the "strong women of the United States" are taking the jobs vacated by "brothers, husbands, or boyfriends" who have gone off to fight overseas. The article features the stories of a number of these women, including Louise Snouffer, who "abandoned her stage career because of her desire to help address the severe shortage of meteorologists." We're told that she trained in the Office of Meteorology in Chicago, and a photograph shows her reading the wind speed on an optical theodolite.

The story also features Aileen Heidgerd, twenty-one years old and a graduate of Cornell University, who previously worked as a dietician but was now working for International Business Machines in Boston. She is pictured at her new job: repairing typewriters.[37]

A similar story, "La defensa femenina civil en Inglaterra," praises young British women for their work in the service of their country during World War II. Illustrations include photographs of women in London registering at the offices of Civil Defense Service of England, a group of British women acquiring the mechanical skills to enter the Women's Defense Service to work in the field ambulance sector, a female British army volunteer wearing a gas mask and "ready for auto transport service," and another group of women "studying maps to become acquainted with the sites to which their company will be assigned in order to defend the homeland." While the article acknowledges that many British women work as nurses, cooks, and typists, it also highlights others who are assigned jobs that require a certain advanced level of technological expertise: "In assigning jobs to the new volunteers, their experience is taken into account, of course, and volunteers are able to express preferences for certain specific occupations. So that while the woman who knows how to handle automobiles is awarded the task of driving a military ambulance, she must also learn to put in place a heavy shield against poison gas and an oxygen tank to facilitate respiration."[38] These two stories give evidence of the complexity of the mutually constitutive relationship of gender and technology. As these last three examples illustrate, women's relationship to technology could not be contained within the walls of the home. I am not arguing that advertising was progressive in its reconceptualization of gender or of domestic practices or that the promotion of new home technologies facilitated Mexican women's agency. What I have attempted to provide is a mediational analysis of a modern gynotechnics of domesticity that understands the interdependence of gender and technology as part of a broad and complex sociopolitical system.

Technologies are not objects or things: they give "material form to dimensions of difference," making gendered power relations visible in the realm of human labor, or as Bray puts it: "Men plow, women weave; men work outside, women inside; those who work with their minds govern, those who work with their hands are governed."[39] Changes within gynotechnics systems generate cultural debate and social anxieties around gendered social roles and relations that are expressed through everyday practices. But social change also opens up new possibilities, and in the case of the introduction of home technologies in the twentieth century, Mexican housewives were offered new ways of being women through the adoption of these technologies.

## Notes

1. Patrice Elizabeth Olsen, *Artifacts of Revolution: Architecture, Society, and Politics in Mexico City, 1920–1940* (Lanham, MD: Rowman and Littlefield Publishers, 2008), 129. Olsen provides a fascinating account of the architectural development of post-revolutionary Mexico City. According to Olsen, contemporary accounts of the city described a situation of "architectural chaos."

2. Olsen, *Artifacts of Revolution*, 55–56.

3. Stephen R. Niblo, *Mexico in the 1940s: Modernity, Politics, and Corruption* (Wilmington, DE: SR Books, 1999), 9.

4. According to one study, "about one-fourth of the population (or some four million people) became actual consumers of electric power" in Mexico by 1920. In a major push to nationalize the electric power industry, Cárdenas established the Comisión Federal de Electricidad (CFE). Miguel S. Wionczek, "Electric Power: The Uneasy Partnership," in *Public Policy and Private Enterprise in Mexico*, ed. Raymond Vernon (Boston: Harvard University Press, 1964), 37, 63.

5. James D. Cockcroft, *Mexico's Hope: An Encounter with Politics and History* (New York: Monthly Review Press, 1998), 129.

6. Robert Redfield, *Tepoztlan, a Mexican Village: A Study of Folk Life* (Chicago: University of Chicago Press, 1930), 39.

7. Niblo, *Mexico in the 1940s*, 2; Stephen H. Haber, *Industry and Development: The Industrialization of Mexico, 1890–1940* (Stanford: Stanford University Press, 1989), 178.

8. See, for example, Susan M. Gauss, "Masculine Bonds and Modern Mothers: The Rationalization of Gender in the Textile Industry in Puebla, 1940–1952," *International Labor and Working-Class History* 63 (Spring 2003), 63–83. In her analysis of the ways in which male-controlled unions participated in "the spread of middle-class ideologies of modern womanhood in order to reinforce the sexual differentiation of labor," Gauss also finds that the state supported these ideologies, writing that "attempts at vigilance over working-class motherhood had their clearest expression in the growth of Social Security, which combined efforts to regulate modern motherhood with benefits that discouraged female labor" (71, 75).

9. "La Madre: Alma de las Sociedades," *Todo*, April 11, 1937, n.p. All translations are mine unless otherwise noted.

10. Patience A. Schell, "Gender, Class, and Anxiety at the Gabriel Mistral Vocational School, Revolutionary Mexico City," in *Sex in Revolution: Gender, Politics, and Power in Modern Mexico*, ed. Jocelyn Olcott (Durham, NC: Duke University Press, 2006), 114.

11. Despite the official discourse and policy that privileged motherhood, thou-

sands of working-class women had to put their children into orphanages or day-care centers. Many of these women worked as paid and unpaid domestic servants in middle- and upper-class households. See Ann S. Blum, "Cleaning the Revolutionary Household: Domestic Servants and Public Welfare in Mexico City, 1900–1935," *Journal of Women's History* 15, no. 4 (2004): 67–90.

12. Joanne Hershfield, *Imagining la Chica Moderna: Women, Nation, and Visual Culture in Mexico, 1917–1936* (Durham, NC: Duke University Press, 2008), 73. See also Amy Kaplan's "Manifest Domesticity," *American Literature* 70, no. 3 (September 1998): 583–84. Kaplan argues that "if domesticity plays a key role in imagining the nation as home, then women, positioned at the center of the home, play a major role in defining the contours of the nation and its shifting borders with the foreign" (184).

13. *Mexican Folkways*, October/December 1928, 192.

14. *Hoy*, April 19, 1937, 55.

15. Judy Wajcman, *Feminism Confronts Technology* (University Park: Pennsylvania State University Press, 1991), 22–23.

16. Ruth Oldenzeil suggests that the contemporary understanding of technology as "male" emerged at the beginning of the twentieth century with the rise of the US engineering industry. See her *Making Technology Masculine: Men, Women, and Modern Machines in America, 1879–1945* (Amsterdam: Amsterdam University Press, 1999).

17. Judy Wajcman, *TechnoFeminism* (Cambridge, UK: Polity Press, 2004), 104.

18. Ibid., 35.

19. Barbara Marshall, "Critical Theory, Feminist Theory, and Technology Studies," in *Modernity and Technology*, ed. Thomas J. Misa (Cambridge, MA: MIT Press, 2004), 108.

20. Francesca Bray, *Technology and Gender: Fabrics of Power in Late Imperial China* (Berkeley: University of California Press, 1997), 4. According to Bray, in order to "understand the part technology plays in supporting a social formation, one must go beyond looking at a single technology or domain of technology (for example, the technologies of economic production), to consider the interplay of sets of technologies, or technological systems" (15–16).

21. Kaplan, "Manifest Domesticity," 583–84.

22. *Hoy*, December 31, 1938, 52.

23. *Mexican Folkways*, October/December 1928, 192.

24. Marshall, "Critical Theory, Feminist Theory, and Technology Studies," 123.

25. *Hoy*, August 13, 1938, 43.

26. Ruth Schwartz Cowan, "The Consumption Junction: A Proposal for Research Strategies in the Sociology of Technology," in *The Social Construction of Tech-

*nological Systems*, ed. Wiebe E. Bijker, Thomas P. Hughes, and Trevor Pinch (Cambridge, MA: MIT Press, 1989), 1–23.

27. "El hogar y la educación," *Hoy*, March 27, 1937, 36.

28. Julie Wosk, *Women and the Machine: Representations from the Spinning Wheel to the Electronic Age* (Baltimore, MD: The Johns Hopkins University Press, 2002), 71.

29. *Todo*, February 26, 1938, 87.

30. *Hoy*, August 13, 1939, 17.

31. Wajcman, *Feminism Confronts Technology*, 14.

32. *Hoy*, August 13, 1938, 43.

33. *Hoy*, February 26, 1938, 4.

34. James D. Norris, *Advertising and the Transformation of American Society, 1865–1920* (New York: Greenwood Press, 1990), 148–49.

35. Wosk, *Women and the Machine*, 120.

36. *Todo*, November 17, 1936, n.p.

37. *Todo*, April 10, 1943, n.p.

38. *Revista de Revistas*, October 22, 1939, n.p.

39. Bray, *Technology and Gender*, 369.

# II
# PHOTOGRAPHY, TELEVISION, AND THE INTERNET

# 4
# Technologies of Seeing
## Photography and Culture

*John Mraz*

Photography and its offspring—cinema, television, video, digital imagery, Internet—have redefined the exchange of information and our very ways of knowing the world. They reflect how we think; but more importantly, they form the structures through which we perceive. The technical images that appear to simply offer windows onto reality are fundamental in teaching people to understand their situations in particular ways. Nonetheless, despite their centrality to modern cognition, when those of us who study human societies think about photography, it is most often as an esoteric expression that is located at the edges of the marginal. The serious subjects come first: politics, economics, history, sociology, energy, health, ecology, and such. Culture is inevitably placed at the end of books, journals, and media of general interest; and photography, if mentioned at all, is subsumed among the arts. In fact, photography ought to be conceived as one of the foundations of today's comprehension of our condition.

In Mexico, photos have served, among other things, to "disappear" the enormous class differences that have been constant, to provide a bucolic and ahistorical vision of what it means to be Mexican (as well as another that embraces modernity), and to proffer celebrities created by modern media as models.[1] Photography got to Mexico quickly: four months after their presentation in Paris, daguerreotype cameras arrived in the port of Veracruz during December of 1839, and the infinitely reproducible *tarjetas de visita* brought the medium to the masses during the 1850s.

By the late Porfiriato, image was intimately linked with power. This was related to the transition from a journalism oriented toward active and informed readers to one that promoted passive consumers. Previously, Mexican periodicals had been connected to certain groups or ideas; they privileged opinions written by experienced essayists. However, the new press was embodied in the title chosen by Rafael Reyes Espíndola for his newspaper, which became the mouthpiece of Porfirian rule, *El Imparcial*. The name asserted that this daily was engaged in an ostensible search for objectivity and truth, in line with the Positivist precepts of the official ideologues and the *científicos* who backed it and in accordance with the "end of ideology" established by Díaz's blurring of liberal and conservative battle lines.[2] It aimed to establish the believability characteristic of modern publications based on information, a journalistic mode that had been stimulated by international news agencies; being up-to-the-minute became increasingly essential to entering the developed world. Following this tendency, *El Imparcial* rejected the old editorialist style of Mexican journalism and adopted the contemporary mode, in which news is collected in the street by reporters and presented to readers as if they are allowed to decide its meaning for themselves. However, its pages were filled with news of disasters, train crashes, streetcar accidents, fires, robberies, homicides, suicides, and assaults, the infamous *nota roja* that planted fear through "news" that demonstrated that the country needed a dictator's firm hand.

Photographs were a fundamental element of the new journalism because they appeared to be windows that opened onto reality. Reyes Espíndola was the first editor in Mexico to utilize the half-tone process (rotogravure) to publish photographs, a key technological breakthrough that became the backbone of the new journalism because it made possible high-quality photographic reproduction in magazines and newspapers. Images were a necessary component to the notion of discovering the news, of encountering reality in the street, because they interlocked with the texts to "prove" the truth of what was placed before the readers. Both the news items and the photos belonged to the same logic of modernity, which was finally based on the inductive principles of the Enlightenment, that "truth" was to be discovered from observation of the world rather than through abstract philosophizing. These ideas were novel in a Mexican society still recovering from the effects of a counter-Reformationist colonialism; and the participation of photography, among the newest of scientific technologies, was no doubt definitive in convincing readers that the publications of Reyes Espíndola were the perfect embodiment of empirical validation and verification.

The sensationalism of *El Imparcial* was designed to play to the lowest instincts, but Reyes Espíndola's magazine, *El Mundo Ilustrado*, was directed toward the refinements of the well-to-do. Hence, while the newspaper pictured the poor as criminals, the magazine essentially ignored them, focusing instead on society congregations: banquets, balls, and fiestas, as well as the sports that

4.1. Guillermo Kahlo, *Train on the Metlac Bridge, 1903.* © Inv. #843539, Fondo Casasola, Sistema Nacional de Fototecas (SINAFO)–Fototeca Nacional del Instituto Nacional de Antropología e Historia (INAH), Mexico City.

developed in the Porfiriato, from basketball to horse races, all of which took place in a stable and progressive culture, relaxed and harmonic in a deserved well-being. The great men who had created Porfirian Mexico were at the magazine's core, and the presence of Díaz was ubiquitous. Among the photographers who documented Porfirian progress for *El Mundo Ilustrado* was the German Guillermo Kahlo (father of the later famous Frida), who provided pictures of modernity such as the train and its passengers, stopped for a portrait on the Metlac Bridge, an image that offered a triple whammy of technological triumph in the railroad, the iron construction, and the camera that recorded them.

Like the Mexican Revolution itself, the photography of the armed struggle had its roots in the Porfiriato since the great majority of the professional photographers who covered it were formed by their work in the illustrated press and the photographic studios of that period.[3] Moreover, although the subjects that appeared in the images became radically other, it seems that the ways of photographing were established prior to the rebellion. It has been common to argue, as did art historian Olivier Debroise, that "to make photographs at the beginning of the twentieth century, it was still necessary to ask history to stop long enough to take the picture. For this reason, photography in the final years of the Porfiriato still largely conformed to the formal, static models established by the studios."[4] However, it is clear that modern, spontaneous forms of photographing had developed during that period. This would become increasingly the case during the revolution, but what changed most was the face of the country. Suddenly, ordinary people appeared in photographs, newspapers, illustrated magazines, and documentary films; they projected a force that overflowed the constrictions of the "popular type" model within which they had been shoehorned in the *tarjetas de visita*.

The *caudillos* who led the different revolutionary factions were as conscious as Porfirio Díaz of photography's power. Francisco I. Madero showed the respect he had for photographic imagery, carefully staging the meeting of the Revolutionary Junta at the Casa de Adobe in Ciudad Juárez, April 30, 1911, to which some fourteen photographers arrived. As the photographic historian Miguel Ángel Berumen remarked, "Madero knew that images of him and his followers would be published, and that the face they put on for the photographers would play a decisive role."[5] Pancho Villa began to understand the importance of photographs in Ciudad Juárez. The watershed event marking his development of this consciousness occurred on April 26, 1911, when Villa received the grade of colonel from Madero in front of dozens of reporters and photographers, as well as almost two thousand spectators from both sides of the border. Before, he dressed in the same dark *ranchero* clothing as his followers, but from then on he changed his uniform to khaki military clothing and his black *charro* sombrero for a beige *americano*-style hat. Villa would end up wearing the uniform designed by the Mutual Film Corporation, but his media projection as a revolutionary leader began in Ciudad Juárez.

Mexico City's photojournalists arrived late to cover the Maderista uprising since their media didn't have much interest in publicizing challenges to the regime; for example, *La Semana Ilustrada* sent Samuel Tinoco, who arrived eight days after the combat had terminated.[6] It was different for the postcard studios because they were motivated by profit rather than by the governmental subsidies that financed the press. Postcards were one of the typical passing fads of modern visual culture; they enjoyed an enormous popularity between 1890 and 1920 because they brought the world and its events to the coffee table, where every family had its precious albums. We have no information about postcard circulation in Mexico, but the numbers in the United States give us an idea of their fashionableness: in 1906, 770,500,000 postcards were mailed; in 1913, the number had reached 968,000,000.[7]

It is ironic that postcards were the visual media that most advertised the Maderista challenge to the Porfiriato because "the first Mexican postcards devoted a disproportionate attention to Porfirio Díaz's administration, glorifying it and praising the country's progress under his rule."[8] Their prior function notwithstanding, they would now serve other masters because Aurelio Escobar Castellanos arrived in the north at the beginning of April and the first photo that registered the presence of the H. J. Gutiérrez agency in the revolution was taken in the middle of that month in Casas Grandes, Chihuahua, documenting a group of Maderistas in front of what must have been a Wells Fargo office of messenger service and transportation.

It seems clear—once we leave behind the myth that Agustín Víctor Casasola was *the* photographer of the revolution—that there were many who covered the long civil war. I believe that specific photographers can be linked to the contending forces in broad strokes—and with strong possibilities of erring—in the following way: the agency of Heliodoro J. Gutiérrez was linked to the Maderista movement both on the northern frontier and in Mexico City, making it the first revolutionary photographic protagonist; Gerónimo Hernández was a Maderista image maker during the truncated presidency, working for *Nueva Era*, the Maderista newspaper; the photographer most engaged with the Orozquista rebellion seems to have been Ignacio Medrano Chávez, "El gran lente"; Amando Salmerón was Emiliano Zapata's photographer, but there were other image makers connected to that movement, among them, Cruz Sánchez and Sara Castrejón (one of the few women to participate with a camera); Eduardo Melhado may have pictured reconstructions of the Tragic Ten Days, and thus could be considered as being a Huertista; the Cachú brothers, Antonio and Juan, were evidently the photographers closest to Pancho Villa; and the Constitutionalists had many image makers, although Jesús H. Abitia has been considered *the* Constitutionalist photographer.

That which is really novel about Mexican photography during the armed struggle is the fact that photographers were committed to revolutionary groups that were at war with each other. Hence, the term "revolutionary photography" is a problem because the images of Abitia are as "revolutionary" as Salmerón's or the Cachú brothers', despite the war that their armies waged against each other. The most direct way to demonstrate commitment and have it paid for was to be subsidized by a *caudillo*; obviously, those who had more money for arms, munitions, and uniforms also had more for photography. Nonetheless, there were other ways to earn a living with imagery and, perhaps, express a point of view: work for the illustrated magazines or in a news photo agency; have a studio or a film business; be employed by an institution with an interest in documenting the war, such as the Cruz Blanca Nacional (National White Cross); sell postcards to publications or in stores or to the very people who therein appeared as souvenirs.

The Mexican Revolution occurred in a moment that photography was being defined as a medium. Just as we find horses next to airplanes and ancient cannons on the same battlefield as the most modern artillery, one sees a great variety of photographic equipment: old and heavy cameras appear together with the new, lighter, more portable ones that emerged as the war advanced. Many photojournalists used Reflex (single lens reflex) cameras, which could be carried and operated in a hand-held fashion, enabling them to go to the news rather than wait for it to come to them and their stationary apparatus.[9] The Reflex cameras offered more mobility and the capacity to capture action; the visor on top enabled photographers to focus on subjects without covering their heads with a dark cloth. The other camera employed by professionals during the revolution was the View, which required the use of a tripod.[10] Some photojournalists, above all the older ones such as Manuel Ramos, utilized View cameras; however, these cameras were generally favored by studio and postcard photographers, probably because they were "an instrument of great sensitivity and precision, provided the subject was immobile."[11] Finally, Kodak Brownies were popular among the wealthy, although it appears that they rarely documented the war; moreover, there is no evidence that revolutionary troops made the quantity of small-format photos that can be found in the albums of soldiers from more prosperous nations during World War I. The differences between the cameras determined, to some extent, whether one captured combat scenes or was limited to posing groups and the degree to which photos were spontaneous or directed.[12] If, however, the variations between the cameras produced different kinds of photographs with their particular aesthetics, in the distribution of photos there was much intermixing between the genres, with photojournalistic images circulating as postcards and vice versa.

According to their means, photographers adopted medium-format cameras (with negatives of 5" × 7", or smaller ones, 4" × 5") whose mobility permitted them to move easily and capture spontaneous scenes. The new cameras, as well as photographic equipment and materials, must have entered the country as easily and constantly as did arms, munitions, uniforms, and supplies. The daily omnipresence of image-making equipment—sometimes as ubiquitous as other arms—created a consciousness that led to the construction of written and visual narratives that incorporated them.

One example is a photo by Eustasio Montoya that offers a story without words (fig. 4.2). Unarmed civilians look at the body of someone who appears to belong to their group. It would seem to be a daily experience for them, to judge

4.2. Eustasio Montoya, *Photographers and a Dead Man in Nuevo León after Being Attacked by Huertistas, 5 July 1913*. © 04.011, Registro 23, Archivo González Garza, Universidad Panamericana, Mexico City.

by their sad but resigned features. The photographer (the other camera present) documented more than their grief: in the photo there are two cameras (and a box to carry another) of the View type that are light and mobile, although they all require tripods. The one on the right is a Kodak Poco, a medium format (5" × 7"), but the one on the left appears to be a more advanced and smaller model (and could be a 4" × 5"). Beyond the evidence of technological advance (and media presence), I wonder about the history behind the image. Was it in homage to a photographer who died in the battle of Colombia, Nuevo León, in 1913, or was the scenario constructed to protest his death?

In the cultural effervescence of post-revolutionary Mexico, battle lines were drawn between two groups of photographers with fundamental differences about what constituted "*lo mexicano.*" The positions represented can be summarized as the picturesque and the anti-picturesque, in which the absence or presence of modern technology was crucial. On one side were Hugo Brehme and Mexicans such as Luis Márquez and Rafael Carrillo, traditionalists who constructed a romantic vision of a bucolic, rural Eden absorbed in its nature and peopled by *charros* and *chinas poblanas*, regional figures who were transformed into national archetypes.[13] Their aesthetic was at one with their content, using their cameras like nineteenth-century paintbrushes to offer what appeared to be transparent, fine-art-inspired images of beautiful, idyllic scenes. On the other side were foreigners Edward Weston and Tina Modotti, accompanied by Mexicans Manuel Álvarez Bravo and Agustín Jiménez, who were experimenters who broke with painterly notions of art and sought to establish photography as a medium in its own right, with its own possibilities and limitations. They rejected the picturesque and focused on modern urban life as found in telegraph lines, typewriters, and toilets.[14]

Contemporary scholars of Mexican photography argue that Hugo Brehme constructed "a graphic system of *lo mexicano*," creating a "visual vocabulary" of *mexicanidad* that constitutes "the base of today's national identity."[15] His photography was influenced by the Pictorialist movement, reflecting traditional principles of composition and idealizing the rural life. More importantly, it embodies classical notions of the picturesque. Nature overwhelms humans and their existence: rugged cliffs of craggy stones and twisted trees hover over shacks embedded in vegetation; in the midst of an open plain, an immense yucca tree renders a burro and rider insignificant; a gargantuan organ cactus dwarfs the *campesino* standing in its shadow; snow-capped volcanoes provide the ultimate backdrop.[16] The pre-Columbian and Colonial buildings on which Brehme focused show the onslaught of rain and sun, the wear and tear of time, as do the dewy-eyed, grandfatherly old men who appear in his portraits. All is made beautiful: ramshackle huts of stone or adobe provide a texture for the camera that is reiterated in rock walls and village markets, where all radiates tranquility and well-being.

The cameras of the traditionalists and the modernists were essentially the same View apparatus that had been utilized prior to and during the revolution, but the photographers focused on distinct subjects and employed them in very different ways. Rubén Gallo has argued convincingly that Tina Modotti was particularly concerned to represent the modernization of Mexico in the 1920s, photographing industrial complexes, construction projects, and machines, as well as telephone and telegraph wires.[17] In speaking of her imagery of telegraph lines, he emphasizes how she taught Mexican culture to see the transformations that were occurring through new eyes. The traditionalists would have considered the tangle of wires "entirely vulgar and unaesthetic—visual pollution marring the pristine beauty of Mexican landscapes."[18] However, Modotti recognized that both the camera and the telegraph had transformed the communication of information: the telegraph extended writing over long distances and the camera allowed a way to write automatically. Modotti shot the telegraph lines from below, rejecting a traditional, painterly perspective; hence, as Gallo affirms, "she used a technological device—the camera—to represent another technological artifact—the telegraph—by means of a technologically informed perspective—the oblique angle. . . . Modotti found a means of representation that was as modern as the subjects it represented."[19]

European modernism emphasized machinery and factories, as well as objects such as tin cans and automobiles. Mexicans who belonged to what has been described as the "Nopal vanguard," which included Manuel Álvarez Bravo, Agustín Jiménez, and Aurora Eugenia Latapí, also looked for novel ways of representing the new.[20] The influences of Weston and Modotti were predominant; after detailing a long list of his teachers, Jiménez noted, "But, when Edward Weston and Tina Modotti were in Mexico, the technique they followed influenced me so that I ended up creating my own way of seeing and feeling things."[21] The Mexican modernity of Jiménez was expressed, among other ways, in the metallic forms of agricultural implements, such as the blades he photographed in 1932. Latapí was Jiménez's outstanding student, and she incorporated rural themes such as close-ups of corn ears.[22] Álvarez Bravo found inspiration in urban themes; one example is offered by advertising images, such as the famous inverted sign outside the "Modern Optical" store and the pair of legs that were painted below an electrical installation shop.[23]

The invention of the 35 mm cameras in Germany during 1924 fundamentally altered the aesthetic of photography, especially photojournalism. The Leica cameras were small and light, compared to the bulky box cameras, whether View or Reflex. They radically increased reporters' mobility, allowing them to move about freely and get into the center of events without being too conspicuous. This was quite a different situation than that of having to stand in one spot, either weighed down by a heavy camera or even more inconvenienced by a tripod, and being immediately identifiable to the forces of repression. The impact of this technological innovation can be appreciated in images taken by the Hermanos Mayo from within the very vortexes of struggles, for example, battles between striking teachers and policemen during 1958 (fig. 4.3).

The pivotal factor in the Mayo Brothers' development was without question their participation in the Spanish Civil War.[24] Their decision to fight on the side of the forces defending democracy in Spain placed them in a very different camp than the supposed objectivity that had characterized prior Mexican photojournalism. For example, their political stance was far from the Positivist eulogies for progress that were typical of Porfirian media or the unconditional acceptance of the post-revolutionary rulers that can be seen in the photos of Enrique Díaz.[25] The Hermanos Mayo's experiences in Spain had given them a

4.3. Hermanos Mayo, *Striking Teachers Battle with Police; Zócalo, Mexico City, 6 September 1958.* Chronological file, no. 12754, Fondo Hermanos Mayo, Archivo General de la Nación, Mexico City.

different way of seeing and focusing on reality; they had something new to offer Mexican photojournalism.

A technical contribution of the Mayo collective was perhaps the most concrete: they introduced Mexican photojournalists to the 35 mm cameras that they had bought in Spain and carried throughout the Civil War.[26] "The arrival of Paco, Cándido, and Faustino with three Leicas was like bringing three Volkswagens into the country fifty years ago," Julio Mayo reflected.[27] However, their road was not quite that smooth: the photo editors in Mexican publications did not believe that publishable images could be made from the Leicas's tiny 35 mm negatives, and they preferred the larger (3¼" × 4¼") negative plates that the bigger cameras held. Despite this, the Mayo Brothers had an enormous advantage in the area of film: the Leicas carried rolls with thirty to forty exposures, while the older cameras had plates that accommodated only twelve exposures. Once it became clear that the 35 mm negatives were perfectly acceptable for newspaper and magazine work, the greater number of photographs the Hermanos Mayo could take gave them more possibilities of covering events thoroughly. The fact that the Leicas could be adapted with different lenses was also important. The telephoto lenses that the Mayo Brothers employed to get nearer to events had not been used previously in Mexico; further, the capacity, especially of Paco

Mayo, to utilize amber and yellow-green filters added yet another distinctive quality to their work. Their technical capacities, combined with their ability to work collectively, allowed for the opportunity to both shoot many more negatives and catalog them, a great problem for photojournalists. The coupling of technological advantage and collective labor accounts for their enormous archive of some five million negatives and explains their "gift of ubiquity," which Carlos Monsiváis has so appropriately recognized.[28]

Although the Hermanos Mayo and other dissenting documentarians such as Nacho López and Héctor García made critical photos, the illustrated magazines offered the only venue in which they infrequently appeared. These publications established themselves from the mid-1930s to the mid-1950s as the newest version of fads such as the *tarjetas de visita* and the picture postcards, sharing the visual culture sphere with more lasting graphic expressions like cinema and comic books. Though there were a number of illustrated magazines, *Hoy*, *Mañana*, and *Siempre!* seem to have been the most important, and they are those that are remembered today. Like their US and European counterparts—*Life*, *Picture Post*, and *Vu*—they enjoyed significant popularity.[29] The Mexican magazines were almost certainly bought by members of the new middle classes born of post-revolutionary economic development.

The illustrated magazines taught Mexicans to love celebrities, such as movie stars Cantinflas, María Félix, Dolores del Río, and Pedro Infante. However, the greatest of all the celebrated was the president. The cornerstone of the magazines' ideology was presidentialism, and it fawned over the occupant of the office, in turn. Tours, banquets, meetings, the inaugurations of public works, decorations received from foreign governments, and any number of other presidential activities filled their pages. In fact, around one-fourth of the illustrated articles published in these magazines dealt with the president in some way, and there were entire issues dedicated to him, which could be up to three times as long as they were normally.[30]

The president served as the great patriarch of a culture still dominated by a traditional family structure. The magazine's portrayal of what a successful man ought to be was encapsulated in this figure, as well as in its infinite replicas: the cabinet members, state governors, and the innumerable functionaries who paid for photo-essays on their activities. The visual message of these men in suits and ties (when not dressed to provoke populist identification)—inaugurating public works, sitting in banquets, appearing in political gatherings, or pictured in visits by photojournalists to their "private lives"—was clear: men should be important, and the clearest path to such public recognition was to be participating members of the PRI (Partido Revolutionario Institucional) dictatorship, finding their place on the ladder of patriarchal dominion. Women's role was defined within this apparatus: they should be wives of important men. The primordial example was offered by the first lady, whose public presence was essentially a reaffirmation of domestic values: she was her husband's shadow at his public appearances, and when she was pictured without him, her role as wife and mother was demonstrated in inaugurating child-care centers, distributing cooking items, and handing out presents to poor children for Christmas or the Day of the Good Kings.

Presidentialism went hand in hand with an unbridled admiration for the ruling classes, but these magazines were a good deal less generous with the *humildes*, who were largely as absent as they had been in Porfirian publications. In relation to the working class, they generally followed the tendency of "industrialist photography," where machines and structures dominated images from which workers were often excluded. Things didn't go any better for *campesinos* and Indians, who were portrayed in the picturesque style of Porfirian photography; Enrique Díaz's images of happy peasants on market day and smiling families of fishermen in Pátzcuaro were a staple fare.[31] The exotic approach also provided the publications with a form of soft-core pornography. For example, the "Indian problem" was depicted by Rafael Carrillo through various images of bare-breasted *indígenas* in "the marvelous landscape of Cosoleacaque," one of them on the cover of *Hoy*.[32] As a symbol of the nation, Indians were important to presidentialism in the media. For example, in a photo-essay, "The President and His People," an Indian is shown shaking hands with President Miguel Alemán. The president is presented with his face to the camera, extending his hand to the Indian, who, photographed from behind, has no personal identity. The cutline states: "This photograph is more eloquent than two hundred words. The spontaneous exaltation of a representative of our humble Indian class to reach out his hand to the First Leader of the Nation who has so profoundly occupied himself with the problems of Mexican agriculture."[33]

The invention of digital imagery would seem to mark a watershed in visual culture because the credibility in the veracity of photographs, what Roland Barthes called the "that-has-been," is threatened by the fact that they can now be easily created using computers without the necessity of the referent having ever really existed.[34] The images of Pedro Meyer, for example, often raise doubts about the act of representation, calling attention to the fact that they are constructed and putting into question the notion of photography as a window onto the real world. They are postmodern in fomenting incredulity in opposition to the believability that was the ideological bedrock of technical images since their invention. Hence, they might be said to initiate the beginnings of a postmodern visual culture, although the appeal to the masses and the manufacture of celebrities that began with modernity continue unabated. A transcultural figure, Meyer was born in Spain (1935) but was brought two years later to Mexico, where he lives, though he also spends time at his US residence. Having devoted himself to photography since 1974, he became the pioneer of digital imagery in Mexico, creating a bilingual site, "Zonezero: From Documentary to Digital Photography," www.Zonezero.com, in 1993.[35]

A major theme of Meyer's digital imagery addresses the interface of the United States and Mexico. In some the digital alteration is clearly signaled. For example, in the image *Mexican Serenade* (1985/92), Meyer created a metaphor for neo-colonialist tourism: a *gringa* sits in a collapsible aluminum chair, reduced to the size of the tiny caricatural figurines of the Mexican musicians to which she listens, in a trailer park that, reproduced on a completely different scale, overwhelms them.[36] Other digital images appear to draw upon documentary credibility. Hence, *Migratory Mexican Farm Workers, California Highway* (1986/90, fig. 4.4) shows men laboring at agricultural tasks, stooped over in a field beneath a billboard advertising "Caesars," an inn that offers "Free Luxury Service From Your Motel"; in the sign, a Roman gladiator opens the door of a fancy private taxi for prospective customers who, presumably, will not include the poor souls straining below. Of this photo, Meyer stated, "I saw the Mexican migratory laborers at some kilometers from the site of the billboard. I had made the association between the two scenes in my mind, but they were separated in space. The photos were taken in pre-digital times, before the existence of instruments to link these two moments." Meyer asserts that his interest was not just that of constructing a discourse about migratory workers, "although it is inevitable that it is ALSO that."[37] But, he would insist, it is more concerned with the experience of observing and the ways of seeing opened up by digitalization, which offers the possibility of constructing a "historical" vision by incorporating the past into the perception of the present.

Of course, the subjectivities of the people behind the camera are a complex reflection of the "climate(s) of opinion" in which they have lived and that have formed both them and their audience.[38] The importance of photographic authorship notwithstanding, where, how, and which photographs appear becomes a determinant factor in shaping our ways of seeing. The Mexican mass media has tended to hide the enormous class differences that characterize the society, filling the void with the powerful, the wealthy, and the celebrities, or turning the class differences into the picturesque. Mexican culture is overwhelmingly visual, but it has not often been an image that revealed the country's reality, but rather, technological advances have made the illusion of objectivity ever more credible. Today, the technical imagery of television carries on the role played

4.4. Pedro Meyer, *Migratory Mexican Farm Workers, California Highway*, 1986/1990. Courtesy of Pedro Meyer.

by Porfirian and post-revolutionary magazines, presenting Mexicans with a "reality" that is ever further from that which they are living. They are being taught to see in ways acceptable to those who rule.

## Notes

1. A more developed version of this essay can be found in John Mraz, *Looking for Mexico: Modern Visual Culture and National Identity* (Durham, NC: Duke University Press, 2009).

2. Alan Knight comments on "the end of ideology" during the Porfiriato. Alan Knight, *The Mexican Revolution*, vol. 1, *Porfirians, Liberals, and Peasants* (Lincoln: University of Nebraska Press, 1990), 15.

3. See John Mraz, *Photographing the Mexican Revolution: Commitments, Testimonies, Icons* (Austin: University of Texas Press, 2012).

4. Olivier Debroise, *Mexican Suite: A History of Photography in Mexico*, trans. Stella de Sá Rego (Austin: University of Texas Press, 2001), 184.

5. Miguel Ángel Berumen, *1911: La batalla de Ciudad Juárez/II; Las imágenes* (Ciudad Juárez, Mexico: Cuadro por Cuadro, 2003), 85.

6. Ibid., 58.

7. George Miller and Dorothy Miller, *Picture Postcards in the United States, 1893–1918* (New York: Clarkson N. Potter, 1976), 22.

8. Gloria Fraser Griffiths, "La postal mexicana," *Artes de México* 48 (1999): 11.

9. See Daniel Escorza Escorza and Heladio Vera, "Las cámaras Graflex en la campaña federal maderista contra Pascual Orozco, 1912," *20/10: Memoria de las Revoluciones* 10 (2010): 254–65.

10. See Arturo Guevara Escobar, "Reflex o View," http:fotografosdelarevolucion.blogspot.com (accessed November 6, 2012).

11. Naomi Rosenblum, *A World History of Photography* (New York: Abbeville Press, 1989), 443.

12. I employ the concept of "directed" to describe a genre of documentary photography that is characterized by the intervention of the photographer in the scene; see John Mraz, "Thinking about Documentary," in *Nacho López: Mexican Photographer* (Minneapolis: University of Minnesota Press, 2003), 169–89; and John Mraz, "What's Documentary about Photography? From Directed to Digital Photojournalism," *Revista Zonezero*, www.zonezero.com, 2002 (accessed November 6, 2012).

13. The clothing of both was adopted as the "typical Mexican image" for the official nationalist campaigns in 1930, and the *china* was named National Archetype in 1941. See Ricardo Pérez Montfort, "Indigenismo, hispanismo y panamericanismo," in *Cultura e identidad nacional*, ed. Roberto Blancarte (Mexico City: UNAM, 1994), 349; Jeanne L. Gillespie, "The Case of the *China Poblana*," in *Imagination beyond Nation: Latin American Popular Culture*, ed. Eva Bueno and Terry Caesar (Pittsburgh: University of Pittsburgh Press, 1998), 19. Luis Márquez is considered to be the "pioneer" among Mexican folklorists. See "El imaginario de Luis Márquez," special issue, *Alquimia* 10 (2000); and José Luis Sánchez Estevez, "Luis Márquez Romay y su obra: Apuntes sobre la búsqueda del nacionalismo cultural en México" (Licenciatura thesis, Centro Universitario de Ciencias Humanas, 1990).

14. Rubén Gallo argues that Modotti was more interested in picturing modernity than Weston, who "refused to photograph scenes relating to Mexico's modernization, which he blamed for 'spoiling' the country." Rubén Gallo, *Mexican Modernity: The Avant-Garde and the Technological Revolution* (Cambridge, MA: MIT Press, 2006), 50. Nonetheless, the toilet that Weston so assiduously photographed was certainly modern.

15. The references to Brehme's importance in constructing a visual nationalism are numerous. See José Antonio Rodríguez, "La construcción de un imaginario," in *Hugo Brehme: Fotograf-Fotógrafo*, ed. Michael Nungesser (Berlin: Verlag Willmuth Arenhövel, 2004), 39; Jesse Lerner, "La exportación de lo mexicano: Hugo Brehme en casa y en el extranjero," *Alquimia* 16 (2002–2003): 30; Mayra Mendoza Áviles, "La colección Hugo Brehme," *Alquimia* 16 (2002–2003): 43; Blanca Garduño Pulido, "Mexico's Enduring Images," in *México: Una nación persistente, Hugo Brehme, fotografías* (Mexico City: Conaculta-INBA, 1995), 15.

16. See Hugo Brehme's photography in these books: *México pintoresco* (1923; repr., Mexico City: Porrúa-INAH, 1990); *Pueblos y paisajes de México* (Mexico City: Porrúa-INAH, 1992); *México: Una nación persistente, Hugo Brehme, fotografías* (Mexico City: Conaculta-INBA, 1995); and in *Hugo Brehme: Fotograf-Fotógrafo* (Berlin: Verlag Willmuth Arenhövel, 2004). Brehme did publish *México pintoresco*, but the other books are compilations of his photography made after his death, although he is listed as the author.

17. Gallo, *Mexican Modernity*, 48.

18. Ibid., 50.

19. Ibid., 55–56.

20. The term "Nopal vanguard" is Córdova's; see Carlos A. Córdova, *Agustín Jiménez y la vanguardia fotográfica mexicana* (Mexico City: Editorial RM, 2005), 202.

21. Jiménez is cited from a 1935 interview; see Córdova, *Agustín Jiménez y la vanguardia fotográfica mexicana*, 34.

22. See this photo in *Agustín Jiménez: Memorias de la vanguardia* (Mexico City: RM and Museo de Arte Moderno, 2008), 84.

23. See these photos in Susan Kismaric, *Manuel Alvarez Bravo* (New York: Museum of Modern Art, 1997), frontispiece and 67.

24. On the Hermanos Mayo, see John Mraz and Jaime Vélez Storey, *Uprooted: Braceros in the Hermanos Mayo Lens* (Houston: Arte Público Press, 1996).

25. On Díaz, see Rebeca Monroy Nasr, *Historias para ver: Enrique Díaz, fotorreportero* (Mexico City: UNAM-INAH, 2003).

26. See my interview with Faustino and Julio Mayo: John Mraz, "Close-up: An Interview with the Hermanos Mayo, Spanish-Mexican Photojournalists (1930s–present)," *Studies in Latin American Popular Culture* 11 (1992): 195–218. After reviewing various archives, among them those of Enrique Díaz and Nacho López, and talking with the photojournalist Héctor García, I am convinced that the Mayo Brothers were the first to use 35 mm cameras in Mexican periodicals. Díaz was one of the best-known graphic reporters during the period 1925–1960 but apparently never used the small-format camera; negatives of his associates indicate that they began to utilize the 35 mm camera in the 1950s. Nacho López began working with a large-format camera in the late 1940s but changed to the smaller camera during the 1950s. This is the same period in which García says he began to use the 35 mm camera. It may be worth mentioning that small-format cameras were advertised in Mexican magazines prior to the arrival of the Hermanos Mayo but were evidently not used in press photography; see *Hoy*, July 23, 1938, 10, where an ad for a Zeiss Ikon appears.

27. Mraz, "Close-up: An Interview with the Hermanos Mayo," 209. Actually, they did not arrive with the cameras because they had been taken away on leaving Spain for France; they bought them after their arrival.

28. See Carlos Monsiváis, "Los Hermanos Mayo: . . . y en una reconquista feliz de otra inocencia," *La Cultura en México*, a supplement of *Siempre!*, August 12, 1981, 2.

29. Circulation figures for Mexican periodicals are notoriously inexact and simply unavailable for the illustrated magazines.

30. See, for example, *Mañana*, June 17, 1950, which is composed of 310 pages on Miguel Alemán's trip to the southeast; issues were normally around 100 pages.

31. *Hoy*, August 13, 1938.

32. *Hoy*, August 6, 1938, 28–33.

33. *Mañana*, September 29, 1951, 14.

34. Roland Barthes, *Camera Lucida*, trans. Richard Howard (New York: Hill and Wang, 1981), 80. This is an enormously complex question: for example, nobody appears to have doubted the authenticity of the digital images made of torture in the Abu Ghraib prison.

35. Among the prolific Pedro Meyer's works are *Truth and Fictions: A Journey from Documentary to Digital Photography* (New York: Aperture, 1995) and a recent retrospective, *Herejías* (Barcelona: Fundación Pedro Meyer and Lundwerg, 2008).

36. P. Meyer, *Truth and Fictions*, 24–25.

37. Pedro Meyer, personal communication, 2001.

38. For a felicitous introduction to the idea of "climates of opinion," see Carl Becker, *The Heavenly City of the Eighteenth-Century Philosophers* (New Haven: Yale University Press, 1932), 1–31.

# 5
# The Early Years of *La Tele*

*Celeste González de Bustamante*

Over the course of the second half of the twentieth century, television became the most significant form of mass communication for the majority of Mexico's citizens. By the beginning of the twenty-first century, 91 percent of the population had at least one television monitor in the home.[1] Today, Grupo Televisa, the nation's largest commercial network, produces more Spanish-language programming than any other company in the world. This is a far cry from over a half century earlier when, on September 1, 1950, the day of the country's first official broadcast, so few citizens owned a television that receivers had to be set up in public places around the capital city for people to watch President Miguel Alemán's fourth *informe*.[2]

With the broadcast of Alemán's address to the nation, Mexico and its citizens were beamed into the television age.[3] On that day, Rómulo O'Farrill Sr., the owner of the country's first TV company and station, attempted to capture the significance of the event, writing in his newspaper *Novedades*: "Today is a day of celebration for Mexico, as today our country will be the first in Latin America to have, for the advantage and benefit of its inhabitants, the most important invention of modern times: television."[4] Even at its inception, O'Farrill grasped the potential power of the medium; and because of the medium's and industry's influence on society over the past six decades, scholars should seek to better understand its development.

Although there is a growing body of literature on television in Mexico, most of the work has focused on the past twenty years.[5] Few scholars have approached the subject from a historical perspective.[6] This chapter examines the interaction between television and culture, while paying particular attention to how Mexicans shaped the trajectory of this new medium and the industry to which it gave birth.

The chapter focuses on the early years of television, specifically from 1950 to 1973. It first examines a series of decisions made early on—the choice to follow a commercial rather than state-run media model, for example—that would significantly impact the development of the industry throughout the rest of the twentieth century. The chapter then turns to some of the first television producers and media magnates who launched the early television industry. Next, it discusses television programming, locating the historical and cultural milieu in which TV emerged, and examines the critical roles played by the news as well as the *telenovela* in fostering a sense of *lo mexicano*. The chapter concludes with an examination of how citizens watched television, paying attention to the complex power dynamics informing the relationship between producers and viewers.

Before turning to some of the first significant decisions regarding TV, it is important to consider that several characteristics unique to television distinguish it from other forms of mass media. Early broadcasters, for example, quickly learned of the dangers of live broadcasting due to its visuality and immediacy. During a televised fashion show sponsored by Salinas y Rocha (a department store chain), a line of beautiful and scantily clad women walked down a runway modeling a new line of lingerie. To the dismay of director Gonzálo Castellot and many viewers, the lighting proved so intense that the picture revealed an improper amount of the women's bodies, forcing him to cut immediately to a more appropriate image.[7]

The immediate nature of live television, as Castellot discovered, was just one of the important differences between electronic and print media. Unlike privately owned newspapers, the airwaves that carried television programming belonged to the public domain, a precedent formally established with the passage of the 1960 Federal Law of Radio and Television. This legislation gave government the right and obligation to regulate the industry through station licensing and the monitoring of broadcast content. Such imperatives allowed—and continue to allow—the government to control the industry much more directly than it could print media.[8]

The limited number of frequencies available restricted the number of individuals and companies that were able to gain broadcasting concessions and thus participate in the production of television content. This had important implications for the types of programming that emerged. Early on the government granted such concessions to an elite group of investors, and as a result, programming, including news, reflected capitalist and elite class interests.

Finally, television had the potential to reach a much larger audience than print media. Because TV programs were broadcast over the air, anyone within a certain range of a signal could watch them for free. In a city the size of Mexico City, this meant that viewership could reach into the millions. Meanwhile, newspapers could at best achieve a readership in the hundreds of thousands. Further complicating matters, newspapers required a nominally literate readership—or a person willing to read out loud to listeners—in addition to the fact that they had to be purchased on a regular basis.

## Early Decisions and the Invention of *Tele-Tradiciones*

One of the most significant early decisions regarding the television industry was reflected in President Miguel Alemán's signing of a decree that called for the establishment of a commercial television model rather than a state-run system similar to the model implemented in the United Kingdom. The commercial model represented one of the first Mexican *tele-tradiciones*.[9]

The state's fairly hands-off attitude toward television was not a *fait accompli*, at least not until the publication of the "Novo Report" in 1948. A year earlier, in 1947, President Alemán called on the director of the Instituto Nacional de Bellas Artes (INBA, National Institute of Performing Arts), Carlos Chávez, to conduct a study of television systems. Chávez selected broadcast engineer Guillermo González Camarena and intellectual Salvador Novo to lead the commission. In October, the two traveled to the United States and Europe, meeting with technical and editorial experts in both countries. Novo praised the content and model of the British Broadcasting Company (BBC) but offered no specific recommendations. González Camarena, on the other hand, proposed that the country adopt a commercial model for technical and financial reasons. He observed that equipment already in place had been modeled after US designs, noting, moreover, that much of the new equipment that was needed to set up a television network would be less expensive if imported from the United States.[10] Following the commission's report, a decree published on February 11, 1950, in the *Diario Oficial de la Federación* announced that the country would follow González Camarena's recommendation and implement a commercial television system.

To understand the government's decision to adopt a commercial model, one must consider radio's impact on the development of television. Early media families already established in both the radio and print media industries represented the most likely investors in television.[11] In the years before World War II, a number of family businesses grew into powerful multimedia empires, which allowed them to invest in the expensive but promising television business.[12] Furthermore, connections between US radio executives and Mexican media barons had already been established, relationships that continued to develop with the advent of television. Indeed, prior to setting up Televicentro, Emilio Azcárraga Vidaurreta toured and trained at NBC.[13]

There is some debate as to whether the establishment of a commercial TV system was a foregone conclusion due to broadcast entrepreneurs' prior experience with radio and the political stance adopted by the Alemán administration. Some scholars argue that the creation of a commission to examine the various TV systems was simply a stalling tactic so that Alemán's front man, Rómulo O'Farrill, could get his business operations in order for the inauguration of the first station, XHTV, Channel 4.[14] Whatever the reasons for the adoption of a commercial instead of a state-run model, the government's decision quickly impacted all levels of the industry.

With a commercial television system in place, the need for program sponsors and advertisers was created. Initially, news and entertainment programs received support from sponsors such as General Motors, Nescafé, and Max Factor. These companies sponsored the early television programs *El Teatro Nescafé*, *General Motors News*, and *El Programa de Max Factor*. Later on revenue would be generated largely through the sale of advertising spots to domestic and foreign companies. Yet the profit motive created a tension between information and entertainment. News had to be "sold," and to do so it had to be entertaining. Therefore, from the earliest days of TV news broadcasting, anchors reported on "hard news" as well as celebrity happenings, such as the latest films starring Mexican starlets like María Félix or Hollywood bombshells like Marilyn Monroe. Similarly, producers embraced sports as a central component of TV news.

While the government tended to allow private interests to direct the budding industry, on occasion it did attempt to exert influence over the airwaves. For example, in 1958, President Adolfo Ruiz Cortines helped establish the country's first state-owned television station, Channel 11, housed at the Instituto Politécnico Nacional (National Polytechnic Institute). Nevertheless, lack of resources prevented the station from competing with programs produced by commercial television.[15]

In addition to direct competition, the government sought to influence the new medium through legislation. At the time of its inauguration, there was no law written to regulate TV, and it therefore fell under a 1940 law concerning radio broadcasting. The Ley de Vías Generales de Comunicaciones contained six basic provisions: It gave the government exclusive regulatory control over the airwaves; established the Secretaría de Comunicaciones de Obras Públicas (SCOP, Ministry of Communications and Public Works) as the agency that would grant concessions and permits; gave the government the right to set advertising rates and establish rules for emergency transmissions; established that government would pay 50 percent less than the rate-card cost to air official programs; gave the SCOP the power to demand the correction or improvement of station services and to suspend any station that did not meet standards, as well as the right to levy fines for violations; and, finally, stipulated that no foreigner could own a radio station, nor could a station change ownership without SCOP approval.[16]

Ten years after the launching of TV, the government finally passed a federal law regulating the industry. The new legislation made it mandatory for stations to grant free airtime to the government, while it also called on the government to participate more directly in the industry. The law stipulated that the president, by way of the Ministry of the Interior, was to act as a watchdog, ensuring that television promoted Mexican cultural and traditional values. Moreover, the law called upon the Ministry of the Interior to produce educational, cultural, and social programs that were to be broadcast free of charge on commercial stations.[17]

This new legislation aimed to counteract the onslaught of foreign, particularly US, programming. One provision, for example, required that only Spanish could be used in broadcasts unless prior permission had been obtained from the Ministry of the Interior. Another prohibited broadcasts that could "corrupt the Spanish language, violate the accepted customs of the community, encourage anti-social behavior, denigrate national heroes, offend commonly-held religious beliefs or discriminate on the grounds of race or color."[18] Such specifications point to the government's increasing concern over television content and illustrate the explicit strategy employed to limit foreign cultural flows.

The government's growing role in television coincided with the expansion of viewers. At the outset of the industry, only one hundred television sets existed in the entire country, yet by 1957 that number had soared to three hundred thousand.[19] During these years, President Adolfo Ruiz Cortines (1952–1958) and members of his administration saw in television a means to implement their anticorruption campaign. As a result, they limited and in some cases banned certain programs. Since the inception of television, *lucha libre* had enjoyed exposure on the air, but in 1954, Ruiz Cortines and then Mexico City mayor Ernesto Uruchurtu outlawed the televising of the sport. Opponents of *lucha libre* characterized it as a barbaric sport and an activity that threatened to further corrupt popular behavior and attitudes, much like the bullfighting during the Porfiriato.[20] The sporting event remained off the air until 1991.[21]

Meanwhile, as the state aimed to shape television content, private investors attempted to consolidate and centralize the industry.[22] By 1955, the owners of Mexico City's three privately owned television stations, XHTV, XEW-TV, and XHGC, understood that although the upper and middle classes were expanding, they had not yet reached a level capable of sustaining three different commercial networks. Each company was forced to compete fiercely over advertising revenue. In 1955, Azcárraga Vidaurreta admitted to the press that the incipient industry was losing millions of pesos a year.[23] He said that the enormous costs associated with television operations were forcing "businessmen down a narrow road without an exit."[24] As a result, on March 26, Rómulo O'Farrill, Emilio Azcárraga Vidaurreta, and Guillermo González Camarena decided that instead of driving each other out of business, they would combine their economic resources.[25]

The consolidation of the national capital's three networks meant that all television producers would now work under one parent organization. The change in company structure had significant implications for the production of news in particular. Since all news production occurred under one company, reporting became increasingly standardized and program diversity diminished.

The formation of this mid-1950s monopoly initiated a tendency that would continue through the 1970s, culminating with the eventual establishment of Televisa in 1973 through the merger of Telesistema Mexicano and Televisión Independiente de México. The merger of these two organizations would produce a commercial television monopoly that would remain unchallenged until 1993, giving the new media conglomerate the ability to expand regionally and internationally.

## Early Producers

Television's pioneering producers represented a group of elite men and very few women.[26] Azcárraga Vidaurreta, the owner of XEW radio, "The Voice of Latin America," quickly emerged as a television luminary. For several years prior to the first inaugural broadcast, he had been planning to establish his own TV network. As the "czar of Mexican radio," Azcárraga Vidaurreta seemed like the natural candidate to receive the first television concession. But President Alemán had another person in mind, and, as we have seen, he granted the first concession to Rómulo O'Farrill.

Less experienced in media production, O'Farrill nonetheless had the right connections, both domestically and internationally, to get his station off the ground. The owner of car dealerships, Mexico City's daily *Novedades*, and the radio station XEX, he learned to skillfully cross-promote his multiple investments. Significantly, O'Farrill also received financial backing from movie theater mogul and Puebla resident William Jenkins. On the international front, O'Farrill worked directly with Meade Brunet, the vice president of RCA, who attended the country's inaugural television broadcast. With the help of these important connections, O'Farrill purchased $2.2 million in television equipment in anticipation of the first broadcast.[27]

Although Guillermo González Camarena's XHGC (Channel 5) had begun transmissions in 1950, in 1952, Azcárraga Vidaurreta became the second major commercial television player to enter the picture when his station, XEW-TV (with the same call letters as his radio station), was inaugurated. With ties to US commercial broadcasting from his radio days and his affiliations with NBC and CBS, Azcárraga Vidaurreta purchased technical equipment from General Electric and DuMont for the first transmissions.[28]

Such connections forged with foreign business owners embody what Fernando Henrique Cardoso termed "associated-dependent development."[29] This relationship was characterized by the mutual goals among foreign and domestic broadcast executives in their efforts to develop the television industry and thus accumulate wealth.[30] The alliances that Azcárraga Vidaurreta and O'Farrill fostered with their North American counterparts played a critical role in shaping the development of the industry.

In the arena of TV news, Jacobo Zabludovsky emerged as the nation's most influential anchor. The role he played in the first twenty years of television remains salient because he served as a model for TV journalists across the country.[31] During his career he interviewed hundreds, perhaps thousands, of politicians, intellectuals, entertainers, and economists. He witnessed and reported on most of the important events of the second half of the twentieth century, including eight presidential administrations, from Alemán to Carlos Salinas de Gortari; Fidel Castro's rise to power; the Cuban missile crisis; the 1968 massacre at the Plaza de las Tres Culturas; the 1985 Mexico City earthquake; and the Zapatista uprising in 1994. A controversial personality, he has been both lauded and lambasted; critics liken him to just another "soldier of the PRI [Institutional Revolutionary Party]."[32]

Like Azcárraga Vidaurreta and O'Farrill, Zabludovsky turned to television after working in radio at XEX with men like Gonzálo Castellot. Once in the new industry, Zabludovsky did much to establish the country's first television news programs. In 1950, as news editor for XHTV, he helped create the first long-running newscast, *Noticiero General Motors*, which aired until 1962. On the program Guillermo Vela served as the news anchor and Pedro Ferríz de Con acted as the commercial narrator.[33] As the news program editor, Zabludovsky wrote the anchor's scripts.[34] At the time, he also worked at the daily *Novedades*, which undoubtedly helped him find content for TV news broadcasts.[35] While Zabludovsky prepared the written copy of the news, television cameramen, who often had experience in film, were sent out to gather the accompanying the images.

Although Zabludovsky was referred to as the editorial director of *Noticiero*

*General Motors*, he wore multiple professional hats. During the period he worked as editor and anchor of newscasts on XHTV, he provided services to the executive branch of government as chief radio and television consultant for presidents Adolfo Ruiz Cortines, Adolfo López Mateos (1958–1964), and Gustavo Díaz Ordaz (1964–1970).[36]

The Mexican Constitution of 1917 theoretically gave television producers like Zabludovsky the right to broadcast newsworthy events. Article 7 of the constitution protected freedom of the press and forbade prior censorship, but what happened in practice was quite different. For example, Article 77 of the law that governed radio and television prohibited the broadcast of news items of a political or religious nature. Aurelio Pérez, who anchored the news in the 1950s, remembered how a representative of the Secretariat of Communications who paid him a visit explained Article 77. The functionary said that everything that had to do with Miguel Henríquez Guzmán (a PRI and pre-candidate for the presidency during the election of 1952) was political and, in accordance with Article 77, was prohibited; but all that had to do with President Ruiz Cortines was civic and, therefore, not prohibited. Article 77, along with the practice of *embutes* (payoffs), diminished the chances of programs critical of the government ever hitting the airwaves.

## Television Programming

The underlying goal of television executives (despite the provisions established in the 1940 and 1960 laws) was to entertain viewers and create a market for commercial goods. They accomplished this through both entertainment and news programming, rarely questioning the dominant ideologies of the time. Through TV newscasts and *telenovelas* and the commercial advertisements that aired alongside them, viewers learned about the latest model of Frigidaire refrigerators, Ford automobiles, and the experience of flying with Aeronaves. Meanwhile, they were presented with instruction on what constituted "acceptable" and "unacceptable" behavior for both public and private spaces. In short, television became a powerful mechanism to promote the values of the petit bourgeois, and the strategy of TV executives fit well with the government's plan to create a modern nation. Much evidence exists to demonstrate that mass media, both in and outside of the country, tend to promote the status quo rather than seek to undermine or criticize it. Yet, as the next section reveals, Mexican TV viewers did not always passively accept these values. Rather, they often developed their own competing interpretations.

Television news became a unifying tool for both media owners and national politicians. TV executives benefited economically from the creation of a mass audience, while government officials benefited politically once their messages could be broadcast across the nation. The ability to disseminate political ideologies over the airwaves resulted in a new form of political communication, called television diplomacy.[37] *Tele*-diplomacy was a more popular form of engagement that involved viewers from all levels of society, in contrast to traditional diplomatic activities that excluded most of the country's citizens. But viewers were only on the receiving end of the *tele*-diplomatic process and therefore could not contribute directly to political discourse over the airwaves.

President Adolfo López Mateos was one of the first adept TV diplomats, as he used the medium to his advantage. Newscasts throughout his presidency featured the president's daily activities. As he traveled throughout the republic, viewers would see him inaugurating various government-sponsored development projects such as new dams and roadways. He topped the news also when heads of state from other nations came for official visits, such as those of US president Dwight D. Eisenhower as well as Soviet Union vice premier Anastas Mikoyan in 1959.

News and entertainment programs helped promote the national government's and media executives' goals by portraying Mexico City as the nation's modern economic, cultural, and political center. Newscasts were filled with the major events transpiring in the capital city, and only on rare occasions did newscasts include stories about events and issues, such as natural disasters, in other states. *Tele-teatros* and, later, *telenovelas* were often set and filmed in the nation's capital. If they were filmed outside of the Ciudad de México, rural areas were often characterized as backwards. These news and entertainment programs helped to shape a sense of *lo mexicano* that was very Mexico City-centric.

Similar to politicians, media moguls used television to promote their own interests.[38] This was particularly evident in news programming. O'Farrill Sr. and Azcárraga Vidaurreta appeared frequently as subjects of news reports as they met for *comidas* and *cócteles* with leaders of various governmental offices and multinational corporations. On February 21, 1955, Rómulo O'Farrill Sr. and Rómulo O'Farrill Jr. served as the focus of a report on *Noticiero PEMEX SOL Novedades* (a television program) that provided details about a meal they hosted for Mr. Vernon Moore and Mr. J. J. McIntire of General Motors, where the O'Farrills demonstrated "their sympathy and understanding" for the work of the GM executives.[39]

Meanwhile, men such as the O'Farrills used television to promote other modern technologies that they hoped would further their own interests and advance national development. An avid automobile enthusiast, O'Farrill Sr. served as the president of the Comité Directivo de los Congresos Panamericanos de Carreteras (Pan American Highway Congress). In 1955, a lead report addressed

O'Farrill's role in the construction of the hemispheric highway.[40] It was probably no coincidence that the story aired in a newscast sponsored by PEMEX, the national oil company. Stories like these demonstrated the important role that television played in the nation's development, as well as in promoting the efforts of the country's leading capitalists.

In the 1960s, as the country geared up for the first Olympic Games to be held in Latin America, media executives and government officials worked in unison to foment their mutual goal of portraying the country as modern and orderly. Newscasts featured various construction projects that, once completed, would be used during the 1968 Olympic Games, including the building of the stadium Estadio Azteca in 1966.[41] News items highlighted the technological feats that made the first worldwide broadcasts possible, such as the inauguration of the satellite station at Tulancingo just days before the opening of the Olympic Games. In general, news broadcasts emphasized the international sporting event and downplayed the unrest among students and other popular groups. Coverage of the government crackdown on students and subsequent massacre at Tlatelolco Plaza on October second was severely limited. As in 1968, major sporting events often sidelined important "hard news" stories.

Beyond sports, citizens learned about their country's precarious position during the Cold War and related space race by watching television. News broadcasts featured stories about Sputnik, the first satellite put into orbit in October 1957, and other Soviet rocket launchings, alongside the coverage of the numerous US space missions. Viewers even learned about Mexico's own space program and the launching of rockets two years after Sputnik. Numerous programs were dedicated to space exploration, including *La verdad en el espacio* (The truth in space), hosted by Jacobo Zabludovsky.

The politics of the Cold War took up hours of airtime at critical moments. News reports, for example, reflected the nation's precarious position as politicians walked a political tightrope in maintaining relations with Cuba and not straining relations with the United States. Tensions reached a high point in October 1963, during the Cuban missile crisis, when viewers saw their president shift positions from supporting Fidel Castro to aligning with the United States after it became apparent that nuclear missiles were found on the island nation.

Despite TV's propensity to reaffirm existing social and political values, news and entertainment programs could not avoid illustrating a country in the midst of dramatic change. During the early years of television, women, for example, gained the right to vote; and two years later they voted for the first time in the congressional elections of 1955. By July 6, 1958, more than 4.5 million women had registered for the presidential election.[42] On the day that women first cast ballots in a presidential election, TV news writers heralded the event, observing:

"Mexican women, women from all social classes have learned to respond to the confidence placed on them when they were given the full political right [to vote]. Almost four million women voted in the whole Republic."[43]

News writers described the event as a positive move for the country's political system but stopped short of giving voice to those citizens affected directly. Viewers heard that at about nine in the morning, "the first citizen of Mexico, Adolfo Ruiz Cortines, and his wife, voted at Pipila School," yet the report referred to María Dolores Izaguirre de Ruiz Cortines as the president's wife, failing to even mention her name.[44] Next, viewers were told that Antonio Carrillo Flores, secretary of the Hacienda, voted, that "his wife Fanny Gamboa de Carrillo and son voted on Nebraska street," and that PRI presidential candidate Licenciado Adolfo López Mateos voted, "accompanied by his wife."[45] Viewers got a chance to see that opposition candidate Luis Álvarez of the Partido Acción Nacional voted, "and just as throughout the campaign, he was accompanied by his distinguished wife."[46] The only woman mentioned by name and as an individual—that is, not as part of a couple—was film star of the Golden Age María Félix.[47] Women were allowed a place on the screen, like the space allocated to them in the movies or in television entertainment programs, but it was a space that was mostly controlled by men. Only under certain circumstances were women allowed to occupy the screen stage by themselves; such was the case of María Félix.[48]

During the first decade of TV, much of the early programming, including advertising spots, catered to the interests of the wealthy as they were the only people able to afford a television set. Yet, from its earliest days, executives strove to cultivate a popular viewership. In an attempt to appeal to a large audience, XHTV, Channel 4, broadcast the first TV musical, *Adelita y sus dorados*, with Rosa de Castillo and Vicente Peña.[49] Soon, however, the *telenovela* emerged as the most profitable and popular form of TV entertainment, enabling Telesistema Mexicano and, later, Televisa to expand viewership at home and abroad. Telenovelas such as *Los ricos también lloran* made actors like Verónica Castro national as well as international stars.

With the rise and international popularity of rock and roll in the 1950s and 1960s, television executives sought to entertain viewers through this new genre of music. Traditional mariachi groups now had competition when programs such as *Hullabaloo, Yeah, Yeah*, and *Discotheque a Go Go* hit the air.[50] There were limits to how many and whose gyrating hips Telesistema Mexicano's conservative owners would allow to be broadcast. Musicians deemed as inappropriate "long hairs" were not afforded a place on screen. Similarly, the controversial Avándaro Music Festival, held outside of Mexico City in 1971, which some have likened to Woodstock, never appeared on television.[51]

Through advancements in satellite technology during the 1970s, Televisa was able to disseminate its programs and cultural aspects that the ownership deemed appropriate to people around the world. Through telenovelas such as *Los ricos también lloran* and Latin America's first international news network, ECO, viewers thousands of miles away could hear about the latest events in Mexico, as well as keep up with cultural trends. Satellite communications allowed news and entertainment programs to be beamed north of the country's border, where the nation's growing diaspora could keep connected to events in their homeland. Spanish-language programming produced in Mexico became significant in communities in US states such as California, Texas, and Florida, where the number of Spanish-speaking residents was on the rise.

By the end of the century, Televisa's news and entertainment programs were being broadcast in more than one hundred countries, bringing millions of people into contact with Mexican culture and society. Of course, the version these viewers encountered—whether in Eastern Europe or East Los Angeles—tended to emphasize the country's male-dominated, urban, non-indigenous, and non-elite characteristics.[52]

## TV Consumers

Mexico, like most post–World War II societies, has been significantly impacted by the power and ubiquity of consumerism. Although consumerism may be tied to the influence of mass media, citizens have often contested and reshaped the ideas and ideologies broadcast on television. Alison Greene, for example, has shown that indigenous viewers in rural Yucatán did not incorporate middle-class and nationalist values embedded in the telenovelas they watched during the 1990s. Instead, some Yucateca Maya used the messages of telenovelas as a way to reaffirm their Mayan identity.[53] The same sort of process of negotiation between TV and lived experience was underway during the earlier stages of the medium. In other words, citizens were not merely "injected" with the messages they watched, but they interpreted TV and mass media in ways that they found meaningful.[54]

Today, television touches the lives of nearly all citizens, but it took decades to reach this point. During the early years, the price of a television set, four thousand pesos in the 1950s, was simply economically out of reach for most citizens. As a result, Mexico City residents often caught bullfights and soccer matches from outside department stores, turning storefronts into stages for public viewings. Bars and restaurants likewise offered alternative venues to catch the news and sporting events. A literal cottage industry of selling television viewing time quickly developed as those who could afford to buy this new device often charged their neighbors twenty to fifty centavos to watch their favorite programs.[55] For TV set owners, the apparatus functioned as a symbol of social status; and they frequently placed their new products, like other forms of conspicuous consumption, in places of high visibility so that friends would be sure to see their possessions.

Less than a decade after the first official TV broadcast, it appeared that the medium was altering the daily lives of citizens, at least those living in urban areas within distance of a broadcast signal. Some early viewers recall that their sleeping habits changed; with the arrival of the small screen, people went to bed later. Mexico City resident Rosa Gomez said she noticed changes: "Girls dressed more in style and more things were bought on the installment plan as a result of televising advertising. People also had new ideas—a neighbor's daughter wanted to become a ballet dancer after she had seen a dance group on TV."[56] Others tended watch TV more and listened less to radio or went less frequently to the movies. The arrival of television was certainly a concern for film producers, who passed measures limiting the broadcast of first-run Mexican films on television for six years after their release dates.[57]

As a new luxury item, the television set increased social tensions among the haves and have-nots while increasing a family's financial liability. Some citizens were forced to pawn items in order to pay their monthly installment on their new set. At least one member of the Gutiérrez family, studied by Oscar Lewis during the 1950s, had to stay home while other family members were away for fear that their television might be stolen.[58]

When they were at home watching television, not all viewers were pleased with what they saw. As early as the mid-1950s, programming content began to produce outrage among some citizens. Viewers formed groups such as the League for Decency to lobby President Ruiz Cortines to prohibit certain types of programming and behavior, particularly *lucha libre* and kissing on television.[59] On October 2, 1953, the president of the Federal Association of Parents of School Families petitioned Ruiz Cortines to "moralize television transmissions" and requested the prohibition of *lucha libre* broadcasts, as well as immoral magazines such as *Pepines* and *Chamacos*.[60] This public outcry against television programming represented part of a larger effort to moralize a city and country in the midst of dramatic social and cultural change.[61]

Public criticism of mainstream media and the government continued into the 1960s. By the summer of 1968, the limits of both the "economic miracle" and the Mexican Revolution prompted thousands of students and concerned citizens to voice their criticisms on the streets of Mexico City. Students criticized the media for having sold out to the government, and they condemned the government for its authoritarian practices.

Large-scale demonstrations worked against both the government's and media executives' mutual goal of creating a positive and orderly image of the city for domestic and international audiences. President Gustavo Díaz Ordaz's warning to protestors during his September 1, 1968, *informe*, stating that the government would take all means necessary to maintain control, did not quell student unrest. Nor did the military takeover of the national university a few weeks later. With the October 12 opening ceremony of the Olympic Games, to be held in Mexico City (the first in Latin America), just ten days away, the government cracked down on students and innocent bystanders in the Plaza de Tlatelolco.[62] Several hundred citizens were killed, including some soldiers, and thousands of so-called *terroristas* were rounded up and taken to Military Camp Number 1.

As mentioned earlier, TV news programs during the run up to the Olympic Games and the massacre demonstrated Telesistema Mexicano's efforts to downplay violence, while it emphasized sports over politics.[63] Jorge Saldaña, a news anchor and host for Telesistema Mexicano in the summer and fall of 1968, argued that limiting coverage of the student movements served the financial interests of the television company and that "for television it was important to have no criticism, problems in the country, or the capital so that the natural course of selling goods could continue. It was a business, and businesses are not interested in alternatives or social disturbances."[64]

Despite Telesistema Mexicano's efforts to limit dissident voices during the summer of 1968, students showed that they had very different interpretations of the events transpiring in the capital city, and they aired their points of view via alternative and low-tech means of communication. Students formed brigades and took to the streets, where they staged mini-plays to inform citizens about their concerns and demands. During demonstrations and throughout the city, students collected money from supporters to print propaganda that could not be aired on television or published in government-supported dailies. Ana Ignacia "La Nacha" Rodríguez was a former student activist and member of el Comité '68.[65] She recalled that *las brigadas* formed a crucial part of the movement, and the messages they disseminated may have been "the simplest medium, but the most effective. We were like mobile newspapers."[66] These "mobile newspapers" offered another version of reality in the growing and tumultuous capital city.

The long-standing practice of exclusionary and politically biased TV news coverage came at a price. As a result of Televisa's efforts to toe the line of the government, television executives began to lose what they needed most, viewers. Studies showed that by the beginning of the twenty-first century, the country's third generation of TV viewers was "tuning-out" in alarming numbers.[67]

## Conclusion

During the first few years of the medium, television executives may have had to dip deeply into their pockets to get the industry off the ground, but it did not take long for TV to grab viewers' attention and for the industry to begin to turn a profit. A growing middle class, advances in technology, and government assistance all worked in favor of network owners. A larger pool of upwardly mobile citizens meant more people could afford to purchase TV sets, as well as the products they saw advertised on TV. In 1959, 120,000 television sets were sold nationwide; most consumers seemed to prefer large screens over smaller portable sets.[68] By 1960, a decade after the first official broadcast, monthly viewership nationwide was estimated at 3,864,122, with about 780,000 television sets in operation.[69] By 1977, Televisa transmitted more than twenty-one thousand hours of programming to an estimated 28 million viewers, and the company's television advertising revenue soared to $144 million.[70]

The industry and those who controlled it marched forward in tandem with major technological advances, including color television in the 1960s, the advent of videotape, cable television, and satellite communications. In 1962, Guillermo González Camarena patented color television technology in Mexico and the United States to be used in various parts of the world.[71] Videotape recording, which first arrived in the late 1950s but began in earnest in the 1960s, allowed live programs to be taped, which facilitated domestic and international distribution. The nation's first satellite station (which allowed satellite transmission from Mexico abroad) arrived in time for the Olympic Games in 1968. Prior to the Tulancingo satellite station, important events were broadcast through microwave signals to satellite stations in the United States and then relayed to other parts of the world.

The government invested heavily in telecommunications infrastructure, mutually benefiting the goals of political officials and media executives. In 1965, the Secretariat of Communications and Transport (SCT) announced it would create a national network of microwave stations. By 1970, sixty-five stations throughout the republic had been established, along with 207 repeaters, covering some 12,800 kilometers of the nation's territory. Although microwave technology allowed programs to be beamed throughout the country, it was limited by line of sight, forcing the construction of high towers, many times in mountainous terrain.

Satellite communications were faster and more efficient because images and sound could be transmitted up and down from a satellite and line of sight was not an issue. While the nation began using satellite communications in the mid-1960s, the government and television executives had to depend on US-owned satellites for transmissions beyond the nation's borders. It was not until 1985, when Morelos I was launched into space, that Mexico established its first telecommunications satellite. Financed by the SCT, Morelos I cost $140 million.

The millions that the government poured into telecommunications infrastructure allowed Televisa to strengthen its commercial monopoly. The company's grip over the TV market, however, began to loosen slightly in the mid-

1990s, after the 1993 establishment of TV Azteca. The new network, owned by Ricardo Salinas Pliego, aired entertainment and news programming that openly criticized the government. This created instant credibility for TV Azteca and cast into sharp relief its competitor's credibility problem. Televisa responded by producing less favorable coverage of the government as a strategy to regain viewers. The company's new president, Emilio Azcárraga Jean, the grandson of Emilio Azcárraga Vidaurreta, conceded: "We are not rooting for the [ruling] Institutional Revolutionary Party (PRI), nor any other party. We are rooting for Mexico."[72] This statement came shortly after the death of Azcárraga Jean's father, Emilio Azcárraga Milmo, who had on occasion called himself a soldier of the PRI.

By the beginning of the twenty-first century, Azcárraga Jean and Salinas Pliego consistently topped the Forbes list as two of the richest men in the world. For decades Televisa has been not only the largest television company in Mexico, but the largest producer of Spanish-language programming in the world. In 2006, Televisa used its prime-time news program to garner support for the so-called Ley Televisa, which established a duopoly between Televisa and TV Azteca for twenty years, until their television concessions go up for renewal.

Since the inception of the medium, television executives have had their eyes on the US market. As early as 1961 Azcárraga Milmo set up a Spanish-language affiliate in San Antonio, Texas, his place of birth. He later renounced his US citizenship so that he could take control of the media empire that his father had established.[73] As new media, such as online publications and social-networking media, entered the media landscape, the owners of "legacy media" (television and print media) searched for ways to compete in a new technological environment. Televisa has attempted to go global through online telecommunications, but its efforts in the United States have been limited. Because of an agreement with Univision, Televisa has been barred from setting up an online channel that is accessible to users in the United States.

In Mexico, the majority of citizens continue to receive information and entertainment via broadcast television. Although satellite and cable television have been available for decades, the cost related to these technologies, as well as online technology, have made them out of reach for millions of potential consumers. As media mogul Salinas Pliego put it, "Television is still the most important medium."[74]

## Notes

1. Comisión Federal de Telecomunicaciones (COFETEL), with data from the Instituto Nacional de Estadística y Geografía (INEGI), 2005.

2. Although 1950 marked the first official broadcast, the country's engineers had been working to develop the medium since the 1930s.

3. Brazil officially inaugurated television on September 18, 1950, and Cuba followed a month later.

4. *Novedades*, September 1, 1950, 1.

5. There are a number of reasons for the lack of research on the history of television in Mexico, the most important being the fact that most television archives are owned by private companies; and in general, it has been difficult for scholars to gain access to the main commercial television network archives.

6. Some significant contributions include Fátima Fernández Christlieb, *Los medios de difusión masiva en México* (Mexico City: J. Pablos, 1982); Claudia Fernández and Andrew Paxman, *El Tigre: Emilio Azcárraga y su imperio Televisa* (Mexico City: Grijalbo, 2000); Fernando Mejía Barquera, *La industria de la radio y la política del Estado Mexicano*, vol. 1, *1920–1960* (Mexico City: Fundación Manuel Buendía, 1989); Marjorie Miller and Juanita Darling, "Emilio Azcárraga and the Televisa Empire," in *A Culture of Collusion: An Inside Look at the Mexican Press*, ed. William A. Orme Jr. (Miami: North-South Center Press, 1997), 59–70; Guillermo Orozco Gómez, "La televisión en México," in *Histórias de la televisión en América Latina: Argentina, Brasil, Colombia, Chile, México, Venezuela*, ed. Guillermo Orozco Gómez and Nora Maziotti (Barcelona: Gedisa, 2002), 59–70; Raúl Trejo Delabre, ed., *Las redes de Televisa* (Mexico City: Claves Latinoamericanas, 1988).

7. Laura Castellot de Ballin, *Historia de la televisión en México: Narrada por sus protagonistas* (Mexico City: Alpe, 1993), 20.

8. Celeste González de Bustamante, *"Muy buenas noches": Mexico, Television, and the Cold War* (Lincoln: University of Nebraska Press, 2012).

9. The English translation would be "tele-traditions." Eric Hobsbawm, "Introduction: Inventing Traditions," in *The Invention of Tradition*, ed. Eric Hobsbawm and Terence Ranger (Cambridge: Cambridge University Press, 1983), 1.

10. Fernando Mejía Barquera, "Del canal 4 a Televisa," in *Apuntes para una historia de la televisión mexicana*, vol. 1, ed. Miguel Ángel Sánchez de Armas (Mexico City: Revista Mexicana de Comunicación, 1998), 19–98.

11. Fátima Fernández Christlieb, *Los medios de difusión masiva en México* (Mexico City: Juan Pablos Editor, 1985), 87–100.

12. Ibid.

13. Joy Elizabeth Hayes, *Radio Nation: Communication, Popular Culture, and Nationalism in Mexico, 1920–1950* (Tucson: University of Arizona Press, 2000), 34.

14. Alex Saragoza, "Azcárraga and the Origins of Mexican Television," in *The State and the Media in Mexico: The Origins of Televisa* (forthcoming).

15. Luis Antonio de Noriega and Frances Leach, *Broadcasting in Mexico* (London: Routledge and Kegan Paul, 1979), 22.

16. Ibid.

17. Ibid., 30–31.

18. Ibid., 32.

19. Miriam Delal Baer, "Television and Political Control in Mexico" (PhD diss., University of Michigan, 1991), 79.

20. José Agustín, *Tragicomedia Mexicana 1: La vida en México de 1940 a 1970* (Mexico City: Planeta, 1990), 140.

21. Heather Levi, *The World of Lucha Libre: Secrets, Revelations, and Mexican National Identity* (Durham: Duke University Press, 2010).

22. Enrique E. Sánchez Ruiz, "Los medios de difusión masiva y la centralización en México," *Mexican Studies/Estudios Mexicanos* 4, no. 1 (1988): 25–54.

23. *Boletín Radiofónico*, num. 62, March 31, 1955.

24. Francisco Hernández Lomeli, "Obstáculos para el establecimiento de la televisión comercial en México (1950–1955)," *Comunicación y Sociedad* 28 (September–December 1996), 147–71.

25. The owners of XEW-TV and XHGC had already joined forces the year before. Fernando Mejía Barquera, "Anexos: Cronología," in *Apuntes para una historia de la televisión mexicana*, vol. 1, ed. Miguel Ángel Sánchez de Armas (Mexico City: Revista Mexicana de Comunicación, 1998), 529.

26. I use the word *producers* inclusively to include media executives such as Emilio Azcárraga Vidaurreta and Rómulo O'Farrill because in Mexico television was perhaps more hierarchical than other television industries. In other words, decisions regarding programming were made from the top.

27. Castellot de Ballin, *Historia de la Televisión en México*, 20.

28. Mejía Barquera, "Del Canal 4 a Televisa," in *Apuntes para un historia de la televisión Mexicana*, ed. Miguel Ángel Sánchez de Armas (Mexico City: Revista Mexicana de Comunicación, Televisa, 1998), 28; Alejandro Olmos, "Algunos protagonistas de la televisión," in *Apuntes para un historia de la televisión Mexicana II*, ed. Miguel Ángel Sánchez de Armas and Maria del Pilar Ramirez (Mexico City: Revista Mexicana de Comunicación, Televisa, 1998), 280.

29. Fernando Henrique Cardoso, "Associated-Dependent Development: Theoretical and Practical Implications," in *Authoritarian Brazil*, ed. Alfred Stephan (New Haven, CT: Yale University Press, 1973).

30. Celeste González de Bustamante, "Dependency and Development: The Importance of TV News in the History of Mexican Television," *Revista Galáxia* (São Paulo) 18 (2009): 247–62.

31. Lolita Ayala, interview by author, Mexico City, July 31, 2008.

32. Fernández and Paxman, *El Tigre*, 325.

33. Sabás Huesca Rebolledo, "La noticia por televisión," in *Apuntes para una historia de la televisión Mexicana*, vol. 2, ed. Miguel Ángel Sánchez de Armas (Mexico City: Revista Mexicana/Televisa, 1998), 71.

34. Elena Poniatowska, *Todo México*, vol. 5 (Mexico City: Diana, 1999), 173.

35. Huesca Rebolledo, "La noticia por televisión," 102.
36. Poniatowska, *Todo México*, 5:184, 185.
37. S. J. Ball, *The Cold War: An International History, 1947–1991* (London: Arnold, 1998), 98.
38. Ibid.
39. *Noticiero PEMEX SOL Novedades*, February 21, 1955, Filmoteca de Noticieros de Televisa, Estadio Azteca, hereafter cited as FNTEA.
40. Ibid.
41. The stadium was owned by Emilio Azcárraga Milmo, who also owned the soccer team Club America that played at the stadium.
42. Philip Taylor Jr., "The Mexican Elections of 1958: Affirmation of Authoritarianism?," *Western Political Quarterly* 13, no. 3 (1960): 722–44.
43. *Noticiero General Motors*, July 6, 1958, FNTEA.
44. Ibid.
45. Ibid.
46. Ibid.
47. Ibid.
48. Celeste González de Bustamante, "From News to Entertainment: Productions of Mexican Femininity in the 1950s," *Studies in Latin American Popular Culture* (in press).
49. Gonzálo Castellot, *La televisión en México, 1950–2000* (Mexico City: Edamex, 1999), 40.
50. Eric Zolov, *Refried Elvis: The Rise of the Mexican Counterculture* (Berkeley: University of California Press, 1999), 103.
51. Ibid., 202.
52. John Sinclair, "Neither West nor Third World: The Mexican Television Industry within the NWICO Debate," *Media, Culture and Society* 12, no. 3 (1990): 343–60.
53. Alison Greene, "Cablevision (Nation) in Rural Yucatán: Performing Modernity and *Mexicanidad* in the Early 1990s," in *Fragments of a Golden Age: The Politics of Culture in Mexico since 1940*, ed. Gilbert M. Joseph, Anne Rubenstein, and Eric Zolov (Durham: Duke University Press, 2001), 415–51.
54. Celeste González de Bustamante, "1968 Olympic Dreams and Tlatelolco Nightmares: Imagining and Imaging Modernity on Television," *Mexican Studies/Estudios Mexicanos* 26, no. 1 (2010): 1–30.
55. Oscar Lewis, *Five Families: Mexican Case Studies in the Culture of Poverty* (New York: Basic Books, 1959), 82.
56. Rosa Gomez, quoted in Lewis, *Five Families*, 83.
57. Francisco Hernández Lomeli, "Obstáculos para el establecimiento de la televisión comercial en México (1950–1955)," *Comunicación y Sociedad* 28 (September–December 1996): 147–71.

58. Lewis, *Five Families*, 147.

59. *Excélsior*, October 28, 1954, 10.

60. Archivo General de la Nación, Mexico City, Fondo presidencial, Adolfo Ruiz Cortines, Caja: 1282, Exp. 704/208.

61. Anne Rubenstein, *Bad Language, Naked Ladies, and Other Threats to the Nation: A Political History of Comic Books in Mexico* (Durham: Duke University Press, 1998); Rachel Kram Villarreal, "Gladiolas for the Children of Sanchez: Ernesto P. Uruchurtu's Mexico City, 1950–1968," (PhD diss., University of Arizona, 2008).

62. The scholarship produced about the 1968 massacre at Tlatelolco is extensive. Some of the most prominent works include Sergio Aguayo, *1968: Los archivos de la violencia* (Mexico City: Grijalbo/Reforma, 1998); Raúl Álvarez Garín, *La estela de Tlatelolco: Una reconstrucción histórica Movimiento estudiantil del 68* (Mexico City: Grijalbo, 1998); Elena Poniatowska, *La noche de Tlatelolco: Testimonios de historia oral* (Mexico City: Ediciones Era, 1971); José Revueltas, *México 68: juventud y revolución* (Mexico City: Ediciones Era, 1978); Julio Scherer García and Carlos Monsiváis, *Parte de Guerra, Tlatelolco 1968: Documentos del General Marcelino García Barragán, los hechos y la historia* (Mexico City: Siglo/Aguilar, 1999). Aside from Elena Poniatowska's *La noche, Massacre in Mexico*, translated by Helen R. Lane (Columbia: University of Missouri Press, 1992), few works in English have emerged. Elaine Carey's *Plaza of Sacrifices* (Albuquerque: University of New Mexico, 2005) is the only scholarly book in English that analyzes the student movements and subsequent massacre at Tlatelolco from a gendered perspective. Focused on the Olympics of 1968, Kevin B. Witherspoon's *Before the Eyes of the World: Mexico and the 1968 Olympics* (DeKalb: Northern Illinois University, 2008) analyzes the myriad of controversies surrounding the XIX Olympiad, including the government crackdown on students. Diana Sorensen examines the work of literary figures such as Octavio Paz and their relationship to politics, modernity, and post-modernity in her book *A Turbulent Decade Remembered: Scenes from the Latin American Sixties* (Palo Alto: Stanford University Press, 2007).

63. Celeste González de Bustamante, "1968 Olympic Dreams and Tlatelolco Nightmares: Imagining and Imaging Modernity on Television," *Mexican Studies/Estudios Mexicanos* 26, no. 1 (2010): 1–30.

64. Jorge Saldaña, telephone interview by author, August 5, 2008.

65. Ana Ignacia Rodríguez was imprisoned three times between 1968 and 1970. She spent almost two years incarcerated in a women's prison, from January 3, 1969, to December 31, 1970. Interview by author, Mexico City, August 5, 2009.

66. Ana Ignacia Rodríguez interview.

67. Manuel Guerrero and Victoria Isabela Corduneanu, "Trust, Credibility, and Relevance in the Consumption of Information among Mexican Youth: Third Generation TV Audiences," in *Empowering Citizenship through Journalism, Information, and Entertainment in Iberoamerica*, ed. Manuel Guerrero and Manuel Chávez

(Mexico City: University of Miami, Michigan State University, and Universidad Iberoamericana, 2009), 157–98.

68. Luis Becerra Celis, "Mexico," in *International Television Almanac: Who, What, Where in Television*, ed. Charles S. Aaronson (New York: Quigley Publications, 1961), 733.

69. Becerra Celis, "Mexico," 733.

70. Luis Antonio de Noriega and Frances Leach, *Broadcasting in Mexico* (London: Routledge and Kegan Paul, 1979), 54.

71. Fernando Mejía Barquera, "Anexos: Cronologías," in *Apuntes para una historia de la televisión Mexicana*, vol. 1, ed. Miguel Ángel Sánchez de Armas (Mexico City: Revista Mexicana/Televisa, 1998), 532.

72. Emilio Azcárraga Jean, quoted in Andrés Oppenheimer, "New Televisa CEO Promises Sweeping Changes: Networks Will Shun Political Biases of the Past, He Says," *Miami Herald*, November 9, 1997.

73. Fernández and Paxman, *El Tigre*.

74. Dom Serafini, "TV in Mexico: Maintaining the Status Quo is a Failure," *Video Age International*, June 1, 2004, http://www.allbusiness.com/legal/laws/168258-1.html (accessed May 1, 2010).

# 6
# And Television Appeared among the Mexicans

*Carlos Monsiváis*
*Translated by Lorna Scott Fox*

### The Audience in the Time of Genesis

The first generation of Mexican television viewers was born, matured, and declined between roughly 1952 and 1960, in a process observed through all of Latin America. Its outstanding feature was a capacity for awe that had been foreshadowed in a *guaracha* song: "Because TV / Will soon be here / And I'll sing for you / And you'll be watching me." The viewers became an instant congregation, much as had already occurred in North America, devoted to this cult of startling or repetitive images, laughter, and tears, gestures designed to be faithfully reproduced, platitudes inescapably drilled in, and politicians parading before the news cameras as if leading a holy procession. This "generation of reverent awe" was galvanized year after year in Mexico by the "Mañanitas," the traditional birthday song broadcast every December 12 from the Basilica of Guadalupe, preferably in the voice of Jorge Negrete, Pedro Infante, or Pedro Vargas.

The first generation of TV addicts could be entertained by any old thing, if "any old thing" meant getting the movies "easy-peasy" into your lounge, bedroom, or one-room shack. Relentless banality was a corollary of awe. Although plenty of films offered more sophisticated plots or settings, people's first reaction to TV was almost unavoidably childlike, given the novelty of the medium. It seems unlikely for any mature adult to be enraptured by lucha libre wrestling, armchair theater, slapstick humor, dubbed US series (*Hopalong Cassidy, The Lone Ranger, I Love Lucy*), sci-fi programs in which the special effects were fairy tales in themselves (*Érase una vez* [Once Upon a Time]) or the clumsiness of maquettes, rustic fantasies (*Así es mi tierra* [That's my country]), and news programs that kowtowed to politicians as if they walked on water. Something you'd never hear at this time: forget "Man Bites Dog"; "Man, Sane, Praises Government"—now that would be an item worth reporting!

In this first phase, what stands out is the learning curve of the faithful television viewers. By and large the "awe generation" swallowed whatever it was fed, indifferent to the tawdry props, the gags that should have stuck in the throat, the absence of televisual pace, the pomposity, the actors and actresses jerking from one movement to the next as though marching in single file through a ravine, the boundless ineptitude, and everything else that, regrettably—to quote the censors' catchphrase—"*is* allowed into your home." Wonder surrounded the very existence of the machine: here was magic domesticated. Even more amazing were the soccer programs, which reinvented the sport as a homely event and shrank the stadium to toy-like proportions. Throw in the international scope of TV newscasts (the Korean War and the Suez crisis fast became familiar pieces of "everyday exoticism"); finally, add for maximum despair the *telenovelas* or soap operas, which discarded any sense of cinematic melodrama by borrowing the magazine-serial technique of piling on the plot twists and climaxes in quick-fire succession. More than the comedies unsuitable for adults or the series set in folksy ranch houses, it was these that "nationalized" every country's output by stating the obvious: the greatest entertainment of all lies in the renovation of melodrama, that sacramental language of the family.

## Of Consequences and Non-sequiturs

Among the many notorious effects of television were the following:

> It interfered abruptly with the routines of the home, undermining conventions of family life, which would never be restored. By this new employment of free time, the family itself was changed, almost without noticing, due to the drastic reinvention of its habits of conversation, leisure, and visual judgment.
> 
> The fortresses of traditionalism were rapidly breached since just by switching on a machine one's isolation from the world (one's "castle of purity") was abolished. Conservative morality claimed to be holding its own to some degree, but in practice the decisive relations were those between convenience (the investment of leisure time in screen hours) and tradition. No prizes for guessing the winner.
> 
> Collective speech was modified by the contraction and adaptation of its vocabulary, advertising slogans acted as the stimulants of the day (advertising as a never very secret utopia), and at work or among friends the talk inevitably revolved around last night's viewing. Where Mexican cinema

provided an overall image of the country, television (private, of course) tragically fragmented experience.

Radio and films had introduced audiences to the broader global environment, but TV consolidated this window on the world in all-encompassing fashion. It announced what was absolutely inescapable (Americanization, the acceptance of a single path to modernity), while treating older customs as relatively elastic. As tolerance for other ways of life gradually became the norm, many hitherto unknown behaviors (self-irony, for instance) also grew more common. Especially in what was still known as "the provinces," TV was the challenger that displaced seven o'clock Mass, family dinners, evening peace, the twilight stroll through the streets, like a nomadic cocktail party. As television took hold, traditional observance of customs receded ever more into devotional memory.

## Newscasts from Mexico

On September 1, 1950, Mexican television was formally unveiled by XHTV–Channel 4's broadcast of President Miguel Alemán's "Fourth Executive Report." On March 21, 1951, Channel 2 appeared, owned by Emilio Azcárraga Vidaurreta. May 10, 1952, saw the launch of XHGC–Channel 5, directed by Guillermo González Camarena. In 1955 the merger of Channels 2, 4, and 5 produced the Telesistema Mexicano, which gradually ramified across the country. The first telenovela, Fernanda Villeli's *Senda prohibida* (Forbidden path), was beamed out in 1958. On January 8, 1972, Televisa was born as the offspring of the union between Telesistema Mexicano and Televisión Independiente, directed by Emilio Azcárraga Milmo, Rómulo O'Farrill, and Miguel Alemán Velasco.

Televisa was gigantic, if you'll forgive the understatement. Besides TV, it owned a crop of radio stations, including XEW and XEQ, several record labels, a movie company, the Azteca stadium, and the América soccer team, plus a stake in many more businesses. In 1965, the group's first satellite, the *Pájaro Madrugador* (Early Bird), swung into operation with a boxing match. By 1968, there were some two million TV sets in Mexico, and advertisers were spending $40 million a year. Thanks mostly to the popularity of its telenovelas and some comedy shows, Televisa expanded throughout Latin America. The other cash cow was soccer.

Azcárraga Vidaurreta, the first airwaves tycoon in Mexico and, indeed, Latin America, was the typical rags-to-riches radio salesman who, almost by default, became a radio and recording magnate and thence, inevitably, a TV baron. He had no need of shrewdness or knowledge, even though he possessed them. His trade was to popularize and sell methods of entertainment, and by *entertainment*, I mean the elevation of domestic passivity to heights of sacred obligation.

His formula was a smart one: invent the family, and the family will sacralize the inventor. Such was its success that this maneuver became entrenched beyond appeal. Whatever else the spectator might yearn for, there would be no escape from the dictatorial definition of *entertainment*: for example, in practice, whatever demanded no effort of concentration and simply invested free time in trivia was promoted to the status of a new custom.

Private television was a fount of daily gratification in the form of jokes, songs, melodramas interrupted or improved by commercials, commercials stressed by the imminent return of melodramas, quiz shows (with questions along the lines of "What color was Napoleon's white horse?" or "Who sponsored the first day of Creation?"), sports roundups, religious services, and news shows that resembled society pages. In those days, the power of the media moguls was crucially dependent on presidential goodwill. They were powerful because they were allowed to do business in peace, and they were expected to repay this trust with loyalty and self-censorship. No sex, so as not to offend the Catholic Church, a prominent ally of the Executive Power; no airtime for the opposition; no subversive rumors. This is not to say that lust was out of bounds (shapely bodies are opportunities for repentance, quipped the comics), and this did not exclude double entendres or the massive subversion that was the opening to modernity itself.

The monopoly was ruthless. Popular singers who did not sign up with a record label linked to Azcárraga Vidaurreta's radio and TV empire found themselves confined to playing village fiestas; exceptions were only made for the insignificant, or else for stars such as Pedro Infante, who could afford to snub the system. In the past, popular tastes had been formed by tacit agreement between the industry and its audience, expressed in box-office receipts; by the 1960s, however, tastes were almost entirely molded by private TV. The government shrugged, as if to say: "Do what you like, provided you do what you must for the state." This pact was ratified in Mexico in 1968, when private TV and radio stations—after ferociously attacking the student movement—skated lightly over the Tlatelolco massacre. Equally obliging was their suppression of protest movements or accurate economic data. Similar pacts were in evidence in Argentina during the "dirty war," in Augusto Pinochet's Chile, and in Alberto Fujimori's Peru. Another brand of wholesale censorship was practiced by Cuban television under Fidel Castro.

## It Doesn't Matter What's on, Be Grateful for What Isn't

From around 1960 to 1968, the "awe generation" was succeeded by what we might call the "treadmill generation." Once the technological shock had been assimilated, viewers still did not feel entitled to any rights from the TV monopolies. They lacked even the choices allowed by cinema: those of not consuming it at all or of selecting a movie according to personal criteria or intuitions. Channel-hopping, the industrialization of Cartesian doubt, had not yet arrived. Television constituted a centralizing force by being the sole leisure occupation shielded from urban violence, "moral contagion," and consumer choice. Moral: if you're not amused by what's on, you'd better change your notion of amusing, because TV isn't about to change. You can either give in or go back to playing board games.

Entrepreneurs, producers, and TV celebrities had nothing but praise for the helplessness of the audience and the effusive resignation with which all offerings were welcomed. The formula goes back to the Spanish Golden Age: "Y pues que paga el vulgo / hablarle en necio para darle gusto" (And since it's the riff-raff who's paying / Please them with some witless saying). Or, to put it another way, nobody ever lost money by underestimating the public's critical capacities. The *philosophy of the spectacle* runs as follows: "Dear spectator, you are unable to stop watching and listening to stupid, clichéd claptrap, patriotic chest-thumping, cheap smut, and jokes so unfunny that only side-splitting guffaws can help you forget them. And you are unable to escape this fate because if you switch off the set, something worse awaits: you'd have to entertain yourself. So what's it to be? Television, or tortured pacing around the room?"

The myth of the power of television was based on the ritual or, more precisely, on the ceremonies of the treadmill routine. In the virtual studio known as "reflections on life," the congregation (the audience) "gratefully accepts whatever is given, however it is given." The motto of private television—the only kind that counts in Latin America—was blunt: repetition satisfies the common thirst for the familiar. Thus "program content" was a euphemism for the spectator's conditioned reflexes, emphasizing copies of copies, serials that reprised earlier serials, and sitcoms already sporting a commemorative plaque. Meanwhile, the devotees put up with the same penance year after year because they had committed, or were about to commit, a sin (reading a book, for example).

If the censorship was implacable, the technology imposed resignation: the corporation knew best what was suitable for the family, after all, and at home there wasn't the moral alibi of a dark cinema auditorium. Technology, as ever, was destiny, so that to switch on the TV meant almost literally to power up a complementary life, the best part of the day. As the passion for telenovelas deepened, the implicit definition of the "televiewer" grew more distinct: the televiewer was someone who accepted what he or she was given because it made him or her feel superior to past generations of people who, in their infinite boredom, would surely have appreciated this box. North American imports (some excellent, like *The Twilight Zone*) filled the bulk of the schedules. The canonical guidelines for the first telenovelas, "classics" such as *Gutierritos, Simplemente María*, and *Ave sin nido* (Bird with no nest), were being defined, while the old rapt amazement had subsided: the technological miracle of television was revered less than the blasphemy of power outages.

Telenovelas became the spectator's other family, not so much in terms of the incidents of their possible or impossible lives but as *the other ideal family*, the one that lives in a convincing décor, suffers with style amidst deluxe or decent furniture, and in the space of an hour undergoes misfortunes that most of the audience would regard, strictly speaking, as small potatoes. The classic melodrama had passed away, unable to be sustained over the four hundred or five hundred episodes of the average telenovela; narrative intensity had also become diluted since no one could possibly remember so much. Conspicuous "indecency" was banned, but there were plenty of "delicate" situations that lent themselves to crude elaboration on the part of the public (after all, it's your "other family"). What didn't take place on screen was spelled out around the water cooler: "Anyone can see that Malena wasn't a virgin!" No wonder telenovelas were on five hours a day. At their height, soaps like *El derecho de nacer* (The right to be born), *María Isabel, Los ricos también lloran* (The wealthy also weep), *El Maleficio* (Evil spell), and *Cuna de lobos* (Cradle of wolves) commanded ratings of 30 or 40 million viewers in Mexico alone.

## Changing the Guard: Emilio Azcárraga Milmo

The political identification of private television ("I am a soldier of the PRI [Institutional Revolutionary Party]," declared Emilio Azcárraga Milmo) was the other great factor of audience infantilization. Not only did a "moralist" censorship oppose every bid for maturity—that is, for an output that trusted the spectator's intelligence and discernment—but the public's own maturity was forever declared to be pending and incomplete. Regardless, the modernization of society through technology owes a good deal to this fragmented, hyper-controlled material; there are always fissures through which new behaviors or notions of family can creep in and take root. Paradoxically, or perhaps typically, isolated societies may swiftly catch up by dint of ritual imitation. We know the formula: "One moment please; our program is about to commence, to continue, to close," whatever—it makes no difference.

From 1960 to 1990, the small screen in Latin America could be described as puttering steadily along: unimaginative, highly censored, and unswervingly derivative. Television was inescapable: everybody had one, and watching it took up all the time that in the past was, well, not usually spent reading either. It proved hard to break up the monopolies, even when different countries came up with more or less acceptable alternatives. At last the transition was made from TV "monotheism" to "polytheism" as choice opened up and channel-hopping encouraged a compulsive, restless curiosity: "Let's see what else is on!" This gave rise to the cult of audience ratings, soon elevated into a bloodthirsty god demanding the sacrifice of anything too boring or serious.

A set of references formed. Because televisual tradition, whose importance must not be underestimated, tends to fade away except around childhood memories, the image of childhood quickly became inseparable from that of TV. To evoke infancy was now to say, "I never once missed that program," followed by, hearers permitting, the enthusiastic re-enactment of gags and catchphrases. This attachment was epitomized by the huge popularity all over Latin America of Roberto Gómez Bolaños, aka "Chespirito," and his two great shows: *El Chavo del Ocho* (The kid in number 8) and *El Chapulín Colorado* (The scarlet cricket). Television became the most fondly recalled nostalgia of each generation. Childhood was also a stage in the development of TV programming. Oddly enough, adults seem grateful for it because the programs they watched as youngsters define the essence of their own childhoods.

During that period, before selective DVD projects made comedy shows and soap operas commercially available, there was no move to embark on any methodical review of the evolution and achievements of Latin American television. Systematic criticism did not yet exist, and—speaking of recall—only a handful of TV stars embedded themselves in the limbo of memory; the rest, the second tier, had to fight to retain visibility on the screen or in studio corridors. "Your face looks familiar. Whatever happened to that real pretty girl in the show I liked so much?"

As late as the year 2000, the strongest tradition in TV was that of transience or oblivion. A picture you are, and a shade of the remote you shall become.

## Media Chronicle

Emilio Azcárraga Milmo, who reigned for decades as the doyen of Mexican TV, once described his responsibility to the public as follows: "Television," he said, "is for losers (*los jodidos*), people who've got no other way to have fun. It's not aimed at rich people like me who have choices, and it's not for people who read political magazines; it's for the losers who don't read and just wait for the entertainment to come to them."[1] This devastating philosophy does much to explain the kind of content that prevailed for so long. Who cares about the minds of clerks and laborers, flopping down on their couch or bed at the end of an exhausting day? If you want to make the viewers think, you'll have to give them scholarships.

Modernity and the pursuit of democracy eventually loosened this dictatorial design. Political debate in Latin America was becoming generalized, albeit with varying degrees of success. But such debate is seldom exciting, apt to trot out the same old slogans, pledges, and insults until the viewer or listener is forced to change channels. This reminds us of a fundamental truth: even in the society of the spectacle, not everything makes good show business, and reliably unspectacular topics are maddening. Since political campaigns tend to be dull as dishwater, boredom acts as a primary depoliticizing force.

The strategies of the consumer society, visually glossy and technologically sophisticated, sought to highlight lack of choice. To this end, the relentless barrage of images of affluence could only frustrate an audience deprived of genuine purchasing power and fed up with dreams that were always demolished in advance. Not in the same words, perhaps, but with the same outcome, such viewers became resigned: their fate was to remain forever as distant observers of mouth-watering landscapes, sleek automobiles, gorgeous specimens of Nordic beauty, and the carefree demeanor of those who are perpetually on vacation, living in mansions into which hundreds of apartments like the spectator's could comfortably be fitted. The whole package was inhibiting enough to make class war look like a bad joke.

While censorship was being increasingly lifted from cinema, it continued to interfere with television. And it did so in a hypocritical, cynical way, well beyond the ornamental criteria of beauty contests. The right was losing its fight against the heresies of *hedonism*, that murky, sexy mist emanating from consumer society. Its manifesto was as follows: chastity removes the need for condoms (known as "sheaths," no matter how awkward to pronounce), so messages about contraception or safe sex could be either eliminated or relegated to the wee hours, accompanied by some lackluster slogan: "AIDS bad. Golden weddings good." But hedonism is sneaky and ubiquitous, and the electorate or television audience began to impose it, taking its cue from commercials and glitzy shows. Suddenly it became possible for late-night broadcasts to approach "edgy" topics, while what used to be considered bad language started featuring—boosted by politics—as mere ambient noise; and the former "smalls," or lingerie, were revealed to be as fundamental as their counterpart: the anxious leering of the spectators. In movies dubbed into Spanish, the famous beep, an admonitory sound if ever there was one, cloaked the "obscenities" but with a curious, unintended effect: people began laying bets on which dirty word had been suppressed in each case.

This patronization of the public went exponential when it came to women, whose intelligence was taken to be either nonexistent or smothered by their feelings. Only the mass arrival of women into higher education tempered the misogynistic onslaughts a little and nudged telenovela plots and dialogues in new directions. And only once it had gone on the record that Latin America was also a continent of single mothers was there a softening of the traditional prejudice against illegitimacy, the great theme of the TV and radio soap that epitomized this era: *El derecho de nacer* (The right to be born). In the same way, it was only the virtually unanimous acknowledgment of a glaring fact—that single women over the age of twenty-six might not still be virgins—that prompted slightly more complex story lines. But this was still far from respecting female rationality, which producers regarded as welded to emotionality. "If she doesn't blub in this scene, she's a transvestite."

As for the poor, they were despised for being deprived of everything, including the right to privacy. The 1990s saw a trend for law-and-order programs copied from North America and Spain, whose crews thought nothing of barging into people's homes unannounced, committing flagrant trespass, whenever they felt like it. "So tell us, ma'am, how do you feel about being raped?" Something very peculiar was in evidence here: the TV magnates had deified their own medium, genuinely convinced of its irresistible omnipotence. "And on the seventh day Yahweh created television, and everything created hitherto fell prostrate before its cameras."

From time to time something is allowed. In 1996, a video recorded by the Guerrero state police was leaked to Ricardo Rocha's current affairs program. The tape showed the murder of seventeen campesinos near the village of Aguas Blancas, the reactions of the survivors, and the movements of police officers meekly exposing their faces to the camera. The upshot was not reassuring: Governor Rubén Figueroa and members of his staff were sacked, but none of the chief culprits went to jail.

In another tragic example, a youth was arrested in Veracruz State in 1995, suspected of raping and murdering a man. An impromptu jury was convened, which sentenced him to be burned alive; someone who owned a video camera was hired to document the event. Lynchings are not unusual in Latin America. The video lasts for forty minutes and includes a moment where the fire goes out and must be relit for the execution to continue. Only three minutes were shown on television. The decision was firm: any more would be too much for the audience to stomach. I don't feel at all certain about this. What was the point of showing such a barbaric event at all? How much *can* the audience stomach? Who evaluates the extremes of humanity and inhumanity?

Sensationalism and mass media–fueled gossip provide another path to the realization that there are more things in heaven and earth than are discreetly dreamt of by traditionalism. This age of chat shows, popular around the world, has made what was unsayable and unthinkable a couple of generations ago into the hot topics of today. The issues discussed form a catalog of weirdnesses that, by being aired on TV, lose much or all of their shock value: women obsessed by the size of their breasts or married to male strippers; gay couples where one partner complains of the other's jealousy; ladies who discover thirty years later that their parents were not incestuous, but that *that* is what goes on between husband and wife; possessive mothers who insist on spending the wedding night with their sons; girls who accuse bosses of asking them to undress out of voyeurism (much worse than sexual harassment); dirty linen galore between couples ("You had the hots for the guy next door, Pepe. Who you trying to fool?"). The conquest of liberal manners over the dead body of privacy.

## He Honors Us Tonight with His Absence

"The Mexicans have the television they deserve," said Emilio Azcárraga Milmo.[2] Does one deserve what cannot be helped, or is one helplessly molded by it? In this as in everything else, communities have felt that their lack of choice was counterbalanced by a vicious circle: the disgraceful narrowing of opportunities necessarily turns into a creative source of entertainment. "Man cannot by nature distinguish what he sees from what he thinks he sees."

Until a few years ago, and to some extent even today, the TV version of Latin America deliberately excised the majority, filling the blank with the president of the republic, assorted leaders, the romantic (and privileged) couple, the anchor, the commercial break, and the voice of government and big business. Ordinary people were never shown trying to meet their real needs, including appetites not sanctioned by advertising, or the legitimate and legal desires forbidden by the Catholic Church. Instead a fantasy of docile, festive communities—and then what happens? A version of The People emerges that, while not necessarily inaccurate on the level of detail, is unreal or false as a whole because it acts in exactly the same way in front of the lens as in front of the screen: the individual performs The People, and this *We* that hungers to appear on TV supplants the *I* that aspires to singularity. Even when there are no cameras around, this individual reacts on cue and does so by his or her own free choice: "I'll pretend to have fun so no one thinks I'm boring. I'll pretend to get excited so no one thinks I'm unfeeling. I'll pretend to be outraged at the politicians they attack on TV so no one thinks I'm apathetic. I'll pretend to enjoy the commercials because they give me a chance to spend more time with my family."

Maybe the word *pretend* is too strong, but how else to describe the submissiveness of the Dear Public (formerly the Republic) to the medium that delivers the pictures? "Unless I'm on TV, my existence will never be registered by others, and though I know there's no reason for me to be on TV, my will to appear is like a prior image that nourishes me and helps me bear my insignificance. Television is my link to the three families I belong to, the world, the nation, and the home, and this inclusion keeps me forever staring at the screen, waiting to spot myself waving in a crowd or being interviewed in the studio. One day it'll happen, and I'll see myself where I never was."

## Characteristics of Recent Years

Here is a list of some recent characteristics of TV:

Endless soccer transmissions have become the axis around which television revolves. Nothing paralyses countries as completely as their national team matches.

Sitcom scripts are sexualized by the use of double entendre (a humor sorely lacking in the distribution of freedoms).

Reality shows, of which the paradigm is *Big Brother*, are affecting our experiences of family life, marital infidelity, and ancestral terrors.

Soap operas have recast the collective psyche. (Very soon we'll all have the right to a telenovela-style life of no fewer than one hundred episodes.)

There is an immoderate appetite for real-life crime: "Unless I'm the corpse, I'm not missing this program." Our rulers make the most of every misdemeanor to pillory the opposition, while claiming that its own scandals have been quoted out of context.

A gender perspective is now common and, indeed, systematic in one sector. Feminism in its various guises offers a critical, complementary view of the world, and traditional displays of machismo are by now insufficiently convincing to seem funny. The increase in female presenters is one token of this phenomenon.

When all is said and done, what historical memory remains is the affair of the left. If anything sums up the right, it's the stubborn attachment to dogma, not the concern for historical memory. When the right wants to praise its heroes, it beatifies or sanctifies them but forgets their names and deeds; it has effaced its cultural and intellectual past and no longer bothers to read its own foundational texts. It is, in a word, an *illiterate* right in both major and minor senses. This means that while it is gratified to have the media empires on its side, it fails to verify the details. The opinions and judgments of the left, by contrast, make an enduring impression in the short, medium, and long terms, thus limiting the power of persuasion and credibility of government discourse.

The emergence of the Internet is the start of the great alternative. Teenagers and young people spend hours at their computers, obtaining what they need straight from the World Wide Web; interactivity exerts a power that ratings never had.

Televiewers are possessed by an undeniable religious spirit, in the sense of a totalizing experience. Not that they take the medium's messages literally or design their lives according to its dictates; it's rather that they cannot conceive of life deprived of their daily psychological fix before the screen. Not so much the content as the very existence of TV regulates their management of time. The particulars of light entertainment or current affairs, the fashion updates, the collection of rumors for spicing social intercourse, these are secondary. (On one level, national news items on TV can only be believed if they are received as gossip.)

The advent of cable TV serials has brought back the notion that televisual products can change your life. *Oz, Six Feet Under, The L-Word, Desperate Housewives, Sex and the City, Queer as Folk*—these all convey a glimpse of new and irrepressible lifestyles.

## The Twenty-First Century, the Power of the Media, and the Struggle over Public Space

In the twenty-first century, a new adjective came to encapsulate the way we think of the predominance of the audiovisual: *lo mediático*.[3] The power of the corporations that trade in this "mediatized" commodity is reinforced whenever a politician buys into their unshakable faith in television—that combination of agora, prophecy, and the miracle of inarticulate speeches turned into slick commercials. Not to be on television, allowing the continuity of one's media presence to lapse, is to enter planned obsolescence (old age being a form of planned childhood). Such is their sacred dread that politicians hesitate to credit television with the possession of visible reality, as if to vanish from the screen would entail a return to the days of house-to-house campaigning and the distribution of smiles along the street.

Besides specific reasons (each illusion of prestige is a world in itself), the power of television is incontrovertible: the medium shapes our overall concepts of global reality, national societies, collective identity, and mass tastes, not to mention the comedy to which children (and adults) are doomed by their helplessness. There is no serious discussion, even today, of the damning certainty that television is the one and only social and governmental mirror we've got. It is the only space in which we see ourselves, intuit ourselves, become hopelessly annoyed with ourselves, enjoy and recognize ourselves. Hence the appeal of reality shows, and our rapturous embrace of the remote as sole proof of the existence of multitudes. Exorbitant outlays on publicity, taking for granted that a politician only comes into existence with his first political ad, enable the miracle to be proclaimed: "After that avalanche of images, I shall be an apparition rather than a politician." Television magics people in and out of sight, and, except on the most climactic occasions, performs much the same trick with the majority of issues. Does anyone stand up against that?

So far we have not been in danger of a homegrown Silvio Berlusconi, thanks to the lack not so much of funds as of a messianic will applied to systematic projects.[4] Yet while we have been spared the sight of decadence propelled to the heights, like Il Cavalieri's, we must note the failure of lawmakers to curb the excesses of private media corporations by, for example, getting them to pay taxes. Whenever some small progress is made in this direction and bills become law, the Executive—backed by a large section of the Legislature and part of the Judiciary—promptly steps in to restore the monopolies' impunity, fiscal exemptions, and radio and TV franchises.

In such circumstances the TV audience meekly submits to the contempt for its abilities, complying with a determinism of our times: there is no such thing as using your brain while watching TV, and it's not the media's job to pose challenges to your imagination or creative fantasy, for these are just abstractions without a sponsor.

In the beginning the sponsor created laughter and weeping, and the face of the audience was formless and void.

## Notes

Regretfully, Carlos Monsiváis died on June 19, 2010, five months after he submitted this chapter.

1. It seems that a lot of people use this quote, but it is unclear if the source is verbatim or if it just has been used and misquoted by Mexican intellectuals; it appears in Jenaro Villamil's article, at http://homozapping.com.mx/2011/03/la-television-mexicana-y-sus-audiencias-el-espejo-roro/ (accessed October 31, 2012). For more information about the discussion of the source and use/misusage of Azcárraga's statement, see Claudia Fernández and Andrew Paxman, *El Tigre: Emilio Azcárraga y su imperio Televisa* (Mexico City: Grijalbo, 2000), 388.

2. See Fernández and Paxman, *El Tigre*.

3. While "mediatic" is not allowable in English, it ought to be, covering everything from media-generated to telegenic.

4. Silvio Berlusconi is an Italian politician whose nickname is "Il Cavalieri."

# 7
# Revolt, Confusion, and the Cult of the Trivial in Mexican Cyberculture

*Naief Yehya*

## Death of Culture: Schizophrenia and the World Wide Web

The printed press is dying, movies are on the brink of collapse, the music business is desperately struggling to survive: in other words, the culture we've known is fading. This planetary manmade cataclysm is the direct consequence of the impact of cyberculture on our daily lives. And yet, simultaneously, new art forms are spawning. Culture can't survive without change. This revolution, in large part, is the result of the obsessive work of a group of hackers (which included Steve Jobs, Steve Wozniak, and Bill Gates, among many others) who, back in the 1980s, were committed to using computers and communications as tools of creation and innovation. Most of these *techies* and self-defined geeks followed a series of unspoken principles, collected by Steven Levy a quarter century ago in his book *Hackers: Heroes of the Computer Revolution*, which guided the progress of computer science and Internet culture in the decades to come with two axioms of particular influence: "Information should be free" and "Mistrust authority, promote decentralization."[1] The first is at least partially responsible for keeping the Internet a mostly open resource of information, data, and tools. The second generated the rebellious spirit characterized by an attitude of defiance against institutions, governments, and corporations that once reigned in cyberspace.

But what started as a hobby and passion evolved into a huge business, a many-headed Hydra. Hackers became web entrepreneurs, and the conflict that arose between their desire to maintain their rebel image and their ambition to enrich themselves brought about, in large part, the confusing nature of today's web.

When the hacker revolution began to take form, no one in the globe possibly could have imagined the repercussions that that primitive soft- and hardware would have on every aspect of life.

## Mexico's First Steps in Cyberspace

Considered to belong to what some have defined as the intermediate zone of technological development (along with Brazil, Chile, and Argentina, among others), Mexico was one of the first Latin American countries to experiment with the Internet.[2] Researchers James Curry (El Colegio de la Frontera Norte, Mexico) and Martin Kenney (University of California, Davis and Berkeley) noted: "The adoption of the communications network did not lag so far behind the USA's, it just took much longer to scale up."[3] Being next door but decades away from the United States in terms of technological development makes Mexico an interesting case study for the impact of and changes brought about by the Internet. The country struggles to balance the pressure to follow the pace of innovation and change imposed by the United States, on the one hand, and its issues and concerns with autonomy, on the other. The huge income gap between the two nations further complicates and limits the assimilation of new and expensive technologies.

At the 1994 Summit of the Americas in Miami, the Organization of American States (OAS) declared Internet connectivity a priority for the region: "A country's information infrastructure—telecommunications, information technology, and broadcasting—is an essential component of political, economic, social and cultural development." The appeal was mainly for imposing free market rules in the communications and information fields, which until then were considered of national strategic priority. Its goals included "encouraging private sector investment to increase participation in the telecommunications and information infrastructure sectors; promoting competition; implementing flexible regulatory regimes." Nevertheless, there were also calls for "stimulating diversity of content, including cultural and linguistic diversity; providing access to information networks for service and information providers; and ensuring universal service, so that the benefits of the information infrastructure will be available to all members of our societies."[4]

In 1981, Ira Fuchs at the City University of New York (CUNY) and Greydon Freeman at Yale University created the BITNET network to facilitate email, Listserv, and file transfer between universities.[5] BITNET plus the connected networks was the embryonic Internet, according to Dublin City University lecturer Mark Humphrys.[6] Curry and Kenney wrote that the first computer communications in Mexico took place in 1987 when the Instituto Tecnológico y de Estudios Superiores of Monterrey (ITESM) and the Universidad Nacional Autónoma de México (UNAM) connected to Fuchs and Freeman's BITNET network. Merit Network, Incorporated, a nonprofit, member-owned organization that provides services to the research and education communities, credits Mexico's first connection to the Internet to when, in February 1989, ITESM employed Internet protocols to communicate with the University of Texas in San Antonio.[7] That same month, Network Information Center–Mexico (NIC-Mexico) was created to manage the name of the territorial domain .mx.[8]

A year later several universities (Universidad de las Americas, Instituto Tecnológico de Estudios Superiores de Oriente, and Universidad de Guadalajara, among others) and government agencies (Secretaría de Educación Pública and the Consejo Nacional de Ciencia y Tecnología—Conacyt) linked to the Internet. "Mexico's network infrastructure consisted of one external backbone, routed through Mexico City, leading to slow, and frequently interrupted service," wrote Curry and Kenney.[9] And, for most users, connection at that time meant dial-up access. In 1990, the Red Tecnológica Nacional (National Technological Network) was created by connecting MexNet, a group of about fifteen universities that used the Internet to share information, send emails, and discuss the direction of communications in Mexico.[10] At that point the connection bandwidth had grown to 2 Mbps (megabits per second), allowing private users to join the Internet.

Before 1997, very few institutions had registered domains, and only thirteen registered .edu domains existed in Mexico according to NIC-Mexico.[11] By 2000 this number had jumped to nearly three thousand. Mexican institutions of higher education were relatively slow to adopt Internet communications, but 60 percent of them had a presence online by 2003. Industry took even longer to follow, leaving Mexico to trail developed countries and some developing nations, like Brazil, in this area.

## Web to Web 2.0

In its infancy the Internet existed as a space with infinite possibilities, but it was extremely confusing to use and, with no adequate search engines, hard to navigate. Digital communications were surrounded by an aura of mystery and a clan mentality. Since then the net, influenced by the hacker ethos, has evolved vertiginously into a space of diversity, openness, and freedom. Many things have been said about the pioneers and architects who shaped the digital world, but it seems paradoxical that a group of introverted geeks created an environment ripe for the development of communities where people with shared interests can find each other and establish links. The World Wide Web turned out to be a social territory, a powerful motor for innovation, and a voracious engine of change. It holds the power to overwhelm users with huge amounts of information and options that, instead of liberating, can drown the imagination. More worrisome is what the virtual-reality pioneer, musician, and programmer Jaron Lanier calls lock-ins. In software, often design decisions, once implemented, turn out to be unalterable, closing the door to better designs. "The consequences of tiny, initially inconsequential decisions," Lanier wrote, "often are amplified to become defining unchangeable rules of our lives."[12] The raw power and speed of computers can morph a minor glitch into a cataclysmic challenge. And the Internet can amplify the impact of software design decisions, turning them into rules, strictures, and norms. This is a factor that gets more prevalent in what is known now as the Web 2.0 (2004), a term coined by the designer and writer Darcy DiNucci but mainly associated with Tim O'Reilly, the publisher, founder, and CEO of O'Reilly Media, who has become a sort of evangelist of its virtues.[13] The original web came to be viewed as a static experience, a place to retrieve information; the future would be interactive. The Web 1.0 followed old software paradigms and depended on desktop applications like the Netscape browser. Designers and programmers were encouraged to create a newer generation of software (or apps) focused on user-centered designs, sharing and participating, and the possibility of leaving behind specific operating systems and considering the web as platform.

Ten years before the decision was made to replace the old Internet with a new web, people globally pondered its power as a resource—not only to create smart online role-playing games and design virtual worlds and even as a platform for dating services, but to actually change the world through action and activism. A group of armed rebels in the southeast jungles of Mexico was among them.

## A Curiously Networked Guerrilla

Nineteen ninety-four was a year of technological transition and radical change in communications. It was also the time when Mexico faced a severe shock to its system in the form of armed rebellion.

In 2005, Steve Coll and Susan B. Glasser published a controversial article in the *Washington Post*, "Terrorists Turn to the Web as Base of Operations," where they proposed that four years after 9-11 (September 11, 2001) "al Qaeda ha[d] become the first guerrilla movement in history to migrate from physical space to cyberspace. With laptops and DVDs, in secret hideouts and at neighborhood Internet cafés, young code-writing jihadists ha[d] sought to replicate the training, communication, planning and preaching facilities they lost in Afghanistan with countless new locations on the Internet."[14]

This widely accepted idea was not exactly precise. The first guerrilla movement that took its message to cyberspace was the Ejército Zapatista de Liberación Nacional (EZLN), or the Zapatista National Liberation Army, a tiny organization of poor, indigenous peasants created around 1993. On January 1, 1994, the official date the country entered the North America Free Trade Agreement (NAFTA), the EZLN became a player on the world stage by declaring war on the Mexican state, creating a severe crisis and forcing a discussion on the unfair living conditions of indigenous peoples throughout the Americas. The EZLN performed an unexpected military feat: underequipped, armed with rudimentary rifles, machetes, and sticks, they were able to take the city of San Cristobal de las Casas and six municipal seats in the southern state of Chiapas. The hold didn't last. The Mexican army counterattacked the next day, and the Zapatistas, who initially numbered around three thousand, suffered heavy casualties and retreated into the jungle. This heavy-handed response immediately aroused dozens of NGOs (nongovernmental organizations) and hundreds of activists to swarm physically into Chiapas and electronically all over the world. They demanded a cease-fire and negotiations. On January 12 the government agreed and halted combat operations.[15] David Ronfeld, John Arquilla, Graham E. Fuller, and Melissa Fuller, in *The Zapatista Social Netwar in Mexico*, a study sponsored by the deputy chief of staff for intelligence and conducted by Rand's Arroyo Center, wrote: "Without the influx of NGO-based social activists, starting hours after the insurrection began, the situation in Chiapas would probably have deteriorated into a conventional insurgency and counterinsurgency, in which the small, poorly equipped EZLN might not have done well and its efforts at 'armed propaganda' would not have seemed out of the ordinary."[16] As Donna Kowal, an expert on political communication from the University of Pittsburgh phrased it, "This group used the Internet to agitate, enact their autonomy, create a collective identity, and ultimately generate worldwide support for their cause all the while maintaining protection against retaliation by the Mexican Government."[17]

The Zapatistas became emblematic of a postmodern guerrilla who fought oppression not only with weapons, but also and mainly with communication and information technology. They had no desire to repeat past experiences of long, bloody guerrilla wars but instead wanted "to exploit the theatre and poetry of a political action," as Graham Meikle wrote.[18] The EZLN was the first indigenous organization to successfully bring its ideology and tactics to the public using digital communications. Its goal was to show that its fight against the government, the World Trade Organization, the International Monetary Fund, and neoliberal ideology was the fight of all indigenous and dispossessed peoples of the world. By spreading their message in this way, the Zapatistas, under the leadership of the charismatic Subcomandante Marcos, struck a powerful, strategic blow against an unprepared government. More importantly, by "hacking" the system and globalizing their message and cause, they won a symbolic battle. The dispossessed gained the technological edge and dominated the media sphere—or so it appeared. In reality, at least in the beginning, the majority of the indigenous peasants involved in the insurrection had little to no knowledge of the digital realm. While the community decided the course of action, the mestizo leadership chose the nature of the media attack.

Deep in the rainforest, the Zapatistas had no laptops or satellite phones, let alone web servers, Internet access, or electrical outlets; still they managed to create a relay system using traditional communication methods and a network of wired volunteers distributed all over the world. So, in a way, their presence in the virtual world was paradoxically virtual. The communiqués from the Lacandón jungle would travel via courier to individuals in Chiapas who faxed or dictated them over phone lines to individuals outside the region with contacts at newspapers (in Mexico, mainly the national newspaper *La Jornada* and *Proceso* magazine) who then printed them within twelve hours of their receipt. Volunteers would post the messages on Internet lists in Spanish, and usually by the next morning they were translated and posted on international lists and web pages.[19]

Sixteen years later, the situation of the indigenous peoples and poor farmers in Chiapas has not improved significantly. Some say it has worsened. Most Zapatista supporters inside and outside Mexico have abandoned the cause. The Zapatistas can no longer count on international NGOs' help and have retreated to their villages, where they concentrate on defending what's left of their autonomy and protecting their people from government forces and former landowners' aggressions. Marcos stopped writing and publishing his communiqués in 2009. His rupture with the left, in particular, with the former presidential candidate from the PRD (Party of the Democratic Revolution), Andrés Manuel López Obrador, and the failure of all negotiations with the government were perceived as the end of his era. Many consider him to have turned into a "vituperative, narcissistic charlatan who is responsible for the depreciation of the Zapatista movement as a national and international player on the Left," as John Ross, a long-time supporter and expert on Zapatismo, wrote.[20] Thousands of pages of reflection and analysis have been devoted to the semi-mythical subcomandante and his admirable talent as a communicator. This characteristic was once considered an asset, but the overexposure generated by an obsessed international media and the lack of tangible victories arguably turned it into a liability. The man who assumed the self-effacing title of "subcommander" (he claimed to be incapable of making any decision without the consent of the people) became, due to the saturation of his image in the media, an *always-on* entertainer, a web celebrity, a decontextualized, masked personality flushed in the vertiginous flow of digital inconsequence. In a sense, the strategy of raising media awareness and intensely focusing attention on political activism on a global scale using the Internet actually degraded and commodified the message. The communications revolution shook the balance of power, but not for long and not enough to create a new world order. The Internet gave the EZLN a stage and the opportunity to make its case, but in the end newer shows distracted the attention of the public and the international pressure gradually disappeared.

## Cyber Mexico

It's very well known that there is a digital divide in Mexico; it couldn't be otherwise in a developing country with huge class differences. Nevertheless, the increase in the number of Internet users over the last twelve years has been remarkable; in 2000 there were 2,712,400 users out of a population of 98,991,200, a 2.70 percent penetration, according to the ITU, the United Nations Agency for information and communications technologies.[21] In 2012 the number of users is 40,600,000 out of a population of 112,336,538, a 36.25 percent penetration, according to AMPICI (Asociación Mexicana de Internet).[22] This data coincides with information from COFETEL (Comisión Federal de Telecomunicaciones), which states that there were 27.6 million Internet users as of May 2008, a 29.80 percent penetration. The same study determined that there were 18.2 million computers in the country (51 percent in private homes) with 11.3 million of those connected to the Internet.[23] Around 2005 most of the users in Mexico started relying on broadband communications (DSL, cable, ISDN, and wireless hot points). In 2006 there were 1.4 million computers using dial-up connections; in 2008 the number decreased to 462,000. During that same period, total broadband connections jumped from 2.6 million to 6.4 million: 93 percent of the total number of Internet accounts.[24]

In a study by the consulting firm Select Mexico (2004), cited by Curry and Kenney, Mexico is divided into three income classes, where the bottom class constitutes 73 percent of the population, 17 percent of whom have access to the Internet. In the AMPICI statistical report from 2009, this is 20 percent. Of the higher income class who make up 13 percent of the Mexican population, 63 percent are Internet users, perhaps a conservative assessment. Nevertheless, in courses, workshops, and seminars I have taught in the last five years to elementary and middle school students in mainly lower-income neighborhoods in Mexico, a surprising number of kids use digital communications constantly. It's just as much a part of their lives as it is for kids from higher-income families. In urban areas, instant messaging among the young is very pervasive. The Internet is accessed through school libraries and Internet cafés, not only for academic reasons but also for playing video games and for social interaction. The highest growth in access to the Internet has been registered in the lowest socioeconomic levels. From 2007 to 2008 the number jumped from 24 to 33 percent.

The main activities of Mexicans on the World Wide Web include downloading music (49 percent), humor websites (32 percent), online gaming (30 to 50 percent of gamers are between twelve and nineteen years old), and sports websites (28 percent). Twenty-two percent of users read newspapers online, compared to 11 percent who listen to the radio and 8 percent who watch TV.[25]

In Mexico, like in most developing countries, most access was originally through Internet cafés because of the prohibitive price of computers, making the growth and specific use difficult to track. While more than half of the users accessed the Internet from cafés in 2000, eight years later that number dropped to 34 percent, while home users went up to 48 percent. In 1999, 63 percent of Internet connections were made through office computers. By 2003 that number had shrunk to 34 percent. By 2008 it had fallen to a low 19 percent. Computer use is becoming more and more a personal and intimate necessity.

A useful way to track Internet development is by considering the increase in domain registrations. The first domain was registered in Mexico in February 1985. By June 2012 the number had reached 579,718, grouped by sub-domains, .com .mx, .gob.mx, .net.mx, .edu.mx, .org.mx, and .mx, according to Network Information Center Mexico.[26] Undoubtedly, the largest category is .com, as almost everywhere else in the world. In Mexico, commercial domains have had the fastest growth. Internet use in Mexico closely follows global trends; still, if the web is mostly a commercial machine, a mega virtual mall, an extremely large number of Mexicans cannot take advantage of the ease and potential of shopping on the web, mainly because only a minority have credit cards and the postal service is not efficient enough to timely and safely deliver goods. In 2007 the total income of Internet sales in Mexico was $955 million, a growth of 78 percent from 2006.[27] Of all Internet sales, 68.0 percent are plane tickets, 8.5 percent are computer goods, 4.0 percent are tickets for shows and entertainment, 3.4 percent are for cell phone accessories, 3.0 percent are for electronic and audio products, 2.2 percent are for hotels, 1.5 percent are for travel packages, and only 1.0 percent are for books.[28]

## La Jornada Virtual and the Cyber Revolution

In 1995, the prestigious Mexican writer Juan Villoro and former *Letras Libres* magazine main editor Ricardo Cayuela took charge of the weekly cultural section *Semanal* of the daily newspaper *La Jornada*, and they invited me to write a column. I had been writing articles on technoculture and cyborgs, and we talked at length about technological innovation and the imminent changes that digital communications were on the verge of offering and imposing on all of us. In the Mexican left, who make up the main readership of *La Jornada*, there existed a prevailing belief that digital culture was elitist and the expression of US cultural colonialism. My personal objective was to create a space in the culture pages (and not in the technology, science, or business sections, where many people believed it belonged) where I could reflect on and discuss the pervasive and unavoidable mediated reality that the Internet and our technological extensions were bringing upon us and the way they impacted the culture at large. La Jornada Virtual, as the column was named, was not focused on showcasing novelties or commenting on products, as many others did, but rather on dispelling myths, exploring and properly addressing technocultural phenomena, and not ignoring it as a passing trend. When Villoro and Cayuela left *Semanal* in 1998, new editor Hugo Gutiérrez Vega and his team introduced radical changes to the section but kept La Jornada Virtual as a biweekly column, and this year will be its fifteenth anniversary.

La Jornada Virtual was by no means the only space devoted to the virtual world and the digital revolution, but it was one of the first such spaces. During the second half of the 1990s the Internet boom was responsible for a worldwide surge of Internet-devoted publications and sections in almost all newspapers; Mexico was no exception. Most devoted themselves to being echo chambers for technology companies, basically printing press releases, industry's hysterical claims, and overhyped product descriptions. A gullible and mostly technologically ignorant media couldn't devote enough coverage to the digital revolution. Some media outlets even published competing sections that eventually cannibalized themselves. Just as fast as they popped up, most of them vanished, more or less around the time the infamous .com bubble busted (1998 to 2000), evaporating huge fortunes and the illusion of the Internet gold rush.[29] La Jornada Virtual chronicled the rise and crash of those euphoric years and covered some of the most relevant events in the digital universe.

The printed media gave a faithful reflection of the confusion, euphoria, and fear that the digital media produced in the world. Instead of the expected convergence, a clash occurred when old media tried to insert itself in cyberspace by using a digital window to showcase at least part of its contents free on the web with the hope of eventually being able to cash in on this presence. Still today the economic mechanisms to make epublications profitable are far from clear. In Mexico most of the newspapers provide their content free on the web, with the exception of the newspaper *Reforma*, which has maintained a for-pay model since its arrival on the web.

With the imminent proliferation of different electronic readers (a variety of ebook formats and devices like the iPad and Kindle) offering electronic versions of major newspapers, the printed media is condemned to change or to become extinct. The new media, though, hasn't offered a better alternative; and, at least for a period, this situation will intensify social segregation and even further limit access to information. By the end of 2009 Amazon accounted for an estimated 80 percent of all ebook sales, establishing the price for most ebooks at $9.99, which put most publishers in a desperate situation. David Young, chairman and CEO of Hachette Book Group USA, said: "If it's allowed to take hold in the consumer's mind that a book is worth ten bucks, to my mind it's game over for this business."[30] These devices will still take some time to be introduced en masse to the Mexican market, but we can count on it happening, and the consequences for an already ailing publishing industry may be calamitous.

## Social Networking

Since the boom of Myspace people all over the world started developing new needs and habits dependent on online communion. The idea of using the web to socialize, link with people we know and meet people we don't, entices. Creating networks of friends and acquaintances is not only fun, but professionally useful.

What started as a pastime has become an authentic planetary obsession that seems to win more converts every day. The social network craze started with the freedom and chaos of Myspace, a website where the user could explore his or her talent as a designer, programmer, and self-promoter while exposing his or her interests and life to anyone interested. From June 2006 to early 2009 Myspace pages were extremely successful. Its visual and audio anarchic aesthetics were eventually displaced by the ordered monotony of Facebook, an amazingly popular social network that, as Jaron Lanier says, has managed to organize people into multiple-choice identities: "If a church or a government were doing these things, it would feel authoritarian, but when technologists are the culprits, we seem hip, fresh and inventive."[31] People seem more than happy to reveal all kinds of personal information and intimate secrets to their "friends," creating the perfect breeding ground for stolen identities and all sorts of scams.

In a very short time Facebook passed from being considered egotistic, frivolous, and extravagant to becoming an extremely popular service, especially but not exclusively among the middle- and higher-income classes all over the world. In Mexico its use followed, for the most part, the international trends, but with an interesting twist. People started to use their smart phones and Internet-accessible devices to alert drivers by tweets or postings not only to traffic jams and closed roads, but also to the location of Breathalyzer checkpoints. This use of micro-blogging services, while a nuisance to law enforcement agencies, seems almost innocuous compared to how kidnappers and drug cartels have used them to spy on and select victims, communicate among associates, plan crimes, threaten rivals, and spread fear through local communities. For a long time criminal gangs have been using *narcocorridos*, *norteño* songs about drug dealers and criminals, to communicate; more recently, they have been using videos posted on YouTube and similar services.

Recently, a cartel in the northern city of Reynosa, in the state of Tamaulipas, during the street gun battles that erupted between cartels and the criminal organization Zetas in February 2010, posted a message on Facebook that read: "The largest scheduled shootout in the history of Reynosa will be tomorrow or Sunday, send this message to people you trust that tomorrow a convoy of 60 trucks full of cartel hit men from the Michoacán Family together with members of the Gulf Cartel are coming to take the city and take everyone out alive or dead!"[32] The message was very successful in creating panic and forcing city officials to close schools, business owners to close shops, and residents to stay home.

Representatives of the leftist Partido de la Revolución Democrática in the senate have proposed creating an online police force in charge of, among other duties, monitoring messages and interactions in social networks.[33] They have drafted a bill that would make sharing information that helps others break or avoid the law a criminal offense. This is another interesting ideological twist where liberal politicians look to create regulation mechanisms to curtail crime that eventually could be used to limit freedom of expression and spy on civilians. Such a law would most likely have little effect on criminals; instead it might prove to be more a success at making any tweet about police activities illegal and in giving corrupt law enforcement agencies new opportunities to extort civilians. In any case, there is no law prohibiting police agencies from creating fake identities to track criminal activities, which makes, in many ways, a law like this redundant and repressive. All this is yet more proof of the extremely poor understanding of this technology.

On the other hand, these services have allowed citizens to report criminal activities of cartels, gangs, and individuals without fear of exposing themselves. According to Reporters Without Borders, Mexico is now the most dangerous place in the world to be a journalist (five journalists were murdered between January 1 and April 20, 2010).[34] It is not a bad idea to conceal one's identity in a time when anyone defying the cartels has good reason to fear retribution. Social networking under these conditions is considered by some as a "safety measure."[35] Citizens have started organizing demonstrations and political actions through Facebook; but on an even more basic level, they use the website to recommend streets and areas to avoid due to criminal or police activities. People have posted videos on YouTube to show how to post anonymous reports on

Twitter and circulate information to wider audiences by using hashtags (#) to denounce cartel operations.[36] It's not surprising that when cartel members go out in public, the first thing they do is confiscate all cell phones.

## A Low-Tech Revolution

While in 2008 there were barely more than 11 million computers connected to the Internet in Mexico, there were 73.6 million cell phones, 25 percent of which had access to the Internet, but only 6 percent of owners used this service. And while close to three-quarters of these cell phones had text-messaging capability, only about half of all owners texted.[37] Cell phones proliferate, and many countries in the developing world have been living a separate technological revolution characterized by its low budget. As Anand Giridharadas points out, America is "spiraling towards fancier, costlier, more network hungry and status giving devices, meanwhile their counterparts in developing nations are innovating to find ever more uses for cheap, basic cell phones."[38] Instead of investing in extending the wired networks or creating new and expensive telecommunication systems, most of the world is busy just adding cell phone towers. In war-torn countries, in regions affected by earthquakes or other natural catastrophes, and in crumbling or devastated states, cell phone technology has proven to be reliable and versatile not only for traditional communication, but also as an effective medium for sending and receiving money wires, paying for services, and even job hunting. Africa is one of the world's fastest-growing mobile phone regions, more than 100 million cell phones are used regularly, not only for calling, messaging, music downloading, and videogame playing, but also as currency devices. Systems like M-Pesa (*pesa* is Swahili for "money"), created in Kenya, allow users to send minutes via SMS (short message service) that are automatically loaded into the phone of the recipient. And it is possible to send money collected at any of the M-Pesa shops all over the country. When shops closed during the violent upheavals that followed the disputed 2007 presidential elections in Kenya, prepaid cell phone cards became one of the most valuable and scarce goods. Charities started distributing cell phone cards, which were used not only to communicate, but also to buy food, fuel, and other essentials. In many countries simple cell phones are used to monitor elections and film demonstrations, repression, and public unrest, as happened recently in Iran.

The situation in Mexico is different from most of the rest of the world because its market is severely controlled by a powerful corporation that restrains its growth and innovation. Mexico followed the trend of other Latin American countries in privatizing state monopolies, as recommended to all nations with huge debts by the International Monetary Fund and the World Bank. The belief was that enterprises under private control would be more efficient, profitable, competitive, and reliable. The result, in many cases, has been that private entrepreneurs favored by the government bought national companies and also inherited badly disguised monopolies. Some of these businessmen did little to improve their service, so the public ended up trapped in a system with fewer options and more expensive services.

In 1990, the Mexican government sold Telmex (Teléfonos de México) to a group of investors lead by Carlos Slim. Since the privatization, other companies started offering phone services in Mexico, but Slim's company remained a semi-monopoly by controlling 80.0 percent of landlines and 72.3 percent of cell phone service. Slim has declared: "Being a monopoly is different from being a dominant player. You're a monopoly if you're alone in the market. If there are five or six wireless competitors, you can't say we're a monopoly."[39] What is clear is that Mexicans today pay some of the highest telephone rates in the world. Slim, the richest man on earth in 2010, according to *Forbes Magazine*, now has operations in all of Latin America, with the exceptions of Cuba, Costa Rica, Bolivia, and Panama.[40] Since the year 2000 Slim has launched a very aggressive campaign, acquiring companies all over Latin America: Telgua from Guatemala, Argentinean cell phone provider CTI and 60 percent of Techtel, 30 percent of the Dominican provider Tricom, the Colombian telephone company Celcaribe, the Brazilian companies BSE and BCP and 52 percent of Embratel, and Conecel from Ecuador, 51 percent of CTE from Salvador, Chilesat, 40 percent of the Empresa Nicaragüense de Telecomunicaciones, Telecomunicaciones de Puerto Rico, and the operations of Verizon Communications in Venezuela. Slim became Bill Gates's partner in the Internet Hispanic web portal T1msn.com and in October 2003 acquired AT&T Latin America. "Slim's monopoly is local and confined to Mexico, but a large part of his wealth comes from his holdings in other countries, where he has no monopoly at all," wrote Felix Salmon.[41]

In 1989, the company Iusacell monopolized the Mexican cell phone market, but its network was limited to Mexico City. A year later Slim launched Telcel, also offering service in the capital. In 1995, the Mexican economy went into a slump. Iusacell kept its strategy of offering expensive plans to mostly wealthy customers. Telcel focused its attention on prepaid mobile plans that were more flexible and affordable, and in two years this policy turned Telcel into the market leader. Other companies offer services in the country, but Mexican mobile telecommunications are mostly controlled by Telcel. "Teledensity," or telephone landline penetration expressed as a percentage of population, in Mexico is still low (18 percent, less than 18 million) with huge disparities between rural and urban areas.[42] At the same time in 2010 there were over 80 million cell phone users, which means that close to 70 percent of Mexicans had a cell phone.

Telmex started providing Internet access in mid-1995 through its branch Uninet. A year later Telmex acquired Prodigy Communications, and with the label Prodigy Infinitum it became the leading broadband provider in the country, holding more than 80 percent of the market. Infinitum is one of the most expensive Internet services in the world: in 2008 it was charging $18.41 as the lowest monthly price per Mbps, while in Japan a similar service was $0.13, and the average cost over thirty countries was $3.77, according to the Information Technology and Innovation Foundation.[43] The brutal stranglehold that Slim has over Mexican telecommunications is a huge burden for its growth, accessibility, and diversity.

## Revolution to Subjectivity

There is no doubt that digital telecommunications is a completely transformative force. The Internet started as a US Defense Department project that eventually was given to the civilian world. In Mexico the Internet gained notoriety for the first time with the majority of the public as an empowering tool for a guerrilla and as a medium with which it was possible to confront the arrogance, corruption, and incompetence of the state. Even with its military origins, the Internet has become a civilian tool with mind-boggling powers to create; it is a liberating and exhilarating medium and gives us the best chance to change our world for the good; it is a participatory and interactive space where we can create, share, discuss, and learn. Lanier asks, "Who would have guessed (at least at first) that millions of people would put so much effort into a project without the presence of advertising, commercial motive, threat of punishment, charismatic figures, identity politics, exploitation of the fear of death, or any other classic motivators of mankind."[44] The Internet is a product of idealism; it is a collective, joyful dream and a resource capable of spreading democracy all over the world. But the net has an uncanny power to amplify subjectivity, and it encourages what Lanier calls the Peter Pan fantasy, where everyone can behave childishly and every indulgency or onanistic behavior is considered acceptable. Computers and telecommunications have changed Mexico, but not so much in the way Marcos could have hoped, and much more in the way Slim had planned.

## Notes

1. Steven Levy, *Hackers: Heroes of the Computer Revolution* (New York: Penguin, 2001), 24–25.

2. Víctor Flores Olea and Rosa Elena Gaspar de Alba, *Internet y la revolución cibernética* (Mexico City: Océano, 1997), 61.

3. James Curry and Martin Kenney, "Digital Divide or Digital Development," *First Monday* 11, nos. 3–6 (March 2006), http://firstmonday.org/htbin/cgiwrap/bin/ojs/index.php/fm/rt/printerFriendly/1318/1238 (accessed October 22, 2012).

4. First Summit of the Americas, Miami, Florida, December 9–11, 1994, Plan of Action, *http://www.summit-americas.org/miamiplan.htm* (accessed October 22, 2012).

5. Listserv was the first electronic mailing list software application, consisting of a set of email addresses. The sender could send one email and it would reach all or some of the people on the list.

6. See http://computing.dcu.ie/~humphrys/net.80s.html (accessed October 22, 2012).

7. See http://www.merit.edu/index.php (accessed October 22, 2012).

8. A geographical domain name determines the country of origin: .us for the United States and .mx for Mexico. Top-level types refer to the group or individual running the website and include .com for commercial websites and .gov for government websites.

9. Curry and Kenney, "Digital Divide," 3.

10. See http://www-cs-faculty.stanford.edu/~eroberts/cs201/projects/2006-07/latin-america/mexicoIntro.html (accessed October 22, 2012).

11. See http://www.nic.mx/es/NicMexico.Historia (accessed November 10, 2009).

12. Jaron Lanier, *You Are Not a Gadget: A Manifesto* (New York: Knopf, Borzoi Books, 2010), 9.

13. Darcy DiNucci, "Fragmented Future," *Print* 53, no. 4 (1999): 32.

14. Steve Coll and Susan B. Glasser, "Terrorists Turn to the Web as Base of Operations," *Washington Post*, August 7, 2005, http://www.washingtonpost.com/wp-dyn/content/article/2005/08/05/AR2005080501138.html (accessed October 22, 2012).

15. David Ronfeldt, John Arquilla, Graham E. Fuller, and Melissa Fuller, *The Zapatista Social Netwar in Mexico* (Washington, DC: RAND Arroyo Center, 1998), 3.

16. Ibid., 23.

17. Donna M. Kowal, "Digitizing and Globalizing Indigenous Voices: The Zapatista Movement," in *Critical Perspectives on the Internet*, ed. Greg Elmer (Lanham, MD: Rowman & Littlefield Publishers, 2002), 106.

18. Graham Meikle, "We Are All Boat People: A Case Study in Internet Activism," *Media International Australia*, no. 107 (May 2003): 9–18.

19. Kowal, "Digitizing," 113.

20. John Ross, "Commodifying the Revolution: Zapatista Villages Become Hot Tourist Destinations," *Counterpunch*, February 17, 2009, http://www.counterpunch.org/ross02172009.html (accessed October 22, 2012).

21. United Nations Economic Commission for Latin America and the Caribbean–ECLAC, "Road Maps Towards an Information Society in Latin America and the Caribbean," December 13, 2002, http://www.itu.int/wsis/docs/rc/bavaro/eclac.pdf (accessed October 22, 2012).

22. AMPICI, Hábitos de los usuarios de internet en México, May 12, 2012, http://www.amipci.org.mx/?P=editomultimediafile&Multimedia=115&Type=1 (accessed October 22, 2012).

23. Ibid.

24. Ibid.

25. Ibid.

26. Network Information Center Mexico, https://www.registry.mx/jsf/domain_statistics/monthly/info.jsf (accessed November 10, 2009).

27. See http://www.razonypalabra.org.mx/N/n67/varia/oislas/emarketer_2000531.pdf (accessed October 22, 2012).

28. AMPICI, Hábitos de los usuarios de internet en México.

29. The ".com bubble" refers to a period in which speculation ran rampant during the period 1998–2000. During this time hundreds of Internet-based companies commonly referred to as "dot coms" were formed and received huge investments in venture capital even though they were unable to prove on many occasions that they were in any way viable. This gold rush was possible due to lax regulation standards and the fact that investors ignored traditional market metrics and, instead, were driven by the promise of huge returns and a blind confidence in the transforming and almost magical power of technology.

30. Ken Auletta, *Publish or Perish: Can the iPad Topple the Kindle, and Save the Book Business?*, *The New Yorker,* April 26, 2010, 24.

31. Lanier, *You Are Not a Gadget*, 48.

32. Alexis Okeowo, "To Battle Cartels, Mexico Weighs Twitter Crackdown," *Time* magazine, April 14, 2010, http://www.time.com/time/world/article/0,8599,1981607,00.html#ixzz0lCBzrEEF (accessed October 22, 2012).

33. See http://www.milenio.com/node/356370 (accessed October 22, 2012).

34. See http://en.rsf.org/ameriques-muscling-in-on-the-media-a-24-02-2011,39608.html (accessed October 22, 2012).

35. Nick Valencia, "Residents Use Social Media to Fight Organized Crime in Mexico," CNN online service, http://www.cnn.com/2010/TECH/03/08/mexico.crime.social.media/index.html (accessed October 22, 2012).

36. A hashtag is a tag embedded in a message posted on the Twitter microblogging service, consisting of a word within the message prefixed with a hash sign. A hashtag is used to narrow down tweets on a particular topic.

37. AMPICI, Hábitos de los usuarios de internet en México.

38. Anand Giridharadas, "Where a Cellphone Is Still Cutting Edge," *New York Times*, "Week in Review," April 11, 2010.

39. Carlos Slim, quoted in Geri Smith, "Carlos Slim on Monopoly, Pemex, and Kids," *Business Week*, March 5, 2007, http://www.businessweek.com/magazine/content/07_10/b4024066.htm (accessed October 22, 2012).

40. See http://www.forbes.com/sites/seankilachand/2012/10/15/worlds-richest-man-and-stefan-persson-get-500-million-dollars-richer/ (accessed October 22, 2012).

41. Felix Salmon, *Carlos Slim, the World's Richest Man*, July 3, 2007, http://www.portfolio.com/views/blogs/market-movers/2007/07/03/carlos-slim-the-worlds-richest-man (accessed October 22, 2012).

42. See http://www.internetworldstats.com/am/mx.htm (accessed October 22, 2012).

43. Robert D. Atkinson, Daniel K. Correa, and Julie A. Hedlund, *Explaining International Broadband Leadership* (Washington, DC: ITIF, 2008), http://www.itif.org/index.php?id=142 (accessed October 22, 2012).

44. Lanier, *You Are Not a Gadget*, 14.

# III
# RADIO AND MUSIC

# 8
# The Race for the Airwaves
## Journalism and the Radio Industry in Modern Mexico

*Viviane Mahieux*

In the spring and summer of 1923, a heated competition took place in Mexico City. The aftershocks of the Mexican Revolution were still being felt throughout the country, and violence remained a part of daily life for many of the capital's dwellers, but this was not a struggle for political territory, nor were the antagonistic camps political parties. This was instead a race for technological and journalistic prominence. Two of the most important periodicals of the capital, *El Universal Ilustrado* and *El Mundo*, competed to be the first to launch a commercial radio station and fill the air with the programming of their choice. At stake was the ability to create an audience of loyal listeners—a much broader one than could be reached through the printed press—and shape the role that the radio industry would play in a nation that was redefining itself after ten years of civil war.

Two very different individuals headlined this race for the airwaves. On the one hand, there was Martín Luis Guzmán, the director, founder, and primary investor of the evening daily *El Mundo*. Guzmán was an established intellectual with a long political trajectory who had recently returned to the country after four years of exile. He was a member of the Ateneo de México, an organization of writers and intellectuals who came together in the final years of Porfirio Díaz's regime, united by a belief in the importance of the humanities and the arts in national education. On the other hand, there was Carlos Noriega Hope, a young cinephile with no political aspirations who frequented avant-garde circles and directed the weekly magazine *El Universal Ilustrado*, which published literary pieces alongside news on fashion, film, theater, and music. Although both were writers with close ties to literary circles, their background and intellectual affiliations were quite different, as was made evident in how the publications they directed promoted the inauguration of their respective broadcasting stations. Their different perspectives would shape how they perceived the role of the press, how they assumed their public responsibilities as journalists and editors, and, consequently, how they envisioned the new technological opportunities offered by the radio. The following pages aim to reconstruct this 1923 race for the airwaves, using it as a prism through which to consider not only how two coexisting intellectual circles—one affiliated with the Ateneo de México and the other with the avant-gardes—imagined the role of new media, but also how a newly industrializing press embraced the potential of the radio.

## Mexico and the Rise of the Radio Industry

In the early 1920s Mexico, like much of the world, was abuzz with the seemingly endless promises of the radio. In September 1921, in the midst of the festivities commemorating the centennial of Mexico's independence, a young doctor named Enrique Gómez Fernández built a transistor radio and installed it in the basement of the Teatro Ideal. He placed a receiver in the theater Bellas Artes, where an international commercial exhibition was taking place. The experiment was a great success. On the evening of September 27, crowds elbowed each other to get a chance to don the headphones that enabled them to listen to singer José Mojica and to María de los Ángeles, Gómez's daughter, who were performing a few streets away in the confines of the Teatro Ideal.[1] In the next months, radio fever continued to spread, and aficionados in Mexico City devoted hours to capturing signals from the United States or Canada, even initiating conversations with some of the other listeners who were scattered across their own city.[2] By the first months of 1923, two small-scale radio stations began experimenting regularly on air. The first was a private station owned and run by an Argentine immigrant who broadcast every Sunday. The second station also sprung from a private initiative, but it would eventually become the official station of the Secretaría de Guerra y Marina.[3]

Because few households owned radio receivers, these first forays into broadcasting only reached a small circle, but this would quickly change. In 1923, Raúl Azcárraga, a businessman with a keen eye for novelty, became intrigued by the possibilities of this new medium. He opened a store, La Casa del Radio, and began selling radios for as little as twelve pesos. At the same time, many newspapers began publishing detailed instructions that explained how to build a receiver from scratch. Although the young poet Salvador Novo would mockingly define a radio receiver as an object of luxury—in its most expensive forms, it became a fashionable must-have of the period—the medium also caught on in more humble social circles.⁴ In November 1923, an article in *El Demócrata* revealed:

> Whomever is a bit observant and circulates among working-class neighborhoods..., if he directs his gaze to the roofs of the innumerable rooming-houses [*vecindades*], especially those of the poorest aspect, he will notice the number of antennae that can be discerned from the street. It can be seen, in contrast with other neighborhoods in the city, that these have not been installed by the most respected businesses nor by an expert electrician, but rather, that they have been constructed according to the strictest rules of thrift.⁵

Antennas began to alter the Mexico City skyline, in poor and rich neighborhoods alike. Not only was the promise of the radio seducing a variety of social sectors in Mexico's capital, it was also catching on in smaller provincial cities. Eventually, it was thought, radio could become a means to breach the distance between urban and rural areas by bringing the culture of the capital to the rest of the country.⁶

The popularity of the radio in the early 1920s manifested itself in many unexpected ways. In June 1923, a radio fair was held in Mexico City's Palacio de Minería. It was organized by private entrepreneurs who sought to pressure President Álvaro Obregón into signing legislation authorizing regular radio broadcasts.⁷ As people milled around stands, such as the one set up by El Buen Tono, a tobacco company that was about to launch its own broadcasting station, they were greeted by young ladies wearing antennae-styled headpieces who distributed the aptly named Radio cigarettes produced by the company.⁸ Other radio-inspired products would be launched then, such as a "radio" soda and "Radiovital," a mysterious health potion that claimed to be prescribed by modern doctors to give strength and energy to "purify the blood."⁹

While Mexico City residents were being bombarded with the imagery of the radio, writers seized this technological novelty as a means to expand literary horizons. In fact, as Rubén Gallo points out, 1920s Mexico lived a "literary infatuation with the radio."¹⁰ The medium particularly influenced members of the Stridentist vanguard group, who in 1923 published a short-lived literary journal called *Irradiador*. In April 1923, Stridentist founder Manuel Maples Arce published his ode to the radio, "TSH" (short for *telefonía sin hilos*), in *El Universal Ilustrado*; and, as we will see, he also read the poem in the magazine's inaugural broadcast. In 1924, fellow Stridentist Luis Quintanilla published a collection of poems entitled *Radio: Poema inalámbrico en trece mensajes*. Other avant-garde writers, later known as the Contemporáneos, also felt the influence of the radio. Francisco Monterde García Icazbalceta founded a short-lived literary journal called *Antena*, to which Salvador Novo and Xavier Villaurrutia would contribute in 1924.

It was in the midst of this effervescence of all things related to the radio that the journalistic race for the airwaves took place.[11] Guzmán was the first to begin advertising the inauguration of *El Mundo*'s station in early 1923, in many ways setting the tone and the pace of Mexican journalism's expansion into broadcasting. But it would be Noriega Hope who launched Mexico's first commercial radio program, inaugurating *El Universal Ilustrado*'s station on May 8, 1923, more than three months before *El Mundo* broadcast for the first time on August 14 of that same year. Noriega Hope's preemptive move would foreshadow the series of setbacks that Guzmán faced in *El Mundo*'s final year of circulation. Although the 1923 race for the airwaves was brief and its consequences seem to have been equally short-lived, it exemplified the complex relationship that intellectuals and journalists had with new media in the 1920s, as well as their fascination with that unfathomable entity, the public.

## Presidential Promises and the Launching of *El Mundo*'s Station

Guzmán inaugurated *El Mundo* on March 18, 1922. From its first issues, the newspaper brazenly advertised itself as "the best evening daily." This was not an empty boast, as its birth coincided with a strike at the well-established *El Universal*, and some of the latter's best chroniclers, such as Marco Aurelio Galindo and Cube Bonifant, soon switched ranks and moved to *El Mundo*. The daily began with a circulation of nine thousand issues, a small but respectable number, and cost five cents. It included sections on politics, sports, film reviews, articles addressed to women, and classified ads. It often gave its leading articles a sensationalist bent, for, as Guzmán once admitted, "crimes are given the cover because our circulation relies on them."[12]

*El Mundo* was not Guzmán's first journalistic venture, nor would it be his last.[13] In 1908, he began working for the daily *El Imparcial*. During the Mexican Revolution, he followed the troops of Pancho Villa, an experience that provided the basis for much of his literary writing and lead to his first exile in New York. By 1919 he was back in Mexico, working briefly as the political columnist for *El Heraldo de México* before launching *El Mundo*. Guzmán's name figured as the daily's director, but, in fact, he was much more than that. The paper was run by an association, the Compañía Editorial Mexicana, of which Guzmán was the principal shareholder. As many of Guzmán's private letters from the period show, he clearly considered *El Mundo* a business venture, and his primary concern was ensuring a sustainable profit to guarantee the continuity of the publication.

*El Mundo* was also a means for Guzmán to reinsert himself into Mexico's political life. It is no coincidence that while he directed the paper he was elected congressman and vice president of the Partido Cooperatista, which supported Adolfo de la Huerta's bid to succeed President Obregón in 1924. Founding and running a newspaper was a means for Guzmán to put into practice his vision of the revolutionary state and to explore the different ways that, as a writer and intellectual, he could contribute to implementing the goals of the revolution. In this aspect, Guzmán's work in *El Mundo* can be read as a means to exercise the ideals of Ateneo de México, the intellectual movement to which he belonged along with Alfonso Reyes and José Vasconcelos, and to transform his experience of the Mexican Revolution into an intellectual project. The members of the Ateneo considered themselves the "theoretical antecedent" of the revolution, often highlighting the parallel between the foundation of their movement and the start of the social struggle.[14] The emphasis on this historical coincidence obscures the careful rhetorical distance that the Ateneístas often kept between their intellectual mission and the popular actors of the revolutionary struggle, but the Ateneo did anticipate, if not the social struggle itself, the formation of a modern, "ethical" state in which intellectuals would acquire a significant role as political and cultural mediators.[15] This role, perhaps best exemplified by José Vasconcelos's work as minister of education under Obregón's presidency, implied fomenting the Ateneo's humanist belief in the transformative potential of arts and letters as a means to incorporate popular circles into the ideal of the nation-state. Yet while the masses were a great concern for the intellectuals of the Ateneo, this entity also signaled one of the elements that they most struggled to understand and that in many ways served as a constant reminder of the inherent limitations of the lettered culture they represented.

Guzmán's *El Mundo* exemplified many of the beliefs and concerns of the Ateneo. The newspaper's various slogans, probably determined by Guzmán himself, indicate the social and intellectual role that he attributed to this enterprise. In March 1922, for example, we can find various slogans headlining the paper, such as these: "EL MUNDO will defend any noble and good idea and attack any one that is contrary to national interests," "Our criteria, our faith, and our ideals are not sidelined by the needs of our department of advertising," and "The culture of a people is measured by the number of newspapers it reads. EL MUNDO will help Mexico read more." A few months later, the paper began using a more constant slogan, "Revolutionary in our essence, constructive in our manner," which speaks to the paternalistic function of the Ateneísta intellectual as an educator who shapes national ideals and builds revolutionary consensus.

The slogans imply that the very existence of *El Mundo* complemented the national project of the revolutionary state by shaping and fomenting a modern, national culture. The fact that *El Mundo* was a private enterprise that relied on its sales to continue in circulation did not hinder its constructive potential: it was fomenting a practice of reading, regardless of whether the text being read was an essay on culture or a sports column on boxing.[16]

Radio broadcasting was a natural extension of Guzmán's pedagogic conceptualization of journalism, especially as it promised to overcome one of the greatest limitations of the press: the fact that a great majority of the Mexican population was illiterate. One of the ways that illiterate citizens could access the information distributed by newspapers was through public readings. The radio offered the possibility of institutionalizing this collective reception by creating an oral, imagined community that shared not only the cultural experience of a given station's programming, but also the repetitive, quasi-ritualistic tradition of gathering to share information and culture on a regular basis.

In February 1923, *El Mundo* announced that it would open a broadcasting station at its headquarters in downtown Mexico City, declaring itself the first Mexican publication to recognize the cultural importance of the radio. The paper also offered a receiver with headphones to those readers willing to buy a subscription before its first anniversary. Around the same time, it inaugurated a new section devoted entirely to the radio, which focused on what the "radio-aficionado" should know: detailed instructions regarding the installation of receivers, technical explanations on how radios worked, and general trivia related to the medium. To encourage more interest in this project, *El Mundo* acquired a receiver that broadcast conferences and concerts from stations in the United States and Cuba. Every day it published coupons announcing these broadcasts and promised that readers would be admitted free of charge by presenting three coupons at the door. The regular practice of reading was, quite literally, the ticket to a new and technologically enhanced cultural experience.

*El Mundo*'s new station was expected to be inaugurated in April 1923, but it was repeatedly delayed by technical and economic problems. The evening of May 8 must have been especially difficult for Guzmán, for that was when *El Universal Ilustrado* inaugurated its own station. At around the same time, the tobacco company El Buen Tono announced its own plans to launch a station in June. It seemed as if *El Mundo*'s station would not be that innovative after all. Despite these setbacks, *El Mundo* doggedly continued to advertise its future station, without relinquishing its self-portrayal as a radiophonic pioneer. Difficulties continued to hamper the projected station throughout the summer, but the mishaps of the days before the official inauguration would reach an almost tragicomic level.

On August 3, the front page of *El Mundo* announced the inauguration of its station for the next evening. Under the headline "Inauguration of the great broadcasting station of *El Mundo*," an anonymous article detailed the program of the transmission. It promised a conference by José Vasconcelos, performances by the singer María Tubau and the composer Manuel M. Ponce, and a poetry reading by Francisco de Icaza. The complete program was to be published the next day. But when August 4 came, the readers of *El Mundo* were instead faced by a surprising headline: "With a message from the President . . . our station will be inaugurated next week." An unsigned article, probably penned by Guzmán himself, announced: "We find ourselves obliged to postpone the event given that a circumstance has risen that fully justifies our decision." This mysterious circumstance was simply that President Obregón had invited himself to speak at the inauguration of the station, where he planned to announce the results of a series of talks that were taking place between the Mexican government and US delegates. The president's decision, this editorial claimed, would propel *El Mundo*'s station to an unprecedented level of importance that was unimaginable even in "the vastest and most flowered capitals of the world":

> This triumph . . . we offer to the public as an attribute of the public itself. To the public belongs entirely the successes, ambitions and conquests of our newspaper. It has only been four months since we celebrated our first anniversary, and today, at the head of all the metropolitan dailies, we inaugurate the first radio station of Mexican newspapers, and the most powerful one in the republic. . . . *El Mundo* is the daily that most sharply reflects national palpitations. By inaugurating this station, not only does it exteriorize its personality as unquestionably the preferred spokesperson of these national palpitations, it also extends through other media its mission of cultural, artistic and patriotic diffusion, taking to all of its subscribers the highest, purest and most noble spiritual vibrations of the most celebrated Mexican and international artists.

This declaration is notable for many reasons. First of all, it ignores *El Universal Ilustrado*'s previous success by reaffirming *El Mundo* as the first daily to launch a station; this was technically true, after all, as the former was an illustrated *weekly*. The passage is also remarkable for its evocation of the sensitivity of *El Mundo* to the country's "national palpitations," an expression that conjures the image of the nation as a living, thriving entity with which the paper developed an intimate, quasi-physical communion—Benedict Anderson's notion of an imagined community presented in one of its most tangible manifestations. Through its radio station, *El Mundo* offered itself as the conductor that brought the "spiritual vibrations" of art and culture to the thriving, palpitating body of the nation. It seemed as if the gap between the spiritual realm of art and the physical world of the masses could finally be breached through the invisible waves of the radio. It was no coincidence that Vasconcelos, then minister of education, was slated as a speaker at this inauguration. *El Mundo* described him as "one of the intellectual voices of noblest hierarchy on the continent" and as "a man of ideas of a torrential and devastating modernity." The radio was the ideal medium to instill in the public the spiritual passion for a modern nation that only an intellectual and educator such as Vasconcelos could conjure. Listening to *El Mundo*'s broadcast was thus an "elevating" experience that would help the masses rise to the intellectual ideals of the Ateneo.

On August 13, *El Mundo* announced once more its inaugural broadcast for the next evening and published the program of this "beautiful social, artistic, intellectual and musical event." It was essentially the same as had been promised ten days before, with the difference that President Obregón would be the first to speak. This time, the event finally took place. On August 15, the front page proudly announced: "*El Mundo*'s radiophonic station was solemnly inaugurated last night." The inauguration was proclaimed a complete success: it was "heard and celebrated in five thousand homes where receivers are already installed, giving the act the national importance, beautifully suggestive, patriotically symbolic, that *El Mundo* intended from the start." The station's first broadcast was not limited only to the subscribers who had earned their radio receivers. Loudspeakers were placed outside *El Mundo*'s headquarters, and crowds gathered on the street to listen to the transmission, where "radiophonic vibrations were launched into the winds."

At the end of the broadcast, according to the August 15 article, Guzmán and his guests celebrated with champagne. But despite the euphoric tone of this narrative, not all had gone according to plan. A small mention toward the end of the piece confirmed an awkward detail: "with motive of the departure of the American delegates, Mr. Warren and Mr. Payne, the President of the Republic, General Álvaro Obregón, who so graciously had promised his presence at this inaugural ceremony, found himself unable to attend." The president's decision to participate in the event, forcing its delay, followed by an abrupt cancellation, reveals to what extent Guzmán was both intimate with political power and at the same time highly vulnerable to the political machinations that surrounded the presidency. Guzmán's close involvement with politics, in fact, would soon lead to the closure of *El Mundo* and to his second exile in Spain. Guzmán was a supporter of Adolfo de la Huerta, and when it became clear that Plutarco Elías Calles would be Obregón's designated successor, he found himself in a precarious situation. On September 24, 1923, *El Mundo* broke the news that de la Huerta had resigned his post as "Secretario de Hacienda," a resignation that the secretary had promised to keep private. De la Huerta accused Guzmán of publishing this information without his consent, and Guzmán found himself effectively caught between two political factions.[17] When de la Huerta began the uprising against the government, Guzmán resigned as director of *El Mundo* and abruptly left the country in December 1923. Guzmán tried to keep the daily in circulation from afar, but it was not an easy task. Companies cancelled their advertisements, the government withdrew its support, and *El Mundo* was forced to close in February 1924.

Guzmán's exile brought about the end of *El Mundo*'s broadcasting station. This project, which had consumed Guzmán's energy and finances for much of 1923, lasted little more than four months, during which music and lectures by figures such as Pedro Henríquez Ureña and Antonio Caso were regularly broadcast on Tuesday and Friday evenings. The Ateneo's most decisive venture into new media thus ended with the disappearance of this short-lived station, and it seemed unlikely that its spiritual ideal of beauty would once again find an ally in the hertz waves of the radio. But as the following pages show, media and aesthetics would also coincide in *El Universal Ilustrado*'s avant-garde radio venture. And although Noriega Hope's project consistently differentiated itself from *El Mundo*'s broadcasting ideals, both initiatives would nonetheless share many journalistic concerns.

## *El Universal Ilustrado* and the Stridentist Imagination

When Noriega Hope began directing *El Universal Ilustrado* in 1920, the twenty-four-year-old writer had just returned to Mexico City after spending a period in Los Angeles, from where he had chronicled the Hollywood silent film industry. Noriega Hope's interest in cinema, especially in the context of a nascent cultural industry, deeply influenced the choices he would make as an editor. From the start, he sought to transform the weekly into a diversified publication comparable to US magazines such as *Vanity Fair* or *Harper's Bazaar*. *El Universal Ilustrado* covered a wide range of topics, such as film, fashion, theater, sports, aviation, and automobiles, interviews with singers and actors, and articles detailing life in New York or Paris, as well as music scores with the latest fox-trot and Charleston compositions. The magazine's covers were usually adorned with glamorous women dressed in the latest flapper fashion, and its pages devoted much space to photographs, caricatures, drawings, and large, illustrated advertisements.

*El Universal Ilustrado* was also a literary publication. Some of the most important texts of the European avant-gardes were reproduced in translation, and

Mexican writers of differing traditions coincided in its pages. This was one of the only venues where young writers could publish and where space was given to experimental poetry and prose. In the early 1920s, Stridentist writers appeared regularly in its pages, especially Arqueles Vela, who was also a contributing editor. By the mid-1920s, when the Stridentist movement moved its headquarters to Xalapa, Veracruz, writers associated with the Contemporáneos group, such as Salvador Novo and Xavier Villaurrutia, would write regularly.

Noriega Hope's aim to maintain a diverse publication was a pragmatic means to ensure a broad reading audience. Only in this manner could various intellectual groups, as well as readers from differing social sectors, find in its pages pieces that would interest them. Like Guzmán, Noriega Hope rarely signed any articles in the publication he directed, but his journalistic vision can be deduced from his editorial choices and the few brief "director's notes" that were occasionally published. These notes stressed the publication's eclecticism and its juxtaposition of frivolous and serious matters. In a September 1921 note Noriega Hope declared: "In these pages we have all done eclectic work, building the only literary newspaper in the country where there are prejudices neither against the old respected writer nor against the young unknown. Work of love, of collegiality, of independence, is that of this newspaper, whose director and whose writers lack grave literary, political or social antecedents."[18]

Unlike Guzmán, Noriega Hope distanced his publication from any overt pedagogic message. On the contrary, the magazine prided itself in being considered a "revista de peluquerías," a magazine to be found at barbershops or beauty salons. The director of *El Universal Ilustrado* justified the importance of his publication by rhetorically diminishing the authority of those who participated in its creation. This was a magazine for all, he implied, because the project was as young as the post-revolutionary nation-state, and those who participated in it had no particular political or cultural influence. Like Mexico, *El Universal Ilustrado* was a work in progress. Noriega Hope thus enhanced the status of his publication through a completely different strategy than the one that had guided Guzmán, which defined itself through the expertise of its regular contributors and its connection to a higher spiritual and intellectual level. While *El Universal Ilustrado* emphasized youth, imagining a scenario where writers and readers collectively discovered new cultural forms, *El Mundo* highlighted experience, proposing the press as a wise guide that oriented the population.

Noriega Hope's anti-solemn attitude would also define his approach to broadcasting. On April 5, 1923, the magazine published an entire issue devoted to all things related to the radio. The director introduced the issue with tongue-in-cheek references to *El Mundo*'s radiophonic ambitions:

The new marvel occupies the pages of honor in this issue. Be convinced that, without being the first, neither are we in second place, because no national weekly has yet had the idea to devote an issue to radiophony. And note that, so as not to be in second place, we avoid all class of technical descriptions. This is left to the newspapers that "have been the first." We barely beat our lyrical drums without the need of putting ourselves in a straightjacket. Mr. Maples Arce leads the way with a stridentist poem in honor of the radio. Who could do it better? Certainly no one. Stridentism is radiophony's blood brother. It's an avant-garde thing![19]

Noriega Hope acknowledges the competitive spirit of the race for the airwaves, but he replaces Guzmán's solemn purposefulness with a nonchalant tone that takes for granted his magazine's connection with new technologies. It is inevitable that *El Universal Ilustrado* expand into radio broadcasting, he suggests, because the magazine's young collaborators are natural allies of modernity. The implication is clear: radiophony comes easy for the creators of the weekly magazine; they don't need to try as hard as Guzmán to catch up with the times.

The texts contained in this April 5, 1923, issue certainly delivered unorthodox meditations on the medium. An article entitled "Radiophony and Love," signed Pepe Rouletabille, includes an image of a young girl, hair curled in the short flapper style of the period, wearing headphones and smiling into the distance. This was obviously not intended to be a very sophisticated reflection. Technology had to be couched in sentimental terms, it is implied, in order to interest women, as is confirmed by the author's aside: "I don't understand stridentism. Universal language is enough for me. It is very sweet to write for women. I would hate to have to write for academics."

"Radiophony and Love" was clearly intended to balance other more experimental texts, such as an article by Arqueles Vela entitled "The Antenna Man." This piece focuses on Manuel Doblado, a radio aficionado known as "El Hombre Antena," who spends hours every day locked up in a little apartment on the highest floor of a tall building, where he has installed his receiver. The article is presented as an interview, but no dialogue takes place because, according to Vela, "to know the Antenna Man one has to sense what he thinks and what he thought. His words are so vague that it is impossible to transcribe them." To interview him, Vela claims, he had to use a mysterious type of intuition: "I had to make him pass through my psychological X Rays. Only this way have I found his silhouette, lost far away from vulgar people." Vela describes a man who has abandoned material life and instead has chosen to live submerged in hertz waves. In this sense, he has created his own cosmopolitan community through the radio, one that eschews frivolity and popular trends: "the Antenna Man is like the lightning rod of crimes, of 'Jazz-Bands,' of the programs that are daily diffused by broadcasting stations." Vela describes this eccentric character in a poetic language that absorbs technical references (X-rays, hertz waves, lightning rods) and associates them with celestial, almost magical qualities. A radiophone, it seems, enables "the Antenna Man" to live connected to his unconscious, to develop an intuitive way of being that is all the more valuable because it keeps him from the mundane distractions of daily life. In this sense, Vela's article recalls Guzmán's description of the radio as a medium associated with a certain "spiritual vibration." But if, for Guzmán, the radio was a conductor that enabled the intangible essence of art and culture to reach a broad public, for Vela, the radio itself embodies the transformative spiritual experience of modernity.

The most notable piece published in *El Universal Ilustrado*'s radio issue was Manuel Maples Arce's poem "TSH." It appeared with an illustration by Bolaños Cacho, featuring lightning bolts striking from the mouths of three disembodied heads that were suggestive of both Greek and Aztec icons. The poem confirmed that for *El Universal Ilustrado*, radio broadcasting was without a doubt "an avant-garde thing." Like Vela's and Rouletabille's articles, "TSH" was specially commissioned by Noriega Hope. In fact, Maples Arce had never before donned the headphones of a radio receiver, and he wrote the poem after listening to the radio for the first time.[20]

"TSH" recreates Maples Arce's experience of tuning in to a radio receiver while searching for a signal. The unexpected sounds he encounters come randomly, as if determined by the shuffling of a pack of cards: static, disembodied voices, the music of a jazz band from New York City. Maples Arce is not interested in scientific details; on the contrary, he seems attracted to the unknown, incomprehensible aspects of the radio. The poem creates an eerie atmosphere, with references to the night, to stars, to the wide and unfathomable ocean. Listening to the radio conjures a state of mind similar to insomnia ("Memory's insomniac antennas / pick up the wireless messages / of some frayed farewell"). The lost and "frayed" sounds captured by the listener are real but without a clear origin; they evoke that confusing in-between state between reality and dreamlike abstraction. Maples Arce appeals to the cosmopolitan reach of broadcasting, yet he describes listening to the radio as a solitary experience. The listener is abstracted from his environment, accompanied only by the misplaced sounds that come from afar. The last verses of the poem, "A golden star / has fallen into the sea," insist on the ephemeral beauty of the radio and the nostalgic sense of loss that accompanies radiophony.[21] The listener knows that what he has just heard is unrepeatable and that he might never meet the others that shared his experience.

"El Hombre Antena" and "TSH" both portray the radio as an individually transformative technology that replaces a local community with an intangible collectivity of dispersed voices. In this regard, both of these works propose the radio as a radically different experience from the one suggested by *El Mundo*, where the medium was instead projected as an instrument to foment the cohesion of a national community. The Stridentist lens privileges the private, sensorial experience of the radio, offering it as a source of artistic inspiration and celebrating the radio as a modern experience that can be lived in multiple forms.

Maples Arce's poem set the tone for *El Universal Ilustrado*'s first radio program, which aired on May 8, 1923. The day was overcast and rainy. Writers and editors feared that a rainstorm would force the cancellation of the broadcast, but the sky cleared and all took place as planned. The program began with Maples Arce reading his ode to radiophony, followed by some words by Noriega Hope and Raúl Azcárraga, the owner of La Casa del Radio and the cosponsor of the station. The speeches were followed by performances by Spanish guitarist Andrés Segovia, revue theater star Celia Montalván, and composer Manuel M. Ponce. The broadcast ended with a few pieces played by pianist Manuel Barajas. The first radio concert of the republic had lasted a little over two hours and, apparently, was hampered by no insurmountable technical difficulties.

The unsigned article on May 10 that chronicled the events of the station's inauguration sealed *El Universal Ilustrado*'s radiophonic victory with these final words:

> Thus ended the concert with which the first great broadcasting station in the nation was inaugurated: with a complete triumph. The Director of *El Universal Ilustrado* and Mr. Raúl Azcárraga were congratulated by the Minister of Communications and by Mr. Miguel Lanz Duret, manager of the Compañía Periodística Nacional. The first made, with a firm voice, the announcement that the broadcasting station and the programming of concerts were inaugurated.
>
> On our way out, like the verse of the "man who was imagined," we saw from afar that:
>
> A golden star
> has fallen into the sea.[22]

The article on the inauguration was not devoid of the official protocol and nationalist sentiment that also characterized the narration of *El Mundo*'s first broadcast in August. However, its formal language is framed by Maples Arce's reading of "TSH." The final lines of the poem, "A golden star / has fallen into the sea," are quoted both at the introduction and the conclusion, reminding all readers that the radio was, above all, a vanguardist adventure and that the "victory" of being the first periodical to launch a commercial broadcasting station could not overshadow the poetic relevance of the event. The station's emphasis on music and poetry differed slightly from the model that Guzmán would espouse a few months later, in which music was accompanied by formal lectures given by the nation's most prominent intellectual figures. *El Universal Ilustrado*'s station would instead privilege music and art from the start, aiming to entertain rather than to straightforwardly educate—although one could argue that exposing a broad public to abstract Stridentist poetry was also a form of aesthetic education.

When looking back at this 1923 race for the airwaves, it is inevitable to wonder how Noriega Hope managed to overcome Guzmán, a more established intellectual with better political connections. One answer lies in the former's alliance with Azcárraga's La Casa del Radio, which offered technical and economic support. Because of this collaboration, *El Universal Ilustrado* did not have to invest in giving away radio receivers, and the station doubled as free advertising for the store. Another factor to Noriega Hope's advantage was that, while *El Mundo* was a relatively independent enterprise financed in great part by Guzmán himself, *El Universal Ilustrado* was part of a greater consortium that included the morning daily *El Universal* and the evening daily *El Universal Gráfico*. With this broader infrastructure came more reliable economic support. As an employee of a large newspaper consortium rather than an investor, Noriega Hope had less at stake than Guzmán in the success of the broadcasting venture and could afford more experimentation.

Despite his initial success, Noriega Hope's participation in the early stages of radio broadcasting would be about as brief as Guzmán's. In September 1923, the collaboration between La Casa del Radio and *El Universal Ilustrado* was replaced by a new alliance between the same business and *El Universal Gráfico*. This meant that Noriega Hope would no longer determine the station's programming. It remains unclear if he was pushed out or, simply, if he had become bored once the novelty of broadcasting wore off. His interest in the radio, however, did not end then. In 1934, it seemed as if Noriga Hope was finally going to step down after directing *El Universal Ilustrado* for fourteen years. A few strategically placed ads in the weekly suggested that he was founding another publication, one that would be entirely devoted to the radio. But this new project never crystallized, for Noriega Hope died suddenly, at the age of thirty-eight, before being able to embark on what would have been his second adventure in radiophony.

## Radio and the Imagined Public

Both Guzmán and Noriega Hope imagined the profoundly transformative effect that radio broadcasting would have on journalism in post-revolutionary Mexico. The promotion and programming of their respective stations clearly indicated that they considered the radio to be a cultural venture more than simply a commercial one. In the early 1920s radio broadcasting relied on an already existing public created by the press, but very little "news" was actually transmitted, partly due to the tight control that the government exercised over information.[23] It was still up to the dailies to inform the public of the latest happenings. Partly for this reason, broadcasts in 1923 were overall quite similar: they consisted of a combination of music, lectures, literary readings, and comic skits. By 1924, the euphoria of the radio subsided amongst artists and educators. It became difficult to convince the best performers and speakers to participate regularly for little, if any, pay, and the quality of the broadcasts decreased.[24] Regardless, the reach of the radio would continue to grow throughout the next decade. In 1926, an estimated twenty-five thousand sets were in operation. By the mid 1930s approximately seventy radio stations were broadcasting in Mexico City, and 250,000 receivers functioned in homes and businesses across the country.[25]

While Guzmán and Noriega Hope shared a belief in the potential of the radio, it was in their conception of modern national culture that they differed. These discrepancies become all the more apparent in how each imagined the reception of their programming. At first glance, Guzmán's approach seems to take for granted the hierarchical distance between an intellectual project and its public, while Noriega Hope's turn toward the irreverent indicates a more horizontal relationship between the providers and the receivers of radio programming. However, Noriega Hope's approach tended to be exclusive in its own way. The relationship he imagined with his public, in fact, left out the many citizens who were not part of a growing white and mestizo middle class and who did not share the same urban referents. Guzmán's and Noriega Hope's approaches should thus not only be read in terms of the difference between the ideals of the Ateneo de México and the proposals of the avant-garde. They also reflected the differences between their respective publications, in terms of their self-definition, the publics they attracted, and the patterns of reading they promoted. While *El Mundo* was an inexpensive and somewhat sensationalist daily, *El Universal Ilustrado* was a cultural and literary weekly that mostly addressed an upper-middle-class population. This might help explain why Noriega Hope felt no need to justify his frivolous editorial choices and radio programming, as did Guzmán.

Early broadcasting in Mexico thus began through the small-scale initiative of a few individuals who saw the radio as a means to expand their intellectual and journalistic projects. It was only as the medium grew during the 1930s that regulations and larger corporations began to shape national broadcasting and the radio lost its individual, almost improvised touch. How these two broadcasting pioneers approached the new medium of the radio in the summer of 1923 began from a very pragmatic basis: they aimed for the continued growth and success of journalistic projects, both in print and on the air. But the journalistic race for the airwaves would also reveal Guzmán's and Noriega Hope's deep concern with the responsibility of intellectuals in post-revolutionary Mexico and with the role that art and culture should play in their nation.

## Notes

1. Gloria Fuentes, *La radiodifusión: Historia de las comunicaciones y los transportes en México* (Mexico City: Secretaría de Comunicaciones y Transportes, 1987), 38.
2. Ibid., 43.
3. Ibid., 46.
4. Salvador Novo, "Radioconferencia sobre el radio," in *Viajes y ensayos I* (Mexico City: Fondo de Cultura Económica, 1996), 39–40.
5. Quoted in Héctor de Mauleón, *El derrumbe de los ídolos: Crónicas de la ciudad* (Mexico City: Cal y Arena, 2010), 50. All translations are mine.
6. For an exploration of radio and rural educational projects in early twentieth-century Mexico, see Joy Elizabeth Hayes, "National Imaginings on the Air: Radio in Mexico, 1920–1950," in *The Eagle and the Virgin: National and Cultural Revolution in Mexico, 1920–1940*, ed. Mary Kay Vaughan and Stephen E. Lewis (Durham, NC: Duke University Press, 2006), 246–58.
7. This event is narrated in Rubén Gallo, *Mexican Modernity: The Avant-Garde and the Technological Revolution* (Cambridge, MA: MIT Press, 2006), 141. Hayes also details government regulations on broadcasting; see Joy Elizabeth Hayes, *Radio Nation: Communication, Popular Culture, and Nationalism in Mexico, 1920–1950* (Tucson: University of Arizona Press, 2000), 35–41.
8. For an analysis of El Buen Tono's forays into broadcasting, see Gallo, *Mexican Modernity*, 292.
9. The advertisement for "Radiovital" appeared in *El Universal Ilustrado* on April 5, 1923.
10. Gallo, *Mexican Modernity*, 120.
11. For more on the history of state and commercial broadcasting from the 1920s to the 1950s, see Jorge Mejía Prieto, *Historia de la Radio y la Televisión en México* (Mexico City: O. Colmenares, 1972).

12. Guzmán, quoted in Fernando Curiel, ed., *Medias Palabras: Correspondencia 1913–1959* (Mexico City: UNAM, 1991), 115.

13. For more biographical information on Guzmán, see Fernando Curiel, *La querella de Martín Luis Guzmán* (Mexico City: Oasis, 1987).

14. Horacio Legrás, "El ateneo y los orígenes del estado ético en México," *LARR* 38, no. 2 (2003): 35.

15. For an analysis of the Ateneo's intellectual relationship with the revolution, see Legrás, "El ateneo y los orígenes del estado ético en México."

16. Although *El Mundo* was a private enterprise that relied on sales and advertising, it received government subsidies; see Pablo Piccato, "Altibajos de la esfera pública en México de la dictadura republicana a la democracia corporativa: La era de la prensa," in *Independencia y Revolución: Pasado, presente y futuro*, ed. Gustavo Leyva et al. (Mexico City: Fondo de Cultura Económica-Universidad Autónoma Metropolitana, 2010), 240.

17. The details surrounding de La Huerta's resignation and Guzmán's second exile remain unclear. For more, see Curiel, *La querella de Martín Luis Guzmán*.

18. Carlos Noriega Hope, *Carlos Noriega Hope: 1896–1934* (Mexico City: Instituto Nacional de Bellas Artes, 1959), 42.

19. *El Universal Ilustrado*, April 5, 1923, 11.

20. Gallo, *Mexican Modernity*, 126–27.

21. *El Universal Ilustrado*, April 5, 1923, 19.

22. Anonymous, "Un gran triunfo de *El Universal Ilustrado* y de la Casa del Radio," *El Universal Ilustrado*, May 10, 1923, 18.

23. Ángel Miquel, *Disolvencias: Literatura Cine y Radio en México (1900–1950)* (Mexico City: Fondo de Cultura Económica, 2005), 164.

24. Ibid., 186.

25. Hayes, "National Imaginings on the Air," 246–47.

# 9
# Music Culture and Resistance in Mexico, 1968–1988
## Popular Music and Mass Media

*Ricardo Pérez Montfort*

In memory of my distant friend Marcial Alejandro

I

During the early 1960s, Mexico's mass media experienced a very particular boom. In radio as well as film and television, the scene was gradually dominated by images and sounds, trends and customs, and, in general, everything that characterized the "American way of life." Mexican cinema experienced one of its many crises, this time caused by the refusal of those who had created the socalled golden age to accept the attempts by a new generation of filmmakers to become part of the industry. While a newly emerging creativity attempted to breathe new life into the old approaches, the mediocrities of several new actors and directors continued to represent the majority of Mexican cinema production, still oscillating between ninety and one hundred films a year.[1] Among the many films featuring El Santo, infidelities and double standards, adventures and ranchero comedies, it is also worth mentioning some films designed for a young audience that had been imposed for some time, with conservative, educational zeal, in response to the nonsense of the "rebels without a cause," and with special emphasis on the Americanization of Mexican post-adolescent stereotypes.

1967 was the first year in which basically all commercial movies made in Mexico were filmed and projected in full color. Only some experimental films and a rare commercial film were still made in black and white. And together with color, there were also attempts to impose stereophonic sound. At any rate, by the late 1960s, films were still characterized by the classic formula of a multiple combination of actors, singers, trendy songs, and rhythms. Consequently, these actors and singers were not only performing the prevailing commercial music on radio, but also working in cinema and, from there, moving to television; they seemed to switch back and forth limitlessly.

It is worth noting here that television, for its part, had not yet become a regular household fixture for the masses. Rather, it could be found in the living rooms of certain upper-middle-class sectors and the country's urban aristocracy. In 1967, there was a total of thirty-four television stations in Mexico, and only one of them was considered to be a cultural station.[2] That year Channel 2 began to broadcast in color, and by the following year, basically because the Olympic Games were held in Mexico, the country became one of the world's ten countries with the best telecommunications facilities.[3]

However, it was a country populated by approximately 37 million inhabitants, while only 1.5 million of them were able to join the category of "TV households" (television owners). Images and trends from the United States penetrated the television sector in a particularly intense way. By the mid-1960s, US television series with Spanish dubbing, including *I Love Lucy, Mr. Ed, Lassie, Combat!, Hopalong Cassidy, Bat Masterson, FBI in Action, Highway Patrol, Dr. Kildare*, and many others, invaded Mexican television channels. These TV series could be watched during prime time and were accompanied by countless commercials advertising Acros stoves, Osterizer blenders, and Frigidaire refrigerators. Also by that time the Teleprogramas Acapulco Company, directed by the Mexican *junior* with the strongest tradition in these circles, Miguel Alemán Jr., became the main producer of national TV soap operas, which would gradually try to compete with the imported TV series and would occasionally include Mexican composers, singers, and musical groups that were in style at the time. Nevertheless, radio continued to be the favorite mass medium for popular music well into the next decade.

By 1965, there were fifty-seven radio stations in Mexico City broadcasting by amplitude modulation (AM).[4] FM (frequency modulation) stations had not yet appeared, and the AM dial was becoming saturated. While a significant number of stations continued to play Mexican or Caribbean music, and even Latin American rhythms or European ballads, nearly 50 percent were dedicated to broadcasting US musical products with a clearly commercial focus.[5] For example, by 1965 the station at 6.20 on the AM dial was broadcasting what was called "music that's here to stay," which basically consisted of melodies interpreted by Doris Day, Connie Francis, Tony Bennett, and Perry Como, to mention a few, and accompanied by the Ray Conniff and Nelson Riddle orchestras. One of the main sponsors was the *Lovable* brand of women's underwear.

However, it was rock and roll, or just *rock* and its multiple derivatives that, from the first appearances in the middle part of the previous decade, remained in constant competition with tropical and ranchero music in the world of music for young people in Mexico.

Groups like Los Hooligans or Los Rocking Devils and singers like Enrique Guzmán, César Costa, and Angélica María demonstrated a clear fondness for the music that was popular among US youth, translating songs like "Jailhouse Rock" and "My Boy Lollypop" into Spanish ("El rock de la cárcel" and "Mi novio eskimal"). The imposed interest in rock and roll expressed by Mexican young people was so strong that one of its initiators, Bill Haley, who sang hits like "Rock around the Clock" and "See You Later, Alligator," spent long periods of time in Mexico City, performing on the airwaves and in movies, night clubs, and dance halls even long after dancing to this music was no longer so much in style. Some groups that actually had more of a "singing-family" style than a rock-and-roll style, like the Castro Brothers and the Carrión Brothers, could switch from playing the romantic tones of young ballad singers and turn into a bastion of quartets or septets ready to rock and roll or perform bossa novas at the drop of a hat.

Nevertheless, during the second half of the 1960s, the writers of a certain "Mexicanistic" repertoire of songs, such as José Alfredo Jiménez, maintained their prestige among both young people and adults who listened to the radio and watched the cinematographic screens and television transmissions newly becoming available to Mexico's middle class.

By that time the discographic productions of companies like RCA Víctor Mexicana, CBS, Peerless, Discos Musart, Orfeón, Capitol, and Gamma were quite prolific. All of these companies were interested in recording and selling Mexican and US music in the national market. They achieved this basically by disseminating their productions through the various radio stations in Mexico City and the rest of the country—all of which had already, for some time, been promoting records and hits in exchange for what had been euphemistically coined *la payola*, which was nothing but payment for positioning a song or artist insistently and ad nauseam within the supposed preferences of the public.[6] In reality the public had absolutely no possibility of influencing the selection of music brought to the airwaves by their favorite radio stations, and thus their capacity to develop their own preferences was, of course, never taken into consideration.

As such, the business of selling records during those times was intimately linked to the radio industry, which allowed itself to be economically influenced, with its pockets generously lined, in exchange for maintaining one artist or another in the preferences of its listening audience. Stores selling LPs or single-play records (33 or 45 rpm) experienced a particular boom in the 1960s and 1970s, not only because of the promotion on radio stations, which unquestionably had enormous influence, but also because devices for playing those records—such as portable record players, stereo consoles, cassette players, amplifiers, and speakers—came much closer to within the economic possibilities of the Mexican middle class. The business of music production companies, controlled by the Mexican Association of Phonographic Producers (Asociación Mexicana de Productores Fonográficos–Amprofon), was also linked to the radio industry, which in turn was linked to other mass communication media. And the most influential magnate in this entire chain was none other than Emilio Azcárraga Vidaurreta.[7]

The popular music heard in national contexts during the second half of the 1960s and the early 1970s oscillated between two basic types: music made in Mexico and music from the English-speaking world. Mexican radio and record companies continued to promote musical genres that were somewhat worn out, although highly popular and therefore economically profitable. These included the bolero as interpreted by veteran Agustín Lara or the Mexican "crooner" Daniel Riolobos; ranchero music in the style of Lucha Villa; the monotonous tropical tunes of Mike Laure; and the overly simple organ music of Juan Torres.

The Mexican romantic and popular music tradition continued to be very much alive. Writers, composers, and singers from times gone by, like Pedro Vargas, and those no longer alive at that time, like Jorge Negrete and Pedro Infante, among so many others, had a place in the record stores and in the increasingly saturated radio. In the many radio stations on the dial with the *X* prefix, which identified them as international radio stations in Mexico, it was possible to hear significant moments from the history of Mexico's popular music on any day of the week and any hour of the day. A review of the list of composers who registered their songs in 1968 makes it clear that the popular and romantic, Caribbean, Latin American, and Mexicanist repertoires continued to enjoy considerable health and did not particularly appear to be on their way out.[8]

Although the majority of these artists belonged to a generation already consolidated in Mexico's music world, some of them demonstrated the way in which Mexican romantic music was being revived with particular determination. Perhaps one of the most prolific young composers in those days was Armando Manzanero from the Yucatan. He had arranged music for and accompanied "Mexico's Sweetheart," Angélica María, who started out singing ballads and rock and roll. Manzanero is recognized for revitalizing the romantic ballad, which in the mid-1960s was in urgent need of a certain modernizing effect. His creativity in this sense was demonstrated in songs winning great commercial success, such as "Esta tarde ví llover" and "No." By 1968 this Yucatan composer seemed to bring a guarantee of success to any emerging artist in commercial music, and he also took advantage of this reality by launching his own solo act, with his first record given the evocative title "A mi amor . . . con mi amor." On that LP's back cover an anonymous commentator introduced Manzanero with the following phrases: "When romantic music was at its worst stage, when frenetic, electronic rhythms were invading the air from north to south and east to west . . . when young people en masse were turning their backs on *spirit* and turning their bodies and souls over to the dizzying, monotonous rhythms from abroad, Armando Manzanero appeared on the scene."[9]

It was not long before this composer-singer entered the world of silver, gold, and platinum records, rapidly becoming a product for exporting to Latin America and the rest of the world. Through Manzanero and his meteoric international success, it appeared that Mexico was reviving its prestige as a breeding ground for Spanish-speaking composers. Unfortunately, it was seemingly more like a shooting star in the late night. Mexico's commercial industry was unable to find another national composer of Manzanero's stature to make this a reality in the 1960s. And other Spanish-speaking singers quickly appeared, with other influences, and they gradually developed a place in Mexico's commercial music world. A phenomenon worth mentioning here was the Spanish ballad singer Raphael, who sang songs written by Manuel Alejandro and found particular success in Mexico. Fortunately, the media in Spain were already promoting someone who would become a symbol of music with a certain tone of protest: Joan Manuel Serrat. This Catalan composer and singer started to develop numerous followers among Mexico's young public, although it would be the 1970s when his promotion on this side of the Atlantic would truly begin.

Another music history was being written at that time, and the always-present commercial interests could certainly not miss the opportunity to participate. What was at stake was a massive international invasion of music initially headed by Elvis Presley, Little Richard, and Jerry Lee Lewis and then leading to the world phenomenon of the Beatles, who unquestionably internationalized music from the English-speaking world. Other groups, English-speaking as well, and with widely differing music, also came to the forefront, ranging from the Rolling Stones to the Beach Boys. And so English became the official language for modern, youth-oriented music.

The theaters and dance halls that had long welcomed fans of *danzón*, swing, cha-cha, and *cumbia* began to open their dance floors and stages to young ballad singers and rockers or twisters. The nightclubs for young existentialist jazz fans occasionally found housed in an old mansion in Mexico City's Roma district or the new Pink Zone experienced an onslaught of rock and rollers.[10] The Lírico theater and the Maxim's and Riviera dance halls in Mexico City and places like Bum Bum and El Tequila in Acapulco witnessed the arrival of several rock groups and soloists. These groups presented both songs for dancing the young rhythms such as the jerk and the surf, or simply go-go, as well as ballads and little love songs.

Appearing shortly thereafter were new versions of cafés featuring singers and, of course, discothèques, which attempted to transform their physical spaces into somewhat more exclusive places for young people with great interest in dancing and minimal knowledge of music. In Mexico City places like Memphis, Harlem, and A Plein Soleil were transformed from rather innocuous cafés into centers for consuming rock-and-roll music.[11] Groups like Los Crazy Birds, which featured at that time a very talkative, theatrical singer, Luis "Vivi" Hernández, and ballad singers like Mona Bell performed on the small stages in these places, which quickly began to receive sanctions from the intolerant, puritanical authorities of that time.

In line with Mexico's Americanized middle-class tendencies, the trendy pastime of ice-skating also had something of an influence on establishing ad hoc centers for young people to get together and listen to music. Perhaps one of the most frequented in Mexico City was the Insurgentes Ice-Skating Rink, where groups like Los Dug-Dugs and Los Yakis brightened the afternoon parties of those who left their adolescence behind as they went round and round a frozen rink or played hockey.

Gradually, a relatively new current was introduced into the music and rock and roll world of Mexican youth, characterized by the psychedelic and hippy counterculture. Long hair, colorfully dyed clothes, the so-called explosion of peace and love, flowers, miniskirts, drugs, and condemnation of wars, especially the Vietnam War, all gained increasing numbers of followers, as demonstrated by the first magazine dedicated exclusively to these topics and based on young people's perspectives and opinions. The publication appeared in the early months of 1968 and was given the synthetic name of *Pop*. Directed by Víctor Blanco Labra, this magazine focused particularly on what was taking place in British and US music and became a Mexican point of reference for the change the world's youth were trying to create through huge musical festivals and protests against segregation and war—of course, with the affirmation that young people were capable of making decisions regarding their lives and actions.[12] In Mexico and precisely in 1968, these young people found themselves clearly at odds with their main detractors, which were not few in number and ranged from the high levels of government and the Catholic Church to parents and schools and from the daily press to occasional commentators.

Various mass media promoted their own commercial preferences—and the conjunction of interests produced by certain veiled alliances with a government of clearly authoritarian roots—by insisting on spreading music and messages appropriate for Mexico's youth. One example in 1968 was that ironic question, broadcast constantly on the radio as a message revealing an attempt at government control, becoming a kind of popular slogan of the time: "It's ten o'clock at night. Do you know where your children are?"

At the very end of so-called Mexican developmentalism and just before the aggressive campaign against the 1968 student movement was unleashed, all of these artists and songs prevailing in the mass media could not drown out the other music that accompanied that emblematic year and culminated in profound national discord.

## II

As with any other event having great social and political impact, the 1968 student movement had its own music. The marches, rallies, assemblies, and even the meetings between activists and society were the scene for a series of musical manifestations known as "protest songs," "social songs," or "committed songs." Their origins were varied, and they were incorporated into a repertoire that included old *corridos* from the Mexican Revolution, parodies of trendy songs from that time, translations into Spanish of international songs about social struggles, and Latin American songs of resistance. Nearly all of them had come to the attention of the singers and fans of this type of music through informal encounters and through records from Cuba, Argentina, Chile, or Peru or, somewhat marginally, from Mexico's musical and lyrical heritage. Personalities like Atahualpa Yupanqui and Violeta Parra were just as highly recognized as Pete Seeger, Bob Dylan, and Joan Baez; and Mexican artists like Oscar Chávez, Judith Reyes, and the Folcloristas were all widely acclaimed by young activists.[13] For example, the somewhat clandestine radio transmissions by the National Strike Council that were broadcast on Radio Universidad, even before the National University was seized by the army in September of that year, began with the classic song by Violeta Parra and interpreted by her son Ángel, "Que vivan los estudiantes" (Long live the students).

Songs like "Las preguntitas sobre Dios" (Questions about God) by Atahualpa Yupanqui or "La zamba del Che" (Zamba for Che) by Víctor Jara were like naturalization papers in the movement, and one of the movement's hymns stated this clearly in its refrain:

> Here remains your clear,
> Heart-warming transparency
> Of your beloved presence
> Comandante Che Guevara.[14]

At rallies Oscar Chávez would often come up to the platform to sing corridos, most notably, "El 30–30" and "El mayor de los dorados." And Enrique Ballesté became famous for his song that said in one of its verses: "Esto de jugar a la vida es algo que a veces duele" (This game called life is something that hurts at times). Also, the Folcloristas group popularized a Chilean *trote* entitled "La Paloma" (The dove) and would frequently end their presentations with this song, saying, "Tráeme de lejanía, Paloma mía, la libertad!" (Bring me back from afar, my dove, freedom!).

Judith Reyes, for her part, composed a significant number of corridos, ranging from parodies to detailed narratives of tragedies resulting from the repression. She wrote, for example, "Corrido de la ocupación militar de la Universidad" (Corrido on the military occupation of the university), and in the first verses she expressed her interest in returning to the musical chronicle that had characterized the corrido since its beginnings:

Ten thousand soldiers left the barracks,
with so many tanks it was horrifying.
It was the month of September, the 18th day,
the year of '68. Very patriotic!

Like beasts in boots, they trampled the patio,
the book, the school, and our dignity.
They pissed on the classrooms,
and turned my precious University into a barracks.[15]

The 1968 student movement also produced some songs that criticized the lack of commitment on the part of some middle-class sectors to support the struggle led by the students. For example, high school students Ismael Colmenares and Armando Vélez of the group Los Nakos wrote a popular ballad entitled "Hippie" that reproached pacifists for their lack of action:

You dream of a world without war
you want people to feel
only love for others . . .
You say that peace is necessary
that mutual understanding
must be our truth . . .
I ask you if this is possible
this harmony, this joy
if all around the world there is hunger.[16]

The repression that ended the 1968 student movement left a deep mark on the musical repertoire of those who participated. October 2 became a necessary reference in subsequent protest songs, both in Mexico and in other parts of the world. Chilean Ángel Parra was one of the Latin American songwriters who used this date as an example of the many dates marking the long chain of massacres in the North American continent's calendar of horror. In the first volume in a series of records dedicated to La Peña de los Parra in the early 1970s, there was a tribute to Mexican students, interpreted by this young songwriter, entitled "Mexico '68." After an introduction very much in line with the style of an old corrido, the lyrics went like this:

So the glorious Olympics
would never be forgotten
the government ordered the death
of four hundred comrades.

Oh, Tlatelolco Plaza
the pain your bullets have brought me!
Four hundred hopes
snatched away in betrayal.[17]

Perhaps the most dramatic musical memory from the end of the Mexican student movement was a song written by Judith Reyes. It had to be recorded outside the country, in France, to be specific, on the Le Chant du Monde recording label. Some of the bitterest lyrics written by this singer and songwriter were on a record that circulated clandestinely among certain Mexican university sectors. In one of them, "Corrido de la Plaza de las Tres Culturas" (Corrido on the Plaza of Three Cultures), the emblematic wound inflicted on October 2 is summed up in these key verses:

On the second of October
we went peacefully
to a rally in Tlatelolco
about 15,000 of us.

In the year of '68
it makes me sad to remember
the jam-packed plaza
at about 6:00 p.m. . . .

Suddenly the sky is pierced
by four flares.

Many men appear,
white gloves and bestial faces.

Bullets zing
panic creeps in.
I look for shelter,
and the troops are everywhere . . .

How bloody was the slaughter.
Even our beautiful young women.
Oh, Plaza of Three Cultures,
you are dripping blood![18]

It is important to mention that years passed before these songs were heard on cultural radio stations in Mexico, which continue to be very few. They have only rarely been heard in Mexican films and even less on Mexican television.

## III

The 1970s witnessed the emergence of new types of forums for bringing attention to music for youth; these were created somewhat outside the influence of mass media. Rock, ballads, rancheros, boleros, *cumbias*, and so-called salsa continued to fill Mexico's radio airwaves and could continually be heard both in cines and on television. However, the opportunities for promoting and listening to the Latin American and protest versions of songs and popular genres were found particularly in what were referred to as *peñas*.[19] Places like La Peña el Cóndor Pasa, La Peña del Nahual, La Peña de los Folcloristas, and Mesón de la Guitarra welcomed many groups of young musicians who interpreted numerous songs from the Ibero-American repertoire to the sound of *quenas, charangos, bombos*, and guitars.

Among the songs presented in the peñas were socially committed songs from the Chilean Nueva Canción and Cuban Nueva Trova movements. These two genres had been consolidated as a result of the support received by both of them from Salvador Allende's socialist government and from Cuba's focus on promoting culture in the early 1970s.[20] In Chile what are now called "social songs" were gradually winning a place in the market and in communication media due to support from some record-producing companies like Odeón, EMI, and Lince. And in Cuba, the Sound Experimentation Group (Grupo de Experimentación Sonora) at the Cuban Institute of Cinematographic Arts and Industries (Instituto Cubano de Artes e Industrias Cinematográficas–ICAIC) was able to capture the attention of a broad public by incorporating key Nueva Trova figures among its ranks, including Silvio Rodríguez, Pablo Milanés, and Noel Nicola, to mention only the most popular at that time.[21] Their songs were never absent at the peñas recently inaugurated in Mexico.

Also incorporated into this particular repertoire were songs from Argentina, Colombia, Brazil, Peru, and Bolivia, with traditional, acoustic instrumentation that over time was identified generically as Andean. Groups like Inti-illimani or Quilapayún from Chile, Anacrusa and Cantaclaro from Argentina, Los Jairas from Bolivia, Jatari from Ecuador, and Los Guaraguao from Venezuela, as well as outstanding singers—like Daniel Viglietti, Alfredo Zitarrosa, Atahualpa Yupanqui, Jorge Cafrune, Mercedes Sosa, Horacio Guarany, the already-mentioned Isabel and Ángel Parra, Víctor Jara, Patricio Manns, Nicomedes Santa Cruz, Soledad Bravo, and Alí Primera—all influenced the popular Latin American trend that was extending gradually among a certain public in Mexico City and some other Mexican cities like Guadalajara, Puebla, Morelia, and Cuernavaca.

Somewhat more limited was the influence from Brazilian music targeting young people. Although the bossa nova had opened the way, becoming very popular in Mexico's urban context since the beginning of the 1960s with personalities like Antonio Carlos Jobim, Joao Gilberto, Luiz Bonfa, Carlos Lira, and Elis Regina, the very committed songs by Chico Buarque de Hollanda, interpreted by María Bethania or Caetano Veloso, were not as well accepted as songs from Spanish-speaking Latin American countries. At any rate, the names of these artists were beginning to be heard on the two cultural radio stations in Mexico, and their records circulated among some bossa nova connoisseurs.

In the early 1970s, several groups of young people from the middle class and from somewhat lower classes appeared; they followed some Latin American examples, turning protest music into their form of expression. While Los Folcloristas must be recognized as having some type of authorship over many of these groups, since members were either their favorite students or at least their dedicated followers, there is no doubt that it was in the peñas that many of the groups that promoted and cultivated this music flourished. The Lacantún and Cuicani groups, Icnocuícatl and Peña Móvil, Cade, La Nopalera, the group formed by exclusively women and called Ihuaye, the Anthar and Margarita duo, and singers Amparo Ochoa, José de Molina, and Gabino Palomares, to mention only some outstanding members of this type of "Mexican new song," consistently demonstrated their inclination to create a type of music that was clearly distanced from commercial, homogeneous currents of music. Most of them began by performing songs that had already become part of the Latin Americanist repertoire, and then gradually groups or singers made an effort to compose something of their own creation.

It is important to mention here that with very few exceptions, the communication media literally ignored this movement. And very few groups had access to the recording studios of Mexican and transnational companies in the record industry. During this time period only Los Folcloristas and Oscar Chávez had the opportunity to make records with labels like Gamma and Polydor. This type of music was basically banned on television, only occasionally granted some minimal space on "cultural" programs. During the 1970s, Radio Universidad and the Secretaría de Educación Pública's (SEP) Radio Educación were the only radio stations including this music in their programming, and on television, only Politécnico's Channel 11 and, very infrequently, Channel 13, still a government channel at that time, considered the pertinence of airing some Latin American songs as long as they were not "protest" songs. Consequently, this fledgling Mexican new song movement became essentially a quasi-clandestine current of music. Very slowly, small companies like Nueva Cultura Latinoamericana, Fotón, and Discos Pueblo appeared and attempted to fill this gap in record production, which had literally marginalized this movement.[22] However, these modest production companies could hardly compete with the major transnational companies and the country's record production associations, which had been dominating the market for decades.

This genre of music was not incorporated into Mexican cinema, although rock and roll, salsa, and boleros had been included in the previous decade and would continue to be integrated in the films produced all throughout the 1970s. In fact, Mexico's cinematographic industry practically ignored the members of this alternative music culture, with only very few exceptions, such as Rafael Corkidi's experimental cinema, specifically, *Auandar Anapu* (1975) and *Pafnucio Santo* (1977), the music for which was composed by Héctor Sánchez, a member of Los Folcloristas, and Jaime Casillas's *Pasajeros en tránsito* (1976), in which La Peña Móvil performed.[23]

As the 1970s progressed, the peñas were no longer dedicated exclusively to performing and listening to Latin American music. In some of them, like the Center for the Study of Latin American Folklore (Centro para el Estudio del Folclore Latinoamericano–CEFOL), efforts were made to expand cultural activity, incorporating expositions, literary workshops, and film festivals. Poetry and music recitals and an occasional theater production were also among the presentations. In addition, it was not unusual for a certain amount of leftist political proselytism to take place in these centers, although membership in a political organization was not typically the focus of these cultural activities. Those who frequented the peñas clearly identified strongly with socialist and national liberation movements; however, the fragmentation and long-standing feuds characterizing Mexican leftist groups made any joint efforts very difficult.

Various alternative means were used to introduce this music to the public, especially concerts at labor unions, rallies, and festivals. The musical groups participated in many protest demonstrations, supporting the causes they considered to be congruent with their political inclinations. These ranged from university labor union struggles to demonstrations in favor of leftist presidential candidate Valentín Campa in 1976 and included demonstrations protesting the military coup in Chile or paying tribute to the heroic people of Vietnam. The musicians were notably present at the so-called opposition festivals organized by the Communist Party and other organizations throughout the second half of the 1970s and the early 1980s, and basically any strike taking place in the Valley of Mexico could count on their presence. At any rate, their music was clearly heard in only marginalized forums, consequently, with only limited dissemination.

Even so, at the end of Luis Echeverría's presidential term and through the initiative of some particularly enthusiastic groups, the Front for Unrestricted Cultural Expression (Frente para la Libre Expresión de la Cultura–FLEC) was created and included groups like Los Folcloristas, La Peña Móvil, Icnocuícatl, and La Nopalera and singers such as Víctor Martínez, Amparo Ochoa, and Gabino Palomares. This Front for Unrestricted Cultural Expression attempted to open up opportunities for disseminating Latin American music and also attempted to become associated with other artist movements in Mexico and the "New Continent." It also worked to develop a strategy for linking its cultural activities with the country's popular and underground sectors as an expression of shared political convictions. Nevertheless, this enthusiasm gradually waned and basically disappeared by the end of the next presidential term.

New initiatives soon emerged at encounters and auditions in Mexico City and around the country. The first to appear was the Independent League of Revolutionary Musicians and Artists (Liga Independiente de Músicos y Artistas Revolucionarios–LIMAR) in 1978, then the Mexican New Song Committee (Comité Mexicano de la Nueva Canción) in 1983.[24] The latter, which grew out of the First Latin American New Song Festival, held in April 1982, was perhaps the most successful organization since it was able to provide continuity for many of the aspirations already manifested in previous initiatives. By the mid-1980s, this committee was able, together with many other cultural organizations that had begun to refer to themselves as "independent," to build upon a significant portion of the social unrest that had emerged in response to the economic and political crises affecting the country. While they were not able to obtain significant coverage of their cultural activities in the communication media, they did outstanding work in seeking alternative forms of mutual support and combining positive experiences in disseminating their musical expressions. This was particularly relevant during the mobilizations prompted by the 1985 earthquake. And the committee also participated as a key player in the Independent Cultural Organizations Encounter (Encuentro de Organizaciones Culturales Independientes) in November 1988. At that forum they seemed to jump ahead to some of the demands that would be featured in the principles identified in the uprising by the Zapatista National Liberation Army in 1994, including the recognition of and demand for respect for Mexico's pluri-cultural, multiethnic wealth.[25]

It is also important to note that the musical trends characterizing the experiences of these groups and singers were gradually transformed in this process. In the beginning the trend toward Latin American music had contributed to increasing awareness of the serious economic, social, and political problems suffered by what was already identified at that time as the Third World. But soon the members of this cultural movement felt intensifying urgency to focus on the dramatic contrasts in Mexico's own particular reality. And gradually there was growing and insistent interest in national music genres, such as *sones huastecos, jarochos*, corridos, and traditional mariachi music. With this interest focused on national popular and folkloric musical expressions, denouncements of local injustices and against the inequality in Mexico's political system were also incorporated into the musical compositions of this time. Perhaps the most important antecedent for this introspective tendency was the performance by a young linguist and anthropologist, Antonio García de León, of verses from "Fandanguito," a song attributed to *jarana* player Arcadio Hidalgo and recorded in a collection, *Sones Jarochos*, produced in 1969 by the National Anthropology Museum. Following are a few lines:

> I joined the revolution
> to fight for the right
> to feel an immense satisfaction
> in my chest.
> But today I'm living in a corner
> paying tribute to my bitterness
> crying out to destiny
> that *campesinos*
> are our hope for the future.[26]

With these popular, socially committed *son jarocho* verses, it was demonstrated that Mexican songs also had the right to transmit messages and denounce the injustices committed within national territory. Many other such expressions followed as musicians attempted to bring their particular political "color" into their communication through music. Some did so with a somewhat nihilistic tone, as evident in "El Huerto," a song that was written by Roberto González and asked the following questions:

> What is the sense of dialectics in History?
> Why go to paradise if we are dead?
> Why long for glory while we are alive
> If glory is so faraway from this land?[27]

Others resorted to a propagandistic tone, responding more to political slogans and commonplace phrases than to creative impulses. A clear example can be seen in the song "La Nopalera":

> A fence is a fence and it means there is an owner,
> The owner of all land, at the limits of your hardship . . .
> Come along brother, your people are calling,
> Come because you are needed.
> Many lives are being snuffed out
> In the bonfire that belongs to the *patrón*.
>
> It will be your sons and daughters who light the people's fire
> To build new men, revolutionary men,
> Without hunger and free from exploitation.[28]

And still others attempted to give a metaphorical semblance to denouncements of what was evident in reality. Such is the case in this song, "Septiembre mentiroso," from the On'ta group:

> I will cut away this deceptive September
> That has been castrating my people
> I will cut away the remaining days of gloom
> That are blinding my people.
> I will declare that independence is a tale
> That has yet to become true:
> I will plant the seed of a new year
> So my people can fasten
> Their new memories
> Their real memories . . .
> When every day is a day of freedom.[29]

Toward the end of the 1970s, the musical trend focused on the conflictive

Latin American situation was intensified due to a significant migration flow from Chile, Argentina, Uruguay, and Brazil of those who had been politically persecuted in their own countries and who began to establish themselves in Mexico. Well-known musical personalities such as Alfredo Zitarrosa and Ángel Parra, as well as the El Galpón theater group, decided to take up residence in Mexico and clearly contributed toward the alternative media, giving them a certain amount of attention. There were some cases of opportunism, with attempts to claim and even displace some of the small achievements made by the Mexican music movement, with arguments of "origins" and "Latin American purity," and there were even petitions to express pity for the struggles lost in the North American continent. Especially notable along these lines were the sorrowful presentations by personalities like Naldo Labrín and Delfor Sombra, two Argentinean members of the Sanampay group, and the embarrassing insistence on remembering the tribulations suffered by guitarist Caito (Carlos Manuel Díaz Alonso) when he fled the dictatorship that devastated his country.

In the mid-1980s, a small opportunity opened up in the mass media for the already acknowledged "Mexican new song." Singer Eugenia León, a former member of the Víctor Jara group, very closely linked to the Mexican Communist Party, began to win some space in the mass media with some presentations in the much-acclaimed programs produced by the Televisa monopoly. Two singers did more or less the same, with very different outcomes. Lupita Pineda managed to position herself in the area of romantic music, within a certain sphere of the television emporium's preferences. And Margie Bermejo was promoted in some culture-oriented spaces sponsored by the government. Nevertheless, it was Eugenia León who attempted to take a bolder jump ahead by taking Marcial Alejandro's song "El Fandango aquí" to the International Festival of the Iberoamerican Television Organization (Organización Televisiva Iberoamericana–OTI) held in Seville. The Mexican edition of this festival presented a competition in which some young people, indulged by commercial media, participated. In addition to Alejandro and León, another musician who was a fresh new talent at that time, Jaime López, participated with an irreverent piece entitled "Blue Demon Blues."

However, just when Eugenia León triumphed in Spain under the monopoly's spectacular brilliance, a stroke of bad luck was to make Mexico shake at its very foundations for another reason: the earthquake of September 19, 1985. While León's triumph was, as commented by Humberto Musaccio, "some pleasant news in the midst of tragedy," it is also true that her accomplishment signified a minimal opening in communication mass media to this genre of young Mexican popular musical creations.[30] "El Fangando aquí" was not a protest song, however, and not even close, since the words of the song merely described a Mexican fiesta. Indeed, it did so in a new way and in the process attempted to give voice and opportunity to a generation of composers and artists who, because of their political position, style, or noncommercial Latin Americanist tendency, had been systematically marginalized by the mass media.

In the end the success enjoyed by Eugenia León and Marcial Alejandro was not followed by any real opening for this particular generation and its creations. One of those responsible for promoting these two young musicians, the media director of the disgusting entertainment program *Siempre en Domingo*, Raúl Velasco, ignored the triumph at the OTI festival and appeared to veto Alejandro, leaving only a minimal opening for the artistic career of León, who gradually won her own terrain. At any rate, the media decided to once again close its doors and go back to the easy winnings from lesser artists, committed only to self-complacency and their third-rate glamour. Mexican new song continued to use its own alternative channels and clearly made a decision to not enter into Televisa's game. In 1988, the apparent unity among the country's leftists and the encouraging mobilization of the opposition around the presidential elections seemed to present a new opportunity for these musicians to link themselves to the political effervescence of the time. But that is another story that deserves to be told on its own.

## Notes

1. Gustavo García and Jose Felipe Coria, *Nuevo cine mexicano* (Mexico City: Clío Publishers, 1997).

2. Francisco J. Martínez Medellín, *Televisa: Siga la huella* (Mexico City: Claves Latinoamericanas Publishers, 1989).

3. Ibid.

4. Fatima Fernández Christlieb, *La radio mexicana: Centro y regiones* (Mexico City: Juan Pablos Editor, 1991).

5. Alma Rosa Alva de la Selva, *Radio e Ideología* (Mexico City: El Caballito Publishers, 1982).

6. Claes af Geijerstam, *Popular Music in Mexico* (Albuquerque: University of New Mexico Press, 1976).

7. Pável Ganados, *XEW 70 años en el aire* (Mexico City: Clío Publishers, 2000).

8. Juan S. Garrido, *Historia de la música popular en México (1896–1973)* (Mexico City: Extemporáneos Publishers, 1974).

9. Author's translation. *Armando Manzanero y sus canciones "A mi amor . . . con mi amor,"* LP record, Víctor, Mexico, 1967, MKL-1760.

10. Alain Derbez, *El jazz en México: Datos para una historia* (Mexico City: Fondo de Cultura Económica, 2001).

11. Yolanda Moreno Rivas, *Historia de la música popular mexicana* (Mexico City: CONACULTA–Alianza Editorial Mexicana, 1979).

12. Victor Blanco Labra, *Rockstalgia: Crónicas rocanroleras, años 50 y 60* (Mexico City: Diana Publishers, 2007).

13. Jorge H. Velasco García, *El Canto de la Tribu* (Mexico City: CONACULTA, 2004).

14. A *guajira* written by Cuban Carlos Puebla, entitled "Hasta siempre, Comandante" (Till forever, Comandante).

15. Judith Reyes, *Mexique: Crónica Mexicana*, Le Nouveau Chansonnier International, Le Chant du Monde, Paris, LDX 74421.

16. *Los Nakos*, Discos Mascarones, Mexico, EPM-06.

17. *La Peña de los Parra–Chile*, vol. 1, América de Hoy publishers, ASFONA, Uruguay, LOF 201.

18. Reyes, *Mexique*.

19. The name *peña* was taken from the connotation used in Spain, Argentina, and Chile in reference to an association of individuals who come together to promote or defend a sports or cultural institution, such as *peñas atléticas*, *peñas tangueras* or *peñas taurinas*. In Mexico in the 1970s and 1980s, peñas were places to listen to music and occasionally enjoy a chamber theater performance while eating some light food.

20. René Largo Farías, *La Nueva Canción Chilena* (Mexico City: Cuadernos de la Casa de Chile No. 9, 1977).

21. Clara Díaz, *La nueva Trova* (La Habana, Cuba: Letras Cubanas Publishers, 1994).

22. René Villanueva, *Cantares de la Memoria: Recuerdos de un folclorista* (Mexico City: Planeta Publishers, 1994).

23. Emilio García Riera, *Historia documental del Cine Sonoro Mexicano* (Mexico City: Era Publishers, 1988).

24. Velasco García, *El Canto de la Tribu*.

25. Ibid.

26. *Sones de Veracruz* record, Museo Nacional de Antropología, MNA, 06, INAH/SEP.

27. *Roberto y Jaime: Sesiones con Emilia* record, Foton, LPF 033.

28. *La Nopalera: Nueva Canción* record, Nueva Cultura Latinoamericana, NCL-LP-0013.

29. *On'ta: Tengo que hablarte* record, Nueva Cultura Latinoamericana, NCL-LP-004.

30. Humberto Musacchio, *Ciudad quebrada* (Mexico City: Océano, 1985).

# 10
# Technology for Cultural Survival
## Indigenous-Language Radio at the End of the Twentieth Century

*Antoni Castells-Talens and José Manuel Ramos Rodríguez*

One of the most overlooked aspects of the cultural history of the media in Mexico is indigenous-language radio, and yet its introduction into indigenous communities has been profoundly transformative. Between 1979 and 1999, the government's branch of indigenous affairs, the Instituto Nacional Indigenista (National Indigenist Institute, INI), installed twenty-one AM radio stations in indigenous regions throughout Mexico. Indigenist broadcasting became, arguably, the most significant public policy that the Mexican state has ever implemented in favor of indigenous languages and cultures.

The network quickly became a space where policy and identity were negotiated. While the government conceived each station as a tool to fight marginalization and implement policies against poverty, they were also intended to allow the state to increase its symbolic presence in communities, present itself as sensitive to the needs of indigenous peoples, and facilitate the co-optation and control of the empty space of indigenous mass communication. In spite of their governmental ownership, though, for indigenous peoples these radio stations became a terrain for cultural survival and resistance to the assimilationist goals of the state.

Languages that had been confined to the private sphere gained unprecedented public visibility and began to be associated with modernity. Radio, a technology that had been introduced from above by the government, witnessed a process of grassroots appropriation. In sum, the very radio stations that the state had envisioned as tools of modernization and assimilation were redeployed by indigenous actors in an effort to make their voices heard.

This chapter places indigenist broadcasting squarely within the history of

Mexican nation and state formation. First, it outlines how the state, through its programs of *indigenismo*, adopted modern communications technologies in an effort to further its policy goals in relation to indigenous people. Second, it traces the origins and growth of indigenist radio and outlines its defining characteristics. Third, it examines the impact of the Zapatista uprising of 1994 on the relationship between the state and the indigenous peoples, scrutinizing the effect of the rebellion on the daily operation of the stations. Finally, the chapter analyzes the processes of negotiation and appropriation that characterize the use of state-owned radio by indigenous peoples.

## *Indigenismo*, Technology, and the Nation-State

*Indigenismo*, or "indigenism," the driving force behind indigenous-language broadcasting, was a central component of nation-state formation in twentieth-century Mexico. If, in the United States, indigenism has been associated with the use of local knowledge to fight for indigenous rights and self-determination, indigenism in Mexico has typically been used in the antithetical sense.[1] Since the 1920s, the country's indigenous peoples have often been studied in order to facilitate de-Indianization—or assimilation into "mainstream" Mexican society—a key aspiration of the post-revolutionary project.[2]

Significantly, the network of radio stations that broadcast in indigenous languages in Mexico are generally referred to as *indigenist* radio, rather than *indigenous* radio, because the stations have been owned by and operated in accordance with INI directives.[3] While an indigenous staff operates the stations, the managers, who are always appointed directly from Mexico City, are often non-indigenous.

The precursors to indigenist broadcasting can be located in the early 1920s, when the state attempted to assimilate and integrate indigenous populations into the nation through a single, mandatory education system and through the promotion of a variety of indigenist cultural policies.[4] In 1926, José Vasconcelos, one of the most influential educational and cultural ideologues of the post-revolutionary period, promoted what were termed *misiones culturales*, temporary teams of educators in charge of bringing education to the most remote regions of the country. For decades, these informal "missions" were the only educational options for peasant and indigenous populations.[5] Over time, the *misiones culturales* incorporated novel media technologies into their pedagogical arsenal—slideshows and sixteen-millimeter film, in particular—likely becoming the first modern communication technologies used for the integration of indigenous peoples in Mexico.

Before launching the first indigenist radio station in 1979, the state had already experimented with radio broadcasts geared toward indigenous peoples. During the late 1950s and early 1960s, the Secretaría de Educación Pública (Department of Public Education) attempted to implement educational shortwave radio. Between 1958 and 1963, selected schools in the Mixtec region of Oaxaca were given shortwave receivers to facilitate the teaching of Spanish, but little documentation has been found on the experiences of these projects.[6] Additionally, it is possible that as early as 1966 INI planned to establish a broadcasting network with the intention of "Westernizing" indigenous peoples.[7]

Once installed, indigenist radio quickly came to constitute an integral part of INI's cultural project. However, the very proximity of the stations to indigenous peoples turned radio into fertile terrain for the negotiation of indigenist public policy, as well as the idea of nation itself. While these interactions took place in the context of asymmetrical power relations, indigenous broadcasters and listeners found a variety of ways to exert a degree of popular agency.

## The Growth of Indigenist Broadcasting

When INI formally began broadcasting in 1979, the dominant indigenist paradigm was still one of indigenous assimilation. Central to this outlook was the Mexican state's desire to *castellanizar* (Hispanicize) or teach Spanish to monolingual indigenous speakers and make Spanish the main language of communication for indigenous peoples. The founding documents of indigenist broadcasting, written by young communication graduates who worked for INI, show clearly that radio, in addition to being used as a tool "to sensitize and motivate communities to adopt a favorable attitude toward the innovations" introduced by the government, would be used to impart Spanish-language courses to the population. Simultaneously, however, a subtle change of paradigm was detectable during these years as policy makers increasingly embraced ideas of cultural sensitivity and respect for indigenous cultures. The same document asserts that radio would also aim to promote indigenous languages and contribute to the "diffusion of local music, art, and literary values to avoid its disappearance and promote its production so as to counterbalance, as far as possible, the penetration of values, norms, and attitudes alien to the region."[8] Internal inconsistencies, such as advocating cultural preservation while promoting the adoption of innovations, reflect contradictions frequently found within indigenism. On the one hand, staff with opposing cultural sensitivities often coexisted on indigenist projects. Some officials advocated assimilation while others advocated the indigenous right to difference. On the other hand, as a nationalist policy, indigenism sought simultaneously to assimilate indigenous peoples and to preserve some of their traits as a proof of Mexican national character.

A defining characteristic of indigenist radio was its refusal to follow the accepted standards set in place by existing commercially oriented stations. Unlike in commercial broadcasting, the majority of the staff of the stations came from local communities and spoke indigenous languages. The first station, Radio XEVZ, La Voz de la Montaña (The voice of the mountain), and others that were established during the following three years shared an innovative philosophy that shook the indigenous areas in which they operated, a philosophy inspired by Paulo Freire's theory of critical pedagogy and the writings of Mario Kaplún on grassroots communication.[9] These theories challenge traditional mass communication boundaries between sender and receiver in a vertical process. Instead, they advocate a horizontal model of communication with grassroots participation. Applied to radio, this philosophy meant creating a fully horizontal medium, a goal that seemed incompatible with the governmental control of the project. However, these stations increasingly adhered to the principles of multilingualism, embraced local issues, engaged in forms of community service, and played indigenous music. By the early 1990s, stations were also engaged in efforts to increase community participation on the airwaves by creating mechanisms such as community correspondents, community advisory boards, and public meetings with the audiences.

The staff of the stations typically included eight to twelve indigenous announcers, producers, and secretaries, whereas, in most occasions, the managers of the stations were not indigenous, did not speak or understand any indigenous language, and came from different states. A number of volunteers also participated in some programs or acted as community reporters. Some stations organized *consejos consultivos*, or "advisory boards," in which traditional authorities, local organizations, and other volunteers representing civil society voiced their opinion on matters of programming and other activities of the stations. Although INI provided the central guiding policies and supervision, many of the programming decisions were, in practice, made locally.

Indigenist radio stations and their staff produced a variety of programs aimed at improving living conditions in indigenous areas. Shows promoted the value of indigenous knowledge, such as traditional medicine or agriculture, and stressed the exercise of indigenous and women's rights while offering public service announcements against cholera and information about federal campaigns to encourage people to obtain voter identification documents. Stations also aired music, news, and messages from individuals, groups, or institutions. In some cases soap operas and other such programs were broadcast, many of which had been produced outside the local area. By contrast, there were strict limitations on the broadcasting of commercial, political, and religious content.

Local indigenous music played an important role in the daily programming. As media scholars have observed, music has the ability to enhance a sense of collective identity and belonging in simple but powerful ways, often more than any

Table 10.1. Indigenous radio stations created between 1979 and 1999

| RadiRadio | RadiYear | RadiAcronym | RadiState | RadiCity | RadiIndigenous languages |
|---|---|---|---|---|---|
| La Voz de la Montaña | 1979 | XEZV | Guerrero | Tlapa | Nahuatl, Ñuu Savi, Me'pha |
| La Voz de los Chontales | 1982 (closed 1990) | XENAC | Tabasco | Nacajuca | Yokot'an |
| La Voz de la Mixteca | 1982 | XETLA | Oaxaca | Tlaxiaco | Ñuu Savi, Driki |
| La Voz de los Purépechas | 1982 | XEPUR | Michoacán | Cherán | Purépecha |
| La Voz de la Sierra Tarahumara | 1982 | XETAR | Chihuahua | Guachochi | Raramuri, O'dam |
| La Voz de los Mayas | 1982 | XEPET | Yucatán | Peto | Yucatec Maya |
| La Voz de la Frontera Sur | 1987 | XEVFS | Chiapas | Las Margaritas | Tojolwinik'otik, Qyool, K'op, Batzil K'op, Abxubal |
| La Voz de la Sierra Juárez | 1990 | XEGLO | Oaxaca | Guelatao de Juárez | Diidzaj, Ayook, Tsa Jujmi |
| La Voz de las Huastecas | 1990 | XEANT | San Luis Potosí | Tancanhuitz de Santos | Nahuatl, Xigüe, Teenek |
| La Voz de la Sierra de Zongolica | 1991 | XEZON | Veracruz | Zongolica | Nahuatl |
| La Voz de la Chinantla | 1991 | XEOJN | Oaxaca | Ojitlán | Ha Shutaenima, Nduudu Yo, Tsa Jujmi |
| La Voz de los Cuatro Pueblos | 1992 | XEJMN | Nayarit | Jesús María | Naayeri, Wirr'árika, O'dam, Mexicanero |
| La Voz del Valle | 1994 | XEQUIN | Baja California | San Quintín | Ñuu Savi, Driki, Diidzaj |
| La Voz de la Costa Chica | 1994 | XEJAM | Oaxaca | Jamiltepec | Ñuu Savi, Tzañcue, Cha'cña |
| La Voz de la Sierra Norte | 1994 | XECTZ | Puebla | Cuetzalan | Nahuatl, Tachihuiin |
| La Voz del Corazón de la Selva | 1996 | XEXPUJ | Campeche | X'pujil | Yucatec Maya, Winik |
| La Voz de los Tres Ríos | 1996 | XEETCH | Sonora | Etchojoa | Mayo, Yoreme, Varogío |
| La Voz de los Vientos | 1997 | XECOPA | Chiapas | Copailaná | O'de püt, Batzil K'op |
| La Voz del Pueblo Hñahñú | 1998 | XECARH | Hidalgo | Cardonal | Hña Hñu, Nahuatl |
| La Voz de la Sierra Oriente | 1998 | XETUMI | Michoacán | Tuxpan | J ñatio, Hña Hñu |
| La Voz del Gran Pueblo | 1999 | XENKA | Quintana Roo | Carrillo Puerto | Yucatec Maya |

other social activity.[10] Similarly, Donald Browne notes that even instrumental music can convey political content as it can express the sensibilities and tastes of marginalized groups.[11] Many stations not only revitalized traditional music, but also encouraged local authors to create indigenous music in new genres. The stations helped, for example, promote rock singers and bands, such as the Sonoran Hamac Casiim, who sang in Konkaak, Sak Tzevul, who performed in Batzil K'op, and Xamoneta, a Purépecha-language band from Michoacán.[12]

The system of indigenist radio stations grew slowly (see table 10.1). In fact, it is not even clear when the stations became a network with a coherent set of policies and a defined vision of the role of broadcasting, although researchers have generally suggested this consolidation took place in the mid-1980s.[13] Nevertheless, by the early 1990s, these stations were at the center of discussions over indigenous people in Mexico.

In 1992, the quincentennial celebrations of the conquest of the Americas reinvigorated the debate over the situation of indigenous peoples in Mexico. That year INI and its radio stations—members of the World Association of Community Broadcasters (AMARC)—hosted the organization's fifth world conference. Through AMARC, indigenist broadcasters established important relationships with indigenous media activists and practitioners from other parts of Latin America, a region that was experiencing an explosion in activities related to indigenous rights. That same year K'iche' Maya intellectual Rigoberta Menchú was awarded the Nobel Peace Prize, the International Labor Organization approved C169 (the Indigenous and Tribal Peoples Convention), and in Mexico indigenist broadcasters began to call for the "free development of the indigenous peoples."[14] Meanwhile, INI planned for the transference of non-administrative functions of the radio stations to indigenous communities.

This sort of indigenist euphoria changed in 1994 when the guerrillas of the Ejército Zapatista de Liberación Nacional (EZLN) took up arms in the state of Chiapas in their fight for indigenous rights. In this context, INI found itself fundamentally transformed. Neither the staff at the institution nor the workers at the stations were prepared for an indigenous armed insurrection. Until then, they had perceived themselves as the exclusive defenders of indigenous peoples. Adding to the confusion was the fact that the Zapatista movement itself embraced many of the INI's own proposals, including indigenous peoples' right to their own media.

## Radio and Zapatismo

The impact of the EZLN rebellion on indigenist radio was twofold. From the state's point of view, indigenist radio became a matter of national security.[15]

Fearing that the stations could become allies of the Zapatistas or inspire new centers of insurgence, the government applied new mechanisms of control and censorship. Meanwhile, the EZLN recognized that these stations played a significant role in indigenous communities and included in its demands that the state transfer the radio network to the communities.[16]

## Radio and National Security

The days that followed the uprising were so exceptional that indigenist radio stations were not prepared to handle the coverage of the conflict. To make matters worse for INI, the Zapatistas had seized the indigenist station in Chiapas, XEVFS, La Voz de la Frontera Sur (The voice of the southern border), for a few hours to broadcast its communiqués. The government reacted with mistrust toward the stations, forcing them to send faxes to INI's headquarters with information on the stories they were broadcasting about the conflict. INI also placed a device on the transmitter of each station that allowed for the signal to be killed by remote control. While officials argued that the device would protect the stations and their staff in case of another rebel takeover, in the stations such a measure was interpreted as an act of censorship.

Each station responded to the conflict in unique ways. In Oaxaca, the INI state delegate reportedly ordered an information blackout on the events of Chiapas. On the night of July 31, 1994, the Guerrero station Radio XEZV experienced one of the most dramatic consequences of the political climate when someone shattered the lights around the station, broke the lock to the main entrance, entered the control booth through a window, and stole the turntables and recording and transmission equipment, thus sending the station off the air.[17] The government spread the rumor that a guerrilla group was responsible for the theft. At the radio station, however, most suspected that the government was involved as local politicians had openly manifested their uneasiness about the content of some programs. Moreover, INI officials had warned the manager to be careful about the extraprofessional political activities of several radio announcers, and the state governor had threatened to shut the station down. The fact that the thieves had also stolen the personal files of the radio staff and were never caught further implicated government officials in the robbery.[18]

In the eyes of the state, indigenist broadcasting became suspicious. In some of the towns where stations were operating, the military went as far as to install training camps, which, over time, became permanent. If 1994 marked a change in the relationship between indigenous peoples and the state in Mexico, it also marked a shift in the state's relationship to communication technologies. While the state was determined to maintain control, the EZLN had shown that such technologies could be used as weapons in the struggle for indigenous rights.

Technology for Cultural Survival / 185

## *The EZLN and the Indigenist Radio Network*

Between 1995 and 1996, the federal government and the Zapatistas negotiated a new indigenous rights agenda. In February 1996, both parties signed the San Andrés Agreements, which included a clause related to radio. It read: "The INI radio stations must be handed over to the communities and indigenous peoples where they operate, so as to become a part of their heritage, and guarantee their right to information, freedom of expression, and development."[19] The agreements further stated that the process would be gradual, beginning with the transfer of stations to the communities upon request, noting that "the rhythm and timing of the appropriation will be decided by the indigenous peoples."[20]

The agreements assumed a collaborative process between governmental institutions and legitimate authorities of indigenous communities, a process that never took place. The treaty was not enforced, and the constitutional and legislative reform that followed did not reflect the demands related to broadcasting. While the law did recognize the indigenous peoples' right to possess their own media, the state remained in control of indigenist radio, and the legal framework made it virtually impossible for indigenous communities and organizations to obtain broadcasting licenses. As a response, by the beginning of the twenty-first century, hundreds of indigenous stations throughout the country were broadcasting without licenses. The growth of this illegal type of broadcasting meant an indigenous appropriation of radio technology for the defense of culture and language. It also meant a defiance to the state. As the state lost the exclusiveness of indigenous-language radio broadcasting, it also lost control over the content of messages. In some cases, the state repressed the new stations and seized the transmitters. In other cases, though, the government has been unwilling or unable to shut them down.[21]

## The Negotiation of Radio Policy

Rather than viewing indigenist radio policy as a succession of unidirectional actions taken by indigenist bureaucrats in Mexico City, it must be understood as a complex process of everyday negotiations between a variety of actors, including radio workers and indigenist officials at the local, state, and federal levels, as well as lawmakers, religious groups, traditional indigenous authorities, the police, the army, nongovernmental organizations (NGOs), academics, local politicians, and even Zapatistas.[22]

Yet negotiation between such actors did not occur on an even playing field since the ultimate power to close down a station rested with INI. In fact, in 1990 INI authorities took Radio XENAC, La Voz de los Chontales (The voice of the Chontal people), the indigenist station in Nacajuca, Tabasco, off the air permanently after a political conflict involving local PRI leaders.[23] While more research is needed to clarify the intricacies of the incident, the message that INI sent to other stations left little room for interpretation: stations could be shut down at the will of the state. By 2010, in Orwellian fashion, Radio XENAC had disappeared from the official history of the indigenist radio network and no mention of it could be found on the indigenist broadcasting website, giving the impression that the eight-year-old station had never existed.

In spite of the ultimate power of the state to exercise a veto over indigenist broadcasting, indigenous workers and listeners maintained and continue to maintain a considerable degree of agency with regard to the everyday functioning of the stations. During the first years of indigenist broadcasting, each station developed its own understanding of what an indigenist station should broadcast, depending, to a large extent, on the managers' ideas and decisions. The central INI offices recognized the inconsistencies and attempted to streamline policies, yet, even then, radio was reinvented on an everyday basis in each indigenous station.[24] Examples of such strategies of resistance vary in their degree of force and in their consequences, but they exemplify some of the limitations of governmental policy.[25]

As mentioned earlier, censorship became common practice during the tensest moments of the Zapatista uprising. Even before the rebellion, however, radio stations had already dealt with pressure from above. Certain LP records in one of the stations, for example, still show the word "NO" next to songs that were considered politically or morally inappropriate for broadcast, such as Violeta Parra's protest song "Gracias a la vida" or Fernando Ubiergo's "Cuando agosto era 21," a song about a teenager who dies from an abortion.[26]

In the 1980s, the stations adopted an unwritten policy of *three no's*: no politics, no religion, no commercials. Rather than externally imposed censorship, the staff saw the measure as an inventive way to avoid trouble. By steering clear of party politics, stations protected themselves from having to open the microphones exclusively to PRI candidates during campaigns, like most commercial and official stations. Religious figures, whether Catholic or Protestant, were also kept from using the stations for proselytization. Similarly, local businesses seeking to buy airtime were turned away so as to guarantee the independence of the stations. The *three no's* policy allowed the stations to retain a relative neutrality in the context of rampant political clientelism.[27]

Meanwhile, unanticipated politic uses of the stations began to appear. In Chihuahua and Guerrero, community reporters of Radio XETAR, La Voz de la Sierra Tarahumara (The voice of the Sierra Tarahumara), and Radio XEVZ reported that local businessmen were involved in clandestine woodcutting. As the director of the radio network from 1989 to 1992 recalls, when Radio XEVZ radicalized its discourse against illegal woodcutting, INI could not intervene: "In [INI's] central offices they saw that this was something that was not easily stopped, something that was not so controllable, especially because it came from radio stations that were so far from the control of the central offices. . . . When in Tlapa the protests started denouncing [what was happening with illegal woodcutting], the central offices were dismayed. [Protesters] created dismay in the central offices but [the central offices] saw that it was impossible to stop."[28]

In Chiapas, guerrilla members seem to have used Radio XEVFS as early as 1992 to send coded messages to one another; and in Oaxaca, members of INI's indigenous video production centers used the official indigenist van to transport Chiapas EZLN activists throughout the state. At night, the Zapatistas would take shelter in the local INI's facilities. Guillermo Monteforte, one of the coordinators of the indigenist video project in Oaxaca in the 1990s, was even invited to represent the Zapatistas in the peace talks with the government. Monteforte asked for vacation time and traveled to Chiapas, where he encountered many of his INI colleagues representing the state, on the other side of the negotiation table.[29] These anecdotes demonstrate what some government officials and radio workers have suggested was an unwritten code of indigenist broadcasting: "Anything can be said, as long as you know how to say it."[30] The sentence, often repeated by INI's officials, managers, and indigenous radio journalists, illustrated the spirit of the negotiation as it expressed a will to be open ("anything can be said"), acknowledged the existence of political pressures and censorship ("as long as"), and stressed the importance of the invisible norms of negotiation ("you know how to say it").

Indigenist radio also offered opportunities for the reinvention of identity. In the Yucatan peninsula, the imagery adopted by local stations (e.g., posters, stickers, logos, murals, pennants, and souvenirs) promoted the figure of the "archaeological Maya."[31] This figure was traditionally used in Mexican nationalism to construct a mestizo identity, as a sign of both Maya history and contemporary regional identity.[32] This imagery contradicted the official narrative of the Mexican nation by establishing a historical link between the archaeological past and the indigenous present without the presence of a Mexican nation.[33] In other cases, the stations promoted the names that indigenous groups preferred over the ones that the state had imposed.[34] Thus, whereas the state used Tlapaneco, Tarahumara, and Otomí, for instance, some stations joined indigenous teachers and intellectuals to favor the names Me´pha, Raramuri, and Hñahñu instead. These examples show how stations challenged hegemonic discourses of the nation by using an institution owned and organized by the state.

Indigenist stations therefore became spaces that simultaneously allowed the state to exert its power and control and offered opportunities for indigenous agendas to be pushed forward and the discourses and practices of nation building to be contested. As Antoni Castells-Talens suggests, the everyday indigenist radio policy is flexible because the state does not expect every rule to be followed, it is malleable because the station knows that the rules can be reshaped to work for its advantage, and it is diffused because the actors involved in the negotiation are scattered at the local, state, and national levels and it is not always clear the kind of agency each actor holds.[35]

## The Appropriation of the Radio Stations

When the first indigenist radio stations were installed, many indigenous communities still lacked electricity and telephone service, dirt roads complicated travel, and radio and television signals from the nearby cities often did not reach such remote areas. Local radio stations became the best way—and, to a certain degree, the only way—to facilitate communications within and outside remote regions. People rapidly learned how to use a medium that permitted the sending and receiving of messages almost instantaneously, free of charge, and in local languages. Apart from its usefulness in resolving daily problems (e.g., finding lost animals, setting appointments, communicating with relatives and friends), radio made it possible for languages, music, and collective memories to circulate beyond the limits of the community and become, in a manner without precedent, part of a "media landscape" that had not existed before then.

With time, digital technologies opened up new communication possibilities; however, radio remained the most popular medium and the only one that continued to use local languages. By the end of the twentieth century, radio, a technology that had reached its peak throughout the world five decades earlier, continued to be of the greatest value in the indigenous areas of Mexico.

The appeal of communication services and the inclusion of the local cultures in programming explain to a large degree why these radio stations became so deeply rooted within the population and why audiences considered them to be "their own." In a study conducted at Radio XEZV, José M. Ramos Rodríguez found that through a process of "sedimentation," occurring over time and in resonance with diverse factors, the station had contributed to activating what Clemencia Rodriguez called "subtle processes of fracture in the social, cultural and power spheres of everyday life."[36] Following Clemencia Rodriguez's proposition, notions about the capacity of community media to induce subtle changes in the symbolic order of the status quo, Ramos Rodríguez found that the use of indigenous languages, music, narratives, and other cultural expressions helped challenge the internalized ethnic stigma that had undervalued their identity. Radio technology allowed public displays of ethnic pride to flourish. When Radio XEZV began broadcasting, Spanish was the only language of communication in downtown Tlapa as indigenous peasants hid their language with embarrassment. A few years later, after thousands of hours of Mepha, Nu Sabi, and Nahuatl broadcasts throughout the region, Spanish lost its hegemony and indigenous languages became a normal part of everyday life in the streets of Tlapa. These fissures in the establishment played a pivotal role in the strengthening of the culture, the identity, and the social cohesion of indigenous peoples in the region. Radio had become an "intangible heritage" of the indigenous peoples and had played a central role in cultural maintenance and reproduction.[37]

## From State Technology to Independent Media

While the state used radio technology to modernize and homogenize the country, indigenous peoples and workers turned to the same technology in an effort to defend their cultures and languages. In some cases, such as during the Zapatista conflict, the state mistrusted its own indigenist network and attempted to develop strategies of control. Yet in the stations, the central offices of INI seemed distant, not only geographically but also culturally, and officials' policies were often bent and reworked. At times, government officials in Mexico City were aware of the limits of their power and allowed for rules to be broken.

Why did the state tolerate and sometimes even encourage critical readings of its own project? On the one hand, as Lucila Vargas observes, the state is not monolithic, nor has it always been efficient enough to achieve its goals through indigenist broadcasting.[38] Moreover, Stephen Harold Riggins notes that when dealing with indigenous media, states tend to adopt approaches that are apparently contradictory.[39] While their outward intention is often to be culturally sensitive, they also aim to co-opt indigenous media and further their larger integrationist goals. By the time Radio XEVZ began operating in 1979, the discourse of indigenism was being transformed through the rhetoric of diversity and multiculturalism, which was in some sense a response to criticism from indigenous peoples and academics.[40] INI's shift toward more culturally sensitive policies suggested a progressive abandonment of *mestizaje* as well as a sign of new attempts to maintain control over indigenous cultural production and to curtail indigenous self-determination.[41]

In some cases, the strategies of indigenous resistance have been too subtle for the state to recognize. It is doubtful, for example, that INI officials were aware that Oaxaca indigenist video workers were using public resources to transport Zapatista operatives. In many such occasions, acts of popular agency have been too small or too secret for them to be noticed by outside observers. It could also be argued that the state allows for dissent as long as it is not perceived as a threat to the integrity of the state. Except for a few instances, like the Zapatista uprising, the attitude of the state toward indigenous peoples has been paternalistic. In some occasions, the state has been unwilling to take action against a particular station due to concerns over the potential for an indigenous response. Negotiation thus acts as a strategy for the state to ensure its survival.

The effects of INI's radio project might be impossible to measure, but some of the consequences of indigenist broadcasting have already become obvious during the first decade of the twenty-first century. In indigenous regions, radio has opened the road toward the legitimacy and normalization of native languages. Languages that, in words of Clemencia Rodriguez, had been treated as "shy languages" are now gaining visibility in the public sphere through the use of media technologies.[42] A generation of indigenous youth has now grown up with its languages broadcast over the airwaves and has begun to establish its own independent media, often inspired by indigenist broadcasting. Meanwhile, indigenous musicians have also found allies in indigenist broadcasting, and INI's stations have recorded and digitized thousands of songs and instrumental pieces, acting as a loudspeaker for new artists by promoting their music.

The influence of the Instituto Nacional Indigenista's communication technologies is also visible in its *video indígena* project, becoming the seed for indigenous video in Mexico, a tool that has contributed to the emergence of broader consciousness about indigenous peoples.[43] In the field of journalism, INI-trained communications workers are also starting their own newspapers and online ventures. Additionally, hundreds of independent stations operate illegally throughout the country in Zapotec, Mixtec, Yaqui, and Tzotzil.

By the beginning of the twenty-first century, the native peoples of Mexico were adopting radio and newer technologies to create their own independent media for cultural survival. Indigenist broadcasting was no longer the keystone of indigenous communication in Mexico.

## Notes

1. Ward Churchill, *Acts of Rebellion: The Ward Churchill Reader* (New York: Routledge, 2003), 275.

2. Guillermo Bonfil Batalla, *México profundo: Una civilización negada* (Mexico City: Grijalbo, 1990), 79.

3. In 2003, INI was replaced by the Comisión Nacional para el Desarrollo de los Pueblos Indígenas (National Commission for the Development of Indigenous Peoples, CDI), the new governmental branch for indigenous affairs. With the CDI, the radio network continued to broadcast under governmental ownership and supervision.

4. Natividad Gutiérrez Chong, *Nationalist Myths and Ethnic Identities: Indigenous Intellectuals and the Mexican State* (Lincoln: University of Nebraska, 1999); Mary Kay Vaughan, *Cultural Politics in Revolution: Teachers, Peasants, and Schools in Mexico, 1930–1940* (Tucson: University of Arizona, 1997).

5. Lloyd H. Hughes, *Las Misiones Culturales Mexicanas y su programa* (Paris: UNESCO, 1951).

6. Maurilio Muñoz, *Mixteca nahua-tlapaneca* (Mexico City: INI, 1963), 116.

7. Rufino Alejandro Gatica García, "Proyecto de un sistema de escuelas radiofónicas para el Instituto Nacional Indigenista" [Project of a radio schools system for the Indigenist National Institute] (BS diss., Instituto Politécnico Nacional, Mexico, 1996).

8. Secretaría de Agricultura y Recursos Hidráulicos–Comisión del Río Balsas, *Anteproyecto para la instalación de una radiodifusora en Tlapa, Gro.* [Project to install a radio station in Tlapa, Gro.], unpublished internal document, Mexico City, 1977, 5–6.

9. Ibid.

10. David Hendy, *Radio in the Global Age* (Cambridge, UK: Polity Press, 2000); Helena Simonett, "Popular Music and the Politics of Identity: The Empowering Sound of Technobanda," *Popular Music and Society* 24, no. 2 (2000): 1–23; Martin Stokes, "Introduction: Ethnicity, Identity, and Music," in *Ethnicity, Identity, and Music: The Musical Construction of Place*, ed. Martin Stokes (Oxford: Berge, 1994), 1–27.

11. Donald R. Browne, *Electronic Media and Indigenous Peoples: A Voice of Our Own?* (Ames: Iowa State University Press, 1996).

12. Instituto Nacional Indigenista, *Encuentro de música indígena: Las nuevas creaciones, de "El costumbre" al Rock* [Indigenous music meeting: new creations, from "El Costumbre," to rock], music CD (Mexico City: INI, 2000).

13. Inés Cornejo Portugal, *Apuntes para una historia de la radio indigenista en México: Las voces del Mayab* [Notes for a history of indigenous radio in Mexico] (Mexico City: Fundación Manuel Buendía, 2002).

14. XEPET, *Consejo consultivo de XEPET "La voz de los mayas": Sexta reunión plenaria* [Advisory Council of XEPET "the voice of the Maya": Sixth general meeting] (Peto, Yucatan, Mexico: Radio XEPET, 1993).

15. Cristina Henríquez and Melba Pría, *Regiones indígenas tradicionales: Un enfoque geopolítico para la seguridad nacional* [Traditional indigenous regions: A geopolitical approach for national security] (Mexico City: Instituto Nacional Indigenista, 2000).

16. EZLN, *EZLN: Documentos y comunicados* [EZLN: Documents and communiqués] (Mexico City: Era, 1995).

17. Juan Angulo Osorio, "Robo a la Voz de la Montaña" [Theft at the voice of the mountain], *El Sur* (Acapulco), "De interés público" section, August 2, 1994.

18. Jorge Raúl Obregón Téllez, personal communication, March 2009, Tlapa,

Guerrero; Carmen García Bermejo, "Silencio institucional ante el robo de la Voz de la Montaña: Ubaldo Segura" [Institutional silence before the theft in the voice of the mountain: Ubaldo Segura], *El Financiero* (Mexico City), "Cultura" section, September 4, 1995.

19. Luis Hernández Navarro and Ramón Vera Herrera, eds., *Acuerdos de San Andrés* (Mexico City: Era, 1998), 151.

20. Ibid., 91.

21. José Manuel Ramos Rodríguez, "La Voz de los sin voz: Emergencia de la radio comunitaria en México," *Revista Iberoamericana de Comunicación* 10 (2006): 13–22.

22. Antoni Castells-Talens, "When Our Media Belong to the State: Policy and Negotiations in Indigenous-Language Radio in Mexico," in *Making Our Media: Global Initiatives toward a Democratic Public Sphere: Creating New Communication Spaces*, ed. Clemencia Rodriguez, Dorothy Kidd, and Laura Stein (Cresskill, NJ: Hampton Press, 2009), 249–70.

23. Corynne McSherry, "Todas las voces: Indigenous Language Radio, State Culturalism, and Everyday Forms of Public Formation," *JILAS–Journal of Iberian and Latin American Studies* 5, no. 2 (December 1999): 99–132.

24. Instituto Nacional Indigenista, *Primeras jornadas de la radiodifusión cultural indigenista* (Mexico City: Instituto Nacional Indigenista, 1996).

25. Schimdt notes the need to distinguish popular agency as *resistance* from popular agency as *cultural creativity*. In the case of indigenist broadcasting, cultural creativity might be an appropriate term, but it is hard to separate both concepts. In many instances, media messages conveyed political meanings that aimed at direct social change or collaboration with resistance movements, such as the EZLN. Arthur Schmidt, "Making It Real Compared to What? Reconceptualizing Mexican History since 1940," in *Fragments of a Golden Age: The Politics of Culture in Mexico since 1940*, ed. Gilbert Joseph, Anne Rubenstein, and Eric Zolov (Durham, NC: Duke University Press, 2001), 23–68.

26. The data were obtained by the authors at the record library of Radio XEPET, in Peto, Yucatán.

27. Castells-Talens, "When Our Media Belong to the State."

28. Eduardo Valenzuela, personal communication via Skype, May 2008.

29. Guillermo Monteforte, personal communication via Skype, May 2008.

30. Eduardo Valenzuela, personal communication via Skype, May 2008.

31. The Yucatan peninsula hosts three indigenist stations: Radio XEPET, La Voz de los Mayas (The voice of the Maya) in Yucatan; Radio XEXPUJ, La Voz del Corazón de la Selva (The voice of the heart of the jungle) in Campeche; and Radio XENKA, La Voz del Gran Pueblo (The voice of the great people) in Quintana Roo.

32. For a thorough analysis of the Yucatec Maya by archaeology, see, e.g., Quetzil

Castañeda, *In the Museum of Mayan Culture: Touring Chichén Itzá* (Minneapolis: University of Minnesota, 1996).

33. Antoni Castells-Talens, "Mexican Nostalgia, Maya Identity: The Reinvention of Iconographic Nationalism in Indigenous-Language Radio," *Journal of Global Mass Communication* 2 (2009): 5–23.

34. Historically, the state had employed the often-derogatory names that the Aztec Empire had imposed throughout the fifteenth century.

35. Castells-Talens, "When Our Media Belong to the State."

36. Clemencia Rodriguez, *Fissures in the Mediascape: An International Study of Citizens' Media* (Cresskill, NJ: Hampton Press, 2001), xiv.

37. José M. Ramos Rodríguez, "Indigenous Radio Stations in Mexico: A Catalyst for Social Cohesion and Cultural Strength," *Radio Journal* 3 (2005), 155–69.

38. Lucila Vargas, *Social Uses and Radio Practices: The Use of Participatory Radio by Ethnic Minorities in Mexico* (Boulder, CO: Westview, 1995), 241–42.

39. Stephen Harold Riggins, *Ethnic Minority Media: An International Perspective* (Newbury Park, CA: SAGE Publications, 1992).

40. Miguel Limón Rojas, "El indigenismo: Un imperativo nacional" [Indigenism: A national imperative], in *INI 40 Años* [INI 40 years], ed. L. Herrasti Maciá (Mexico City: Instituto Nacional Indigenista, 1988), 81–101.

41. Chong, *Nationalist Myths and Ethnic Identities*.

42. Rodriguez, *Fissures in the Mediascape*.

43. Claudia Magallanes Blanco, *The Use of Video for Political Consciousness-Raising in Mexico: An Analysis of Independent Videos about the Zapatistas* (Lampeter, UK: Mellen Press, 2008); Alexandra Halkin, "Outside the Indigenous Lens: Zapatistas and Autonomous Videomaking," in *Global Indigenous Media: Cultures, Poetics, and Politics*, ed. Pamela Wilson and Michelle Stewart (Durham, NC: Duke University Press, 2008), 160–80; Laurel C. Smith, "Mobilizing Indigenous Video: The Mexican Case," *Journal of Latin American Geography* 5, no. 1 (2006): 113–28.

# IV
# RAILROADS, AUTOMOBILES, AND THE METRO

# 11
# Film, Time, and the Railway in Porfirian and Revolutionary Mexico

*David M. J. Wood*

In January 1907, Mexican president Porfirio Díaz traveled to Salina Cruz, a major industrial port on the Pacific coast of the Isthmus of Tehuantepec, for a ceremonial reopening of the interoceanic railway that covered the 310 kilometers separating Salina Cruz from the isthmus's Atlantic port of Coatzacoalcos. As had been customary for some years on Díaz's official engagements, he was accompanied by a contingent of photographers and cameramen whose still and moving images were destined to provide a positive and propagandistic public record of the occasion. For Mexican cinemagoers and readers of the illustrated press, the inauguration of the interoceanic line across the Tehuantepec railway was the latest installment of an ongoing discourse in early actuality film that associated the figure of Díaz with the railway, the symbol par excellence in Porfirian Mexico of technological progress and modernity.[1] The railway also symbolized international economic integration.[2]

The visual association between the dynamic technologies of the cinema and the railway was by no means specific to early Mexican film, dating right back to films such as the Lumières' *Arrivée d'un train en gare de La Ciotat* (1897), in which the advance of the locomotive toward the camera illustrates its novel capacity to register and reproduce movement. Despite the immobility of the early cinematograph, filmmakers soon began to experiment with ways of enriching their films with a further level of movement; by mounting the film camera on a moving train, the necessarily fixed camera position itself became mobile. As a result, the "phantom ride" film offered viewers an unprecedented visual and sensory experience, enabling them to sense the train's velocity as it cut through the unfolding space ahead, observed from the privileged vantage point of the front of the engine from the driver's perspective (rather than the side view to which passengers were accustomed).[3]

Many spectators of the first phantom rides would have experienced something akin to the "annihilation of space and time" produced by rail travel itself from the early nineteenth century.[4] As Wolfgang Schivelbusch has observed, while traditional forms of transport such as horseback and stagecoach had a "mimetic relationship with the space traversed, permit[ting] the traveler to perceive that space as a living entity," the velocity of the railway brought closer newly accessible, topographically distant spaces, even as the intermediate spaces through which the locomotive cut its path were effectively destroyed.[5] The old temporal relationship to space—the time taken to travel through a given space—was thus radically reconfigured. Although the development of railway infrastructure came later to Mexico than to the western European countries to which Schivelbusch refers, it expanded greatly during the final two decades of the nineteenth century and the first decade of the twentieth century (coinciding roughly with Díaz's rule).[6] This expansion ensured that by the time phantom rides came to Mexico, for many Mexicans—or at least those living within reach of the rail network on which itinerant film exhibitors themselves depended to reach their audiences—this spatiotemporal pulverization was familiar, although in all likelihood less ingrained than for their European counterparts.[7]

For Frank Gray, one of the attractions of the phantom ride was that "the invisibility of the engine and the lack of sounds from it enabled the viewer to become disconnected from the presence of a real train": the film viewer became a *vicarious* train passenger, experiencing a bodily displacement that separated spatiotemporal annihilation from the material technological apparatus—the train—that brought it about.[8] In the edited films (made from 1907) that we will discuss here, this heightened sense of spatial and temporal compression would be accentuated by the cinema's own temporal ellipsis: just as the railway does away with traditional transportation's mimetic relationship to space, the temporal ellipses implied in the editing together of shots occurring at distinct moments create a new, non-mimetic relationship to lived time, thus further separating the viewer from the actual experience of riding a train. Even in later phantom rides in which parts of the engine and train carriages do become visible, I would then argue, the sense of abstraction from the actual sensation of being physically aboard a train persisted, aided by the wider range of viewpoints and spatial and temporal transitions characteristic of later actuality and fiction film. The exhibitionist drive in at least early phantom ride films thus distills the experience of travel in modernity, disdainfully defined by Siegfried Kracauer as "a pure experience of space," "a shift . . . away from movement that refers to a meaning and toward movement that is solely self-referential." This was symptomatic of a society in which "technology becomes an end in itself."[9]

The remainder of this chapter will focus on the shift that Kracauer observes, away from meaning and toward self-referentiality, that, for him, is deeply rooted in the ambiguous presence of contingency in modernity. It will consider the dialogue established between film, mechanized transport, particularly rail transport, and modernity's reconfiguration of temporal and spatial relations at various points in the silent period of Mexican cinema, corresponding roughly to the first three decades of the twentieth century. I will begin by returning to Porfirio Díaz's trip to Tehuantepec, discussed briefly in the opening paragraph.

## Porfirio Díaz in Tehuantepec: Modernity, Folklore, and Contingency

*Inauguración del tráfico internacional de Tehuantepec* (Salvador Toscano, 1907) is an actuality film of roughly fifteen minutes duration that narrates the arrival of the steamboat *Arizonian* at Salina Cruz.[10] It features the transferal of the ship's cargo with electric cranes onto a freight train; the train's passage across the isthmus, accompanied by Díaz and his contingent, including picturesque scenes of local life; and the convoy's arrival at Coatzacoalcos (referred to in the film by its former name of Puerto México), where the cargo is transferred to the northbound *Lewis Luchembach* steamer.[11] As Ángel Miquel rightly points out, the film's two narrative premises—technical progress, on the one hand, and the exotic scenery and inhabitants of the isthmus, on the other—are skillfully bound together around the figure of Porfirio Díaz, who is thus seen to articulate modernity with tradition.[12] Although devices of modernity (railway rolling stock and infrastructure, steamboats, the electrical crane) feature almost constantly, they are on the whole filmed from a static position, giving the impression that modernity emanates from the external world depicted in the film—hinging upon the prominent and domineering presence of Porfirio Díaz—and not from the filmic sphere through which the viewer experiences it.

Aurelio de los Reyes has, though, commented on *Inauguración*'s formal embracement of the notion of progress in at least one particularly energetic shot at the film's close, which highlights its own dynamism in depicting and embodying movement. As the *Lewis Luchembach* is setting sail from Coatzacoalcos, de los Reyes identifies four simultaneous levels of motion: the movement of the boat, upon which the camera is mounted, producing the effect of a tracking shot; a slight panning movement by the camera; the external movement of the current against the ship's own momentum; and the ship's own movement as it heads for the open sea.[13] One earlier shot in this actuality film depicts movement by means of the same device, as de los Reyes also points out: as the presidential train pushes on across the isthmus, the camera looks out sideways as we cross a bridge over a river, identified with the point of view of the passenger gazing through the train window—indeed, at least symbolically, from the perspective of Díaz himself. Here the camera's own movement is dependent not on the natural swaying of the ocean tide, as in the Coatzacoalcos shot, but on the mechanized motion of the train. By contrast, the actual scene that the camera depicts here is wholly static: the train's heady velocity is far too great to allow us to perceive any slight movement of the river's current or of the small group of people standing outside a rudimentary thatched hut, reduced to a blur as they watch the train thunder past, or of their paddle boat, moored to the river's edge.

These anonymous onlookers, of whose lives we catch the briefest of glimpses from the interoceanic express, occupy the "space in-between" of which Schivelbusch writes: a fleeting presence pulverized as the train strikes its way through the landscape they inhabit and represent.[14] To follow Schivelbusch's argument, for those fortunate passengers able to buy tickets to board the Tehuantepec railway, this spectacle of local life is roughly equivalent to a theatrical spectacle (the period on which Schivelbusch writes predates the cinema): "the landscape thus purchased becomes imaginary."[15] The landscapes traversed—all intermediate points between the Pacific and the Atlantic ports—lose their local identity, "their traditional here-and-now," or their Benjaminian aura, becoming converted, in Marxian terms, into circulating commodities.[16] There is in this brief shot, though, one *apparent* movement external to the camera/train itself that Schivelbusch does not contemplate here in likening the train's window spectacle to the theatrical spectacle: that of the diagonal iron struts of the railway bridge in extreme close-up, each blurred mass of metal obscuring our view of the river for the briefest of instants before being rapidly replaced by the next.

It does not seem fortuitous that of the many hours of unobstructed views available to the filmmaker/editor on the long haul across the isthmus, he should choose precisely (and only) this one to illustrate the train's passage through the hinterland: the struts, of course, are stationary, but the movement of the train lends them the appearance of dynamism, heightening the velocity at which the train, the president, and the valuable international cargo cut through the countryside. The struts are also a powerful and material reminder of the way in which mechanized locomotion mediates the train passenger/film viewer's experience of the landscape he or she traverses.[17] If we are to conceive of the Tehuantepec landscape as a theatrical spectacle, there is at play in this particular experience of modern spatiotemporality something of a Brechtian breaking down, however unintentional, of the fourth wall that separates it from its audience. The passing paraphernalia of the railway punctuate any mimetic "representation" of the passing landscape that the traveler/viewer might perceive through the train window, like so many "shocks" that, for Walter Benjamin, characterize the experience of modernity.[18]

11.1. Frame enlargement from *Inauguración del tráfico internacional de Tehuantepec*, Salvador Toscano, Mexico, 1907. Frame reproduced by Filmoteca UNAM; authorized by Filmoteca UNAM and Fundación Carmen Toscano.

When, in the film's following sequence, the train stops to observe a leisurely group of people leaving mass, scenes from a local market, and children bathing in the river, the relative stasis of these scenes might seem to be little more than a rhetorical, folkloric counterpoint to the dynamism of the train or, building on Miquel's point cited above, a patrimonial vestige of tradition protected by the benevolent presidential figure. Yet such a reading of the editing strategy of *Inauguración del tráfico internacional de Tehuantepec* would be misleading, for this actuality film was made at an intermediate stage in the transition from the static view to the logic of montage.[19] Far now from the bustling pomp of the official ceremonies surrounding the inauguration on which the film is centered, or from the busy loading and unloading of cargo, or from the exciting phantom ride across the Tehuantepec isthmus, the present-day viewer might be mildly unsettled by the long duration of these folkloric and largely uneventful shots, some running to a full minute.[20] The lengthy duration of the shots described here, as was the general rule in early films, allows the eye to roam throughout the entire frame undirected, or directed by the unpredictable, random, and idiosyncratic movement of the people, animals, or objects that wander in and out of the frame. This capacity of the early cinematograph to register movement—and, therefore, its conversion of the whole of reality into potentially filmable space—has been interpreted by Tom Gunning as a key aspect of what he terms the "cinema of attractions": an "exhibitionist" (as opposed to the later voyeuristic) mode of filmmaking concerned with the display of its own illusory power and showing a "lack of concern with creating a self-sufficient narrative world upon the screen," and which, for Gunning, dominated US cinema until around 1906 or 1907.[21] Mary Ann Doane has similarly described the pre-1907 period (in the United States) as a time in which "the cinema seemed to be preoccupied with the minute examination of the realm of the contingent, persistently displaying the camera's aptitude for recording."[22] Conversely, later filmmakers would view shots as distinct forms of semantic units, giving rise to the continuity system of editing in the United States, which aimed to forge a seamless narrative out of fragments of filmed reality, and to the montage editing, popularized in the Soviet Union, that sought to produce, among other reactions, an emotive dialectical analysis by the spectator through the juxtaposition of diverse images.

For Doane, the contingency captured by the early film camera was not merely a by-product of its capacity to record actuality (its indexicality), but rather the two functions were intimately linked as part of a wider tension in modernity between, on the one hand, the indexical and the instantaneous, which tend toward meaninglessness and, on the other hand, the elliptical and the legible, which tend toward meaning. Cinematic movement (and therefore time) is, in fact, based on an illusion of movement produced by the rapid projection of successive still images; late nineteenth- and early twentieth-century experiments found time itself (and therefore movement) to be ultimately unrepresentable in any legible form. The project of narrative film is thus the structuring and rendering readable of time and contingency.[23]

The anxiety produced by the visibility of contingency in early cinema—the apparently (although illusory) pure experience of time and the consequent descent into meaninglessness—is linked to the anxiety expressed by Kracauer in the quotation above over the search for inexorable modernity (here, Porfirian progress): the pure experience of *space* and the consequent descent into meaninglessness. They must, then, be tamed by an appeal to narrative, in the case of contingency; or, in Kracauer's discussion of modern travel, to the eternal, to the absolute, and to art, which "by giving shape to the phenomenal, . . . provides a form that enables it to be touched by a meaning which is not simply given along with it."[24] While temporal contingency in *Inauguración* is far from fully tamed, any anxiety over inexorable progress and the resulting societal change would appear to be resolved precisely by this sequence of scenes of traditional life, whose ultimate significance cannot be fully grasped or absorbed into the main technological and economic premise of the film and which persists with great vitality even as the express train shakes the economic and social foundations of modern Mexico. This sequence, then, serves at once as a *formal* display of modernity's contingency and as a *cultural* defense against it.

## Madero's Triumphal Journey: Revolution as Movement

By May 1911, the revolutionary leader Francisco I. Madero had ousted Porfirio Díaz, following Díaz's fraudulent reelection in June 1910 and Madero's uprising against the regime in November of the same year. In the ensuing months *maderismo* gathered considerable momentum with revolts across the country, and by May Madero, who counted such military strongmen as Pascual Orozco and Francisco Villa among his numbers, had taken the border city of Ciudad Juárez. Díaz, realizing by now that his position was untenable, finally resigned, and an interim government was installed under the presidency of Francisco León de la Barra, following an agreement between Madero and the ancien régime. Madero, riding high on his tide of popularity, embarked on a triumphal journey by train from Juárez to Mexico City via Piedras Negras, accompanied, as Díaz had been four years earlier on his trip through Tehuantepec, by the filmmaker Salvador Toscano.[25] Toscano, egged on by his *maderista* partner Antonio Ocañas, came to see great commercial potential in filming the new revolutionary figurehead as he traveled victoriously to the capital. By Toscano's account at least, the resulting film, *La toma de Ciudad Juárez y el viaje del héroe de la Revolución D. Francisco I. Madero* (The capture of Ciudad Juárez and the journey of the hero of the revolution Francisco I. Madero), comprised "views" that were the "only authentic" ones to have been filmed since they were "authorized by Mr. Madero on board his train, on which this company's representative also traveled."[26] In this discussion we will concentrate on the triumphal journey itself, which occupies three of the film's five parts.[27]

The recently reconstructed tour de force that is *La toma de Ciudad Juárez* has a far greater sense of dynamism than the Tehuantepec film, using a wide array of camera positions, angles, and movements to associate Madero's revolutionary momentum with the rapid and inexorable drive forward of the train. The railway between the frontier and the capital was itself, of course, a product of the Porfirian railway-building project, but what is on display here is not the virility of modernity itself but the revolution's literal *and* symbolic appropriation of the nation, for which the railway stands as synecdoche. The Porfiriato had voiced the need to *forjar patria* (to forge the fatherland) partly through the development of transport infrastructure, a sentiment later echoed by the revolutionary regimes. But while the growth of the railways under Díaz symbolized the growing military reach (and limitations) of the state in hitherto remote regions of the country, that power was relatively fragile, as the frequent battles over railways and their changing hands between insurgent and government control during the Mexican Revolution would come to prove.[28] It was thus crucial that Madero should demonstrate not only his control of the strategic railway but also the enormous and enthusiastic grassroots support that he enjoyed at every stopping-off place on the way. Much of the triumphal journey section of the film thus shows the masses of common people cheering Madero on his way.

Although the triumphal journey sections of *La toma de Ciudad Juárez* display a wide variety of phantom ride shots, it seems significant that, at least in the surviving footage included in the 2010 reconstruction, there are relatively few picturesque views of the passing landscape-turned-spectacle, the focus instead falling strongly on the crowds of supporters at stations on the way.[29] For the most part Toscano thus elides the pulverization of the in-between topographical spaces traversed and rendered irrelevant by Madero's train, instead depicting the stations and the people who fill them as active intermediate subjects in revolutionary modernity. Two significant exceptions are the crossing of the bridge over the River Nazas between Gómez Palacio and Torreón and a panoramic view of the outskirts of Zacatecas as the train presses on toward Mexico City.

The shot of Madero's train crossing the River Nazas bears a striking contrast in its composition to the bridge shot that it echoes in *Inauguración del tráfico internacional de Tehuantepec*, cited above. In *La toma de Ciudad Juárez*, the bridge over the River Nazas is shot by a static camera mounted on the back of the train and pointing diagonally rightward, thus enhancing the perceived dynamism of the train's movement, rather than the flat sideways view in *Inauguración*. The frame is dominated by the railway track and the striking iron struts of the bridge that loom large in sharp focus in the foreground and middle distance; the bland scenery, consisting of arid land and a few trees, is relegated to the flat, inert, out-of-focus background of the frame. The modern, dynamic railway bridge that privileges movement over stasis is thus not an irruption of contingency in the placid contemplation of a picturesque scene, but rather the very point—the message—of the shot. This shot, which foregrounds the predictable and the mechanized over the contingency of nature or of the moment, is closer to the syntactic unit of both continuity and montage editing than to the persistent exhibitionism of earlier views.

Contingency, though, is far from absent in *La toma de Ciudad Juárez*. As we saw in the previous section, both contingency and inexorable modernity are seen as a potential menace to a preexisting (representational or societal) order: "contingency introduces the element of life and the concrete, but *too much* contingency threatens the crucial representational concept of totality, wholeness"—a feeling of representational excess that both cinema's transition toward narrative and Kracauer's appeal to art and to the absolute strove to correct.[30] Much of the

11.2. Frame enlargement from *La toma de Ciudad Juárez y el viaje del héroe de la Revolución D. Francisco I. Madero*, Salvador Toscano, Mexico, 1911. Courtesy of Fundación Carmen Toscano.

visual and aesthetic pleasure that the spectator might derive from viewing the triumphal journey sections of *La toma de Ciudad Juárez* resides in the absorbing compositions of the shots as the train, the railway lines and their peripheral furniture, and the crowds of onlookers converge and diverge, and in the heady combinations of different planes of motion within a given shot as the train, the camera itself, now far more given to panning than a few years earlier, and the variously homogeneous and chaotic movements of the crowds accentuate our feeling of participation in this irresistible propulsion of revolutionary force toward the capital.

In nearly all of these shots we are able to perceive the horizon or fixed elements such as railway tracks, buildings, trees, or even just the ground, enabling us to find the perspective of the scene. But one stunning moment breaks out of this perspectival logic: as the train pulls slowly out of Gómez Palacio, the camera, probably mounted on the back of the train, pans to the right while the crowd of admirers on horseback and foot move vaguely and unevenly toward the train, *wholly* filling the frame so as to exclude the slightest hint of any stable spatial referent that might enable us to orient ourselves comfortably within the filmed space. There is no sense of representational wholeness here: we are rather confronted with a pure experience of movement, that is to say, of time. Any vestige of perspective or distance that the other crowd shots might maintain is here entirely shut out. Such moments of contingency, when the promise of narrative or visual fulfillment is momentarily lost, work mildly against the narrative "efficiency" that, for Miquel, broadly characterizes the structure of Toscano's journey films, from *Inauguración* to *La toma*.[31]

## Leisure Time on the Post-revolutionary Railway

With the construction of the corporatist post-revolutionary state underway in the 1920s, the railway stood in a somewhat ambiguous relationship with modernity and nation-(re)building. For Álvaro Obregón's 1920–1924 government, intent on promoting national reconciliation on a political level alongside the economic business of reconstruction, the extensive national railway infrastructure—the object of countless acts of sabotage during the revolution—continued to be of strategic importance. But the 1920s signaled a shift of the center of infrastructural gravity from rail to road and, to a lesser extent, to air transportation. This transition enabled the post-revolutionary state, on the one hand, to subcontract investment in construction to the private sector in tune with its broader liberal economic policies, in contrast to the greater centralization of railway building, and, on the other hand, to consolidate and extend its own military power to areas of the country inaccessible by rail.[32] The railways themselves switched back and forth between state and private control as their respective administrators struggled to reorganize them and deal with their general decline, exacerbated by a lengthy strike of the traditionally militant railway men in late 1926 and early 1927 and by frequent rebel attacks on the railway infrastructure during the Cristero rebellion of 1927–1929.[33] A whole range of both government and privately produced pictures of the era suggest a desire to impress the cinema-going public with diverse manifestations of modern Mexico's domination of both aviation and automotive transport technologies. Even so, the railway, unwilling to surrender its place in the cultural and visual imaginary of post-revolutionary Mexico, also figured in numerous official films.[34]

In 1925, the Chilean journalist Miguel Chejade made *México ante los ojos del mundo* (Mexico before the world), a documentary produced by National Railways of Mexico with bilingual Spanish and English intertitles, destined to be exhibited across the Americas.[35] The surviving footage narrates two separate journeys from Mexico City, to Cuernavaca and to the Nevado de Toluca volcano, and aims to attract both local and foreign tourists to avail themselves of Ferrocarriles Nacionales de México's comfortable, rapid, and modern railway service in order to enjoy the touristic delights of Mexico City's environs.

In a broader study of stylistic conventions in nineteenth-century photography and early actuality films in Mexico, John Fullerton and Elaine King rightly point out that a sequence of *México ante los ojos del mundo* set in the Jardín Borda in Cuernavaca uses visual devices such as composition in depth, perspective, receding planes, and framing devices that evoke nineteenth-century melancholic and picturesque conventions. These devices work to close down contingency, in stark contrast to the mobilized views identified in other films of the period.[36] What Fullerton and King omit to mention, though, is that the entire scene filmed and set in the Jardín Borda is framed not as a direct narration of the rail passengers' sojourn in the gardens, but as an imaginary "flash-forward" imagined in the mind's eye of two young men who peruse the publicity literature provided by the train company. This becomes clear when the men's gaze as they look and point at the railway guide is cross-cut with our view of the gardens; and an intertitle announces, "The traveler looks at the Time-Table an takes information that 'Jardín Borda' in Cuernavaca is one of the most beautiful gardens in the country [sic]."

The insertion of the Jardín Borda sequence as a parenthesis within the main narration of the journey indicates that such attractive scenarios appear not in their own right but as so many aspects of the theatrical "spectacle" purchased by the railway traveler.[37] The ultimate aim of Ferrocarriles Nacionales' propaganda film is, of course, precisely to encourage travel as an end in itself: the same self-referential system that Kracauer decried as a symbol of the meaninglessness of spatiotemporal contraction in modernity. Meanwhile, the cross-cutting described above is also symptomatic of continuity editing, of which Chejade's film makes extensive use, indicating a far more synthetic spatiotemporality than that of Toscano's earlier, linear documentaries discussed above—as might be expected of a film produced at this juncture in the development of film style.[38] *México ante los ojos del mundo* is thus clearly inserted in cinema's drive toward narrative and away from contingency: the effort not to mimic objective and contingent time—the necessarily frustrated project of the early views—but to forge a new, synthetic, self-contained, and rationalized narrative time associated with the "attempt to structure contingency."[39] It is further worth noting that the passengers with whom Chejade's film asks us to identify are tourists enjoying their leisure time away from the jobs at which we imagine they toil on a daily basis: another aspect of the rationalization of lived time in capitalist modernity to which the post-revolutionary Mexican regime aspired.

We return, then, to the so-called narrative efficiency of which Miquel detects a glimmer in Salvador Toscano's much earlier actuality films. The frequent views from the side or back of the train of the passing scenery and railway tracks in *México ante los ojos del mundo* tend to stand less as the exhibitionist phantom rides of early cinema than as syntactical units that denote, for instance, the well-heeled passengers' pleasure as they glimpse out of the train's side windows to admire a particularly spectacular view. A great many of these tracking shots use framing devices (narrow gorges are a dramatic example) that guide the eye toward a particular aspect of each shot (as Fullerton and King discuss in the Jardín Borda sequence), thus working further to structure contingency. These devices, which contribute to the "efficiency" of film style—the closing down of contingency to the extent that *nothing* is meaningless, there is *no* time outside the synthetic narrative time—are clearly associated with the efficiency of the mechanized transport, which is the real protagonist of this film, and with the electricity upon which the cinematic apparatus itself depends. For Doane, "the cinema participates in the rationalization of time characterizing the industrial age. 'Economy' is a fundamental value of the developed narrative film, and the efficiency of electricity is paralleled by the efficiency of narrative."[40]

In another brief sequence, we encounter an editing strategy that seems further in intent from US continuity editing, which forecloses contingency in order to produce a self-contained spatiotemporal sphere, than it does from Sergei Eisenstein's montage of attractions, which, by contrast, charges previously existing fragments of reality (film frames as syntactic units) with new meaning, thus referring the activated viewer not to a hermetic, escapist narrative but to the external world. As the Mexico City–bound train prepares to depart from Cuernavaca, a medium shot shows the train dispatcher sitting at his desk; there is a cut to the timetable over which he pores with a pencil and then back to the dispatcher, then a close-up of his hand as he activates the telegraph mechanism on his desk: the efficiency of the railway bureaucrat is bound up in the systematized procedures that enable communication. The dispatcher glances up; the clock on his wall thus denotes urgency and precision. The dispatcher reflects on what the clock has told him; elsewhere the telegraph operator receives the dispatcher's message and transmits it with great speed to the conductor.[41] Then there is a double exposure: screen right and in the background, the telegraph operator continues his labor, while screen left and in the foreground, the conductor writes out the message. Finally, the conductor hands the message to the engineer toiling over the train engine. This dynamic sequence of images visually conveys the abstract notions of the spatial and temporal simultaneity and the irreversibility of time produced by the related technologies of wireless communication, the railway, and the cinema.

The material associations thrown up by this brief montage sequence, beyond the temporal framework of any enclosed narrative system that might frame the rest of the film, would seem to approach Kracauer's "meaning which is not simply given along with" the film: a meaning residing rather in the material world available and known to the spectator. The fleeting use of this editing strategy also indicates that *México ante los ojos del mundo* stands as something of a hybrid

# Film, Time, and the Railway / 209

11.3. Frame enlargement from *México ante los ojos del mundo*, Miguel Chejade, Mexico, 1925. Filmoteca UNAM; authorized by Filmoteca UNAM.

of various representational systems. In addition to the continuity and montage systems we have discussed here, some shots onboard the train recall the contingency of earlier views, albeit due, perhaps, to what Chejade might have considered errors in framing and exposure, to which I will finally turn. As the train presses on toward Cuernavaca, a side-view, tracking shot of the passing countryside, interspersed with passing telegraph wires, cuts to a close-up near-silhouette of the engineer cast against the bright light of the scenery behind; the man's proximity to the camera exaggerates the shaky movement of the train as he jolts about the frame. This has a giddying effect that enables us to distinguish the scene depicted but not to focus steadily on any part of the shot.

This does not appear to be the result of a mere technical or editing accident since the same technique crops up a number of times, converting it almost into a narrative trope. A further scene begins with a picturesque view of the passing countryside, followed by another montage sequence: a shaky phantom ride shot as the camera is seemingly dangled from the front of the train with the engine and railway tracks in full view, an equally shaky shot in silhouette of the engineer in action inside his cabin, and the train's whistle blowing off steam. We cut back to the railway man; then there is a stark transition to the bourgeois passengers smoking and chatting as they enjoy the ride: the smoothness of *their* ride is palpable in the steadiness of the camera. Remarkably, for a simple industrial propaganda film, Chejade appears to be casting a whole set of associations: not only the simple connection between the engineer's action inside his cabin and the sound of the whistle on top of the train, but also the connection between the precarious conditions of blue-collar labor, the placid lives of the bourgeoisie, and the ability of mechanized transportation to convert atomized space into a spectacle. Doubtless, Chejade's intention was far removed from the revolutionary consciousness-raising intended by the pioneers of Soviet montage; indeed, this may be an ideologically reactionary attempt to derive spectatorial pleasure from an identification with the tourists' ability to enjoy such trappings of privilege.

The railway, as this brief discussion suggests, bears an ambiguous relationship with inexorable progress in silent Mexican actuality and documentary films. Modernity is, on the one hand, the efficiency and rationalization that allows for the expansion of both the railway and the revolution; but it is also the chaos, the contingency, and the continual uprooting of a process marked by perpetual change. The three films considered here reflect cinema's only partially successful attempt to circumscribe this contingency through narrative efficiency, as part of a broader set of debates in the artistic and media spheres over the possibilities and limits of technological progress. Late 1910s and 1920s fiction films bear witness to post-revolutionary Mexico's continued accommodation to rapid change in relation to the twin technologies of transport (with the focus shifting toward road and air transport in tune with economic reality) and cinematic representation. The question of how such films articulate the altering relationships that post-revolutionary spectators experienced with time and space as a result of such technologies might be addressed in a longer study.

## Notes

1. Aurelio de los Reyes, *Cine y sociedad en México, 1896–1930*, vol. 1, *Vivir de sueños (1896–1920)* (Mexico City: UNAM/Cineteca Nacional, 1981), 95–96; Gustavo Casasola, *Historia gráfica de la revolución mexicana, 1900–1960* (Mexico City: Trillas, 1960), 1:96–97.

2. The Salina Cruz–Coatzacoalcos line, first opened in 1894, was one of the key strategic axes of Porfirian railway building, connecting Atlantic and Pacific commercial shipping routes. See Francisco R. Calderón, "Los ferrocarriles," in *Historia moderna de México*, vol. 7, part 1, ed. Daniel Cosío Villegas (Mexico City: Editorial Hermes, 1974). 1907 marked the opening of new drainage and coastal defense works in both ports and the reconstruction of the railway itself. See Casasola, *Historia gráfica*, 1:96–97.

3. Frank Gray, "*The Kiss in the Tunnel* (1899), G. A. Smith, and the Emergence of the Edited Film in England," in *The Silent Cinema Reader*, ed. Lee Grieveson and Peter Krämer (London: Routledge, 2004), 55.

4. Wolfgang Schivelbusch, "Railroad Space and Railroad Time," *New German Critique* 14 (Spring 1978): 31.

5. Ibid., 33.

6. In 1880 Mexico had 1,073 kilometers of railway tracks; in 1898, 12,801 kilometers. By 1910 the network had grown to 19,280 kilometers. Calderón, "Los ferrocarriles," 540, 567, 628.

7. De los Reyes, *Cine y sociedad*, 42–44; Aurelio de los Reyes, "El cine en el noroeste de México," in *Regionalización en el arte: Teoría y praxis*, ed. José Guadalupe Victoria, Elisa Vargas Lugo, and María Teresa Uriarte (Mexico City: Gobierno del Estado de Sinaloa/Instituto de Investigaciones Estéticas–UNAM, 1992), 165–79. However, as de los Reyes points out, competition pushed many early exhibitors beyond the railway routes, sometimes traveling to remote locations by horse, mule, or donkey, or even on foot.

8. Gray, "*The Kiss in the Tunnel*," 55.

9. Siegfried Kracauer, "Travel and Dance," in *The Mass Ornament: Weimar Essays*, ed. and trans. Thomas Y. Levin (Cambridge, MA: Harvard University Press, 1995), 66–70.

10. The Filmoteca UNAM's print runs to just over fifteen minutes; for Miquel, it is a rare example of a Mexican actuality film of the era to be preserved "as it was exhibited." Ángel Miquel, *Acercamientos al cine silente mexicano* (Cuernavaca: Universidad Autónoma del Estado de Morelos/Ediciones Sin Nombre, 2005), 12; Ángel Miquel, *Salvador Toscano* (Mexico City: Universidad de Guadalajara/Gobierno del Estado de Puebla/Universidad Veracruzana/UNAM, 1997), 41.

11. The *Arizonian*'s cargo consisted of sugar on its way from Hawaii to New York; Miquel, *Salvador Toscano*, 41–42.

12. Miquel, *Salvador Toscano*, 42.

13. See his voiceover commentary on *Inauguración*, included on the DVD *Oaxaca* (*Imágenes de México*), Universidad Nacional Autónoma de México, 2005, compiled and with voiceover commentary by Aurelio de los Reyes. I would add that as well as the horizontal tracking effect provided by the motion of the boat upon which the camera is mounted, the forceful tide to which the boat is subjected toward the end of the shot produces a strong, rhythmic, tilting effect on the vertical axis.

14. Schivelbusch, "Railroad Space," 34.

15. Ibid., 36.

16. Ibid., 37–40.

17. Charles Musser, "Moving towards Fictional Narratives: Story Films Become the Dominant Product, 1903–1904," in *The Silent Cinema Reader*, ed. Lee Grieveson and Peter Krämer (London: Routledge, 2004), 93.

18. Walter Benjamin, "On Some Motifs in Baudelaire [1939]," in *Illuminations*, trans. Harry Zorn (London: Pimlico, 1999), 152–96.

19. See Miquel, *Salvador Toscano*, 42; de los Reyes, *Cine y sociedad*, 97.

20. By the late 1910s the average shot duration in US cinema was around five seconds; David Bordwell and Kristin Thompson, *Film Art: An Introduction*, 7th ed. (New York: McGraw-Hill, 2004), 285.

21. Tom Gunning, "The Cinema of Attractions: Early Film, Its Spectator, and the Avant-Garde," in *Early Cinema: Space, Frame, Narrative*, ed. Thomas Elsaesser with Adam Barker (London: British Film Institute, 1990), 57–58.

22. Mary Ann Doane, *The Emergence of Cinematic Time: Modernity, Contingency, the Archive* (Cambridge, MA: Harvard University Press, 2002), 142.

23. Ibid., 29, 67.

24. Kracauer, "Travel and Dance," 69.

25. Jesús Silva Herzog, *Breve historia de la revolución Mexicana*, vol. 1, *Los antecedentes y la etapa maderista* (1960; repr., Mexico City: Fondo de Cultura Económica, 2007), 231.

26. Miquel, *Salvador Toscano*, 58. For other films depicting the journey, see Aurelio de los Reyes, *Filmografía del cine mudo mexicano*, vol. 1, *1896–1920* (Mexico City: Filmoteca UNAM, 1986), 65–71.

27. Toscano's film lasted around an hour. Ángel Miquel's 2010 reconstruction, running to just over twenty-five minutes, was edited according to the order of sequences listed in detailed exhibition posters; little of Toscano's original editing was found to have survived. Email correspondence with Miquel, May 2010.

28. Alan Knight, "The Weight of the State in Modern Mexico," in *Studies in the Formation of the Nation State in Latin America*, ed. James Dunkerley (London: Institute of Latin American Studies, 2002), 212–53.

29. The exhibition posters seem to confirm this impression; Miquel, *Acercamientos*, 137–40.

30. Doane, *The Emergence*, 12 (original italics).

31. Ángel Miquel, "Documentales de la revolución maderista," paper delivered at the conference Cine mudo en Iberoamérica: naciones, narraciones, centenarios, Universidad Nacional Autónoma de México, Mexico City, April 2010.

32. Knight, "The Weight of the State"; Jean Meyer, "Revolution and Reconstruction in the 1920s," in *Mexico since Independence*, ed. Leslie Bethell (Cambridge: Cambridge University Press, 1991), 201–40.

33. Meyer, "Revolution and Reconstruction."

34. For titles, see de los Reyes, *Filmografía*, 2:34, 67–68, 70, 134–35, 160–61, 196–216.

35. De los Reyes, *Filmografía*, 3:22. The toned print preserved in the Filmoteca UNAM runs to some twenty-two minutes; reel 3 of the original four is missing. Email correspondence with Ángel Martínez, Filmoteca UNAM, June 2010. The closing title announces the film to be the first of a series.

36. John Fullerton and Elaine King, "Local Views, Distant Scenes: Registering Affect in Surviving Mexican Actuality Films of the 1920s," *Film History* 17, no. 1 (2005): 66–87.

37. Schivelbusch, "Railroad Space." We later witness a spectacular hike on the Nevado de Toluca.

38. On the development of continuity editing in this regard, see David Bordwell, *On the History of Film Style* (Cambridge, MA: Harvard University Press, 1997).

39. Doane, *The Emergence*, 29.

40. Ibid., 162.

41. The psychological effect of the montage is disturbed at this transition by an intertitle that, somewhat needlessly, narrates this fact.

# 12
# "Los Hijos de Ford"
## Mexico in the Automobile Age, 1900–1930

*J. Brian Freeman*

Contemplating the many changes in everyday life since the end of the Mexican Revolution (1910–1920), writer and critic Salvador Novo observed that due to the proliferation of automobiles—*los hijos de Ford*, or "the children of Ford," as he labeled them—crossing the street in Mexico City had become a struggle of epic proportions.[1] By the end of the 1920s, the use of private cars, taxis, jitneys, and buses by residents, from the eminently wealthy to the working class, had altered the spatiality of the city, given birth to new sounds and smells on streets, produced novel dangers, and led to the appearance of new occupational opportunities.

Novo was by no means alone in his attention to this "other" revolution that had begun during the waning years of the dictatorship of Porfirio Díaz (1876–1911), grown alongside the armed struggle, and finally exploded during the early post-revolutionary era. The presence of automobiles soon captured the attention of countless observers, from writers and artists to government officials, motoring advocates, and foreign visitors. In the context of a changing intellectual and political landscape shaped by the nation's sweeping social revolution, these many and diverse onlookers projected onto the automobile their varying levels of enthusiasm and concern over the striking process of technological modernization that was taking place in the capital and slowly but surely filtering into the countryside.

As the armed revolution came to an end and the new post-revolutionary government set about fostering a common national identity by drawing heavily on rural and indigenous aesthetics, the proliferation of automobiles struck many people—citizens and foreigner visitors alike—as a force of cultural corruption and the embodiment of the "soulless" materialism that seemed to characterize the United States, not Mexico.² Yet such a view was not universally held, and others, like Mexico City's iconoclastic avant-garde writers and artists, took inspiration from automobiles and used them to produce new forms of art and literature and actively studied their impact on urban life. Meanwhile, state builders and business interests became active advocates of automobility as they embraced motorized movement as an important component of post-revolutionary modernization.

Alongside these debates over the appropriateness of cars in Mexico, an increasing number of motorists took part in an early form of automobile tourism as they sought out a variety of rural, indigenous, and colonial-era manifestations of Mexican culture. In so doing, they did much to promote and popularize knowledge about the provincial nation, even as they employed a certain condescending gaze from behind the steering wheels of their modern machines. As they mapped passable roads and catalogued cultural attractions, they established the basis for a massive expansion in both tourist and everyday motoring during the 1930s, 1940s, and beyond.

By the end of the 1920s, then, Mexico City was well on its way to becoming the bustling, automobile-age metropolis it would be known as during the second half of the twentieth century while a growing amount of the nation had been brought into the orbit of the modern motorist. What emerged was a discursive duality, which contrasted a machine-age Mexico City with a rural world of folklore and tradition that was distinct but accessible to the modern automobile tourist. Over the course of the middle decades of the twentieth century, a period of accelerated modernization and increasingly rigid folkloric nationalism, this spatialized vision of national culture would fundamentally shape both domestic and international narratives of modern Mexico.³

This chapter begins by tracing the contours of an early car culture that emerged during the last decade of the Porfirian regime and examines how it was transformed by the revolutionary process that followed. It then turns to the development of an incipient form of urban mass automobile transportation in the aftermath of the 1916 Mexico City general strike, which was consolidated over the next decade and a half. Finally, the chapter outlines the characteristics of the growing practice of automobile touring in rural areas during the early post-revolutionary period and concludes by contemplating the state's embrace of automobility as a tool for the promotion of both modernization and traditional Mexican culture.

## Teatime at the Automóvil Club

The economic prosperity and political stability associated with the rule of authoritarian president Porfirio Díaz (1876–1910) provided new opportunities for the wealthy to publicly perform their elite status through conspicuous consumption, and during the last decade of the era the automobile came to represent an ideal item to fulfill that desire.[4] Motoring formed part of a larger effort by Mexican political and economic elites to transform the national capital into a modern, cosmopolitan city; and during the first decade of the twentieth century the privileged few sped up and down the capital's Champs-Élysées, the Paseo de la Reforma, they had tea at the newly established Automobile Club on the shores of Chapultepec Lake, and they soon began to go on excursions to nearby haciendas.[5] Meanwhile, automobiles began to appear in an assortment of other activities from parades and festivals to commercial and sporting events.[6]

Although the first self-propelled car to arrive in Mexico had initially dismayed both residents and roaming dogs in 1895—the daily *El Universal* even dubbed it "El coche del Diablo" (the Devil's car)—within a year bicycle mechanics Alexander Byron Mohler and William P. De Gress had built the first domestically designed automobile, a tiny two-person machine that rode atop four bicycle wheels.[7] By 1902, Alonso Fernández Castelló had organized the first tour into the countryside as he and five friends made their way to the Mariscala hacienda, passing through Tlalnepantla and Cuautitlán. Days later, a group of five cars traveled to the Soltepec hacienda, while that same year, H. Menel and Pedro Z. Méndez drove to Pachuca and back in one day.[8] Over the course of the decade cars continued to arrive, and they soon became a fixture of everyday life in the city center.

Meanwhile, automobile and parts suppliers aimed to garner public attention and prove the value of their products by staging spectacular trips out into the hinterland of the capital. In January of 1910, Billy Knipper, hailed as "the famous chauffeur" by local newspapers, traveled to Puebla to demonstrate the latest "Chalmers" model car imported by dealers Mohler and De Gress.[9] In March of that same year, a Packard "30," owned by distributor Kenneth Walter, made a trip from Mexico City to the western city of Guadalajara in under thirty-three hours.[10] Boasting of their accomplishments, automobile agencies published the results of such events in the pages of the capital's periodicals. When a Flanders "20" arrived in Mexico, having departed from Quebec, Canada, Mohler and De Gress took out a full page in *El Diario* to announce that they had the model in stock.[11] Meanwhile, that summer a "White" model truck also garnered attention by hauling two tons of cement from Mexico City to Toluca in three hours and fifteen minutes.[12] Forming part of the growing consumer culture that came to characterize early twentieth-century urban Mexico, these daring drives emphasized to buyers that such models could handle the rugged terrain of a country that still lacked a system of good roads.[13]

Although local inventors had attempted to build their own vehicles, virtually all automobiles had to be imported from abroad, and foreign manufacturers, particularly those located in the United States, carefully watched the growing acceptance of cars among Mexico's urban elites. In a 1908 study of the global demand for motor vehicles, the US Bureau of Manufactures announced that Mexico had become the third largest market for US makes. The trade had nearly doubled in two years, reaching $812,639 in 1907. To advance the trade in motorcars, consular studies closely tracked the tastes of Porfirian consumers, and reports noted the general preference among the city's affluent residents for fanciful designs of French and Italian origin. In order to expand the trade abroad, US producers determined they would have to appeal to the tastes of domestic buyers. Yet with prices ranging from 2,000 to 20,000 pesos apiece, motoring remained firmly confined to the wealthiest members of society.[14] Consequently, by 1910 there were only about two thousand automobiles in the national capital, a city of five hundred thousand people.[15]

Yet just as the bicycle had initially met resistance from residents and municipal governments (they were banned from the center of town for a few months during 1891), the automobile had its detractors.[16] The poor, for example, now had to navigate new obstacles, as cars produced conflicts over how the street ought to be used, while some of the city's well-to-do residents also questioned the benefits of such technologies.[17] As early as 1900, journalists at *El Universal* called for police action in order to ease the circulation of people and vehicles and to "put things in order" during the hours of transit.[18] Over the course of the decade accidents plagued the city, and periodicals began to report on the gruesome details while slipping in their own editorial commentary. *El Diario* lamented the "frequent automobile accidents . . . that day by day are becoming increasingly disgraceful."[19] Dr. Salvador Quevedo y Zubieta characterized automobile travel as detrimental to one's health, arguing, "The automobile can be medically considered to be a cause of an impulsive neurosis."[20] *Diario del Hogar*, on the other hand, reported on the deleterious effects of exhaust fumes that filled the air along San Francisco Avenue during the "daily procession of the elegant world," demanding action to address what the paper understood to be a public health threat.[21]

Modernista writers, in particular, disconcerted by the effects of the machine age, published their own attacks.[22] In "El automóvil de la muerte" (The automobile of death), poet Amado Nervo told a presumably fictional tale of a group of peasants plagued by the crazed speed of automobilists who, in their reckless excursions through the countryside, ran over geese and into cows. Seeking revenge, the peasants strung a piece of wire across the road, positioned at just the right height. When an open-topped car came along, the peasants watched as "the wire, taut and rigid, cut through, with the same ease with which a wire cuts a block of butter, first two heads, then three." "Oh, automobile of the decapitated!" Nervo lamented, "the frightening automobile of death, with its five torsos leaning slightly back and slowly spilling their blood!"[23] Likewise, José Juan Tablada expressed his own misgivings in the poem "El automóvil en México," written after a short stint working for Pepe and Andrés Sánchez Juárez, owners of the major auto dealership Garage Internacional.[24] Unlike the Italian futurist F. T. Marinetti, who believed auto races to be "more beautiful than the Victory of Samothrace," for Tablada the automobile was instead a "dragon made by cubists," a "mechanical caricature of an apocalyptic beast," and a "dynamic coffin" that left behind "carbon flatulence."[25]

By the end of the Porfirian era an incipient car culture, characterized by its restriction to the wealthiest of residents, a dependence on foreign manufacturers and parts suppliers, and its association with the conflict-ridden modernizing project of the era had taken hold of Mexico City. When revolutionary writer and intellectual José Vasconcelos looked back on the period, he took a dim view of the elite fascination with automobiles, noting how they embodied the superficiality and unevenness of modernization and technological change during the era. The Porfirian governing elite, according to Vasconcelos, seemed to have believed that progress had been achieved simply "because an automobile had arrived in Mexico" even as most city residents "continued cooking with charcoal like in the time of Moctezuma."[26] Yet with the onset of the revolution in 1910, the world of the wealthy motorist soon gave way to a more broad-based and not entirely elite car culture.

## The Automobile Revolution

The outbreak of the Mexican Revolution and the collapse of the Porfirian regime brought about a significant redistribution of automobiles as military forces crisscrossed the country, helping themselves to wealthy motorists' stock of vehicles, even as additional cars continued to pour into the country. The revolution itself took place during the decade in which Henry Ford and other manufacturers perfected the techniques of mass production and distribution. As a result, the price of a Model T declined throughout the decade, and by 1916 a runabout could be acquired for $345 in the United States, while a touring car cost $360. That same year production hit 738,811 units, half the world market.[27] Many of these vehicles made their way to Mexico, and in 1918 the *Los Angeles Times* asserted the country was "a land of used automobiles," noting that while a few new models had been introduced over the previous eight years, used cars had been imported "in large numbers, not only from the United States, but from Central and South American countries."[28] The political, industrial, and technological transformations of the 1910s thus encouraged a redefinition of the automobile as non-elites increasingly learned to use cars as both tools of war and everyday means of movement. As a result, by 1919 the number of cars in Mexico City had nearly tripled.[29]

An event that took place during the waning months of the Porfirian period is illustrative of the way in which the outbreak of revolution challenged the meaning of the automobile in Mexico. In January 1911, *La Iberia* reported on a strange occurrence in the state of Guerrero. A group of "rebels" had commandeered a vehicle, owned by a local *hacendado*, and packed it with dynamite. Late at night, they lit the explosives and sent the car running toward an encampment of federal forces with whom they had been skirmishing. But a kilometer before reaching the base, the charge detonated without injuring anyone. The anticlimactic event, nevertheless, may have been one of the world's first car bombings.[30] Throughout the course of the armed phase of the revolution, rebel forces would continue to confiscate automobiles; and cars soon came to represent highly visible symbols of the new distribution of power that had emerged since the collapse of the Porfirian regime.

Indeed, the many revolutionary generals who took up residence in the nation's capital represented a significant market for cars. Nearly every major figure of the period owned one, if not multiple, automobiles, and even the rural-oriented rebel leader Emiliano Zapata reportedly traveled around Mexico City in a large touring car.[31] In 1915, General Venustiano Carranza reportedly purchased six train carloads of automobiles from the US side of the border, and by 1919 there were around fifty-five hundred in the capital and surrounding suburbs.[32] In some cases the new revolutionary officials—who were not always discreet consumers and were, in fact, often criticized for the luxurious cars they owned—hosted auto races and went on excursions to nearby attractions.[33] On a trip to the city during the late 1910s, Harvey Middleton was surprised to witness in Chapultepec park "a parade of automobiles . . . four lines deep, two lines going in each direction, the cars being so numerous that they can only go at a walking rate."[34]

Above and beyond the new automobilist revolutionaries, the most profound transformation in Mexico City's "motorscape" of the 1910s came with the appearance of a new class of workers and entrepreneurs: bus and taxi drivers. Recalling the birth of the *chafirete* (professional chauffeur) and his able assistant, the *cobrador* (fare collector), Salvador Novo observed that many of these workers had initially been drivers for an assortment of generals; but over time they purchased their own cars and began to provide "the newly fashionable speed to a greater number of citizens at a moderate price." To Novo, these young chafiretes and cobradores were none other than the "first sons of the Revolution."[35]

Indeed, taxi and bus drivers benefited indirectly from the powerful labor movement that had emerged in Mexico City following the collapse of the Porfirian regime. During a general strike that hit the Federal District on July 31, 1916, and brought tram services to a halt, entrepreneurial chauffeurs began to transport multiple passengers for a nominal fee, thus giving rise to the new industry. After moving people around the capital in their own cars during the first day of the strike, by the second day they had begun to transform the bodies of vehicles into makeshift buses or jitneys by positioning wooden planks atop old chassis and stringing up cloth around rod canopies.[36] Shortly after the initial, improvised vehicles hit the streets, chauffeurs began to experiment with new design and service innovations. On mainly Ford chassis, they crafted wooden bodies that contained laterally running benches to fit eight, sturdy roofs, sides to act as backrests, and curtains to keep out rain. Costing between 1,200 and 1,400 pesos, they were eminently affordable and quickly came to represent a permanent competitor of the trams. The vehicles charged ten centavos a person, roaming the city like taxis, at all hours, and keeping to no permanent routes.[37] To attract passengers, chauffeurs displayed a diversity of cardboard signs listing the tramway routes they claimed to serve, though they did not always do so faithfully, and on every stop along the trajectory, the cobradores announced the various parts of the city through which they would pass.[38]

During the late 1910s and early 1920s, drivers and cobradores established the first permanent and predictable routes. In 1917, a group of bus owners from the same neighborhood assembled and formed the first such line, which by the early 1940s carried the name Santa María Mixcalco y Anexas. Quickly, new associations of owners emerged to service other parts of the city, and between 1918 and 1923, over 90 percent of the lines functioning in the early 1940s had been established.[39] Yet with no official policy addressing matters of permits and routes, an excessive number of vehicles led to constant conflict on the streets. To defend their interests, bus line owners and workers established the Centro Social de Chóferes in 1921, an affiliate of the Confederación Regional Obrera Mexicana (CROM). Through the Centro Social and the CROM, the budding industry attempted to limit the granting of permits by the municipal government of Mexico City and the other municipalities of the Federal District.[40] On February 27, 1922, bus line workers halted all services and marched to the Palacio del Ayuntamiento (City Hall), where they exchanged gunfire with authorities. Negotiations with President Álvaro Obregón ended favorably for the nascent industry, as officials reportedly met all demands of strikers.[41] The Department of Transit was shifted to the government of the Federal District, away from the municipalities, while members of the Centro Social would henceforth advise that department and eventually take up high-level positions within its hierarchy.⁴² Chauffeurs thereafter represented key allies of the Obregón administration, and during the failed de la Huerta rebellion, La Alianza de Camioneros de México mobilized three hundred cars for the president, transporting troops to the western battlefront.⁴³

By the end of the 1910s, changes in both domestic politics and international automobile production had fostered a new car culture characterized by its extension beyond the confines of the wealthy. During the early 1920s, as the country entered a period of post-revolutionary reconstruction and cultural revival, the popularization of the automobile would encourage authorities to embrace road building as a means to unify the country both economically and culturally. Meanwhile, however, the proliferation of cars struck many, particularly foreign visitors, as the embodiment of a destructive machine age that ought not be exported to a "traditional" nation like Mexico.

## Foreign Visitors and Machine-Age Mexicans

The increasing presence of automobiles elicited an almost uniform disdain from the countless foreigners who arrived in the nation's capital during the first decades of post-revolutionary reconstruction. This period forms part of what Mauricio Tenorio Trillo has called the "cosmopolitan summer," a moment of profound fascination for nearly all things Mexican. During these years, foreign artists, writers, intellectuals, and tourists flocked to take part in a national cultural renaissance championed by such figures as artist Diego Rivera, intellectual José Vasconcelos, and anthropologist Manuel Gamio.⁴⁴ Many, if not all, foreigners arrived in search of "traditional" and "authentic" expressions of Mexicanness that were being promoted by the new regime as part of its effort to forge a common national identity. Yet visitors were consistently shocked by the proliferation of motorized vehicles and the generalized process of technological modernization in the nation's capital. To such observers, machines, particularly automobiles, were inconsistent with what they understood to be an idyllic, if primitive, land of peasants and Indians. Worse yet, the automobile, and the machine more generally, appeared to be corrupting "authentic" Mexico.

One of the more vociferous opponents to the globalization of the machine, a form of "Americanism" as he saw it, was English writer D. H. Lawrence. In his novel *The Plumed Serpent*, Lawrence populates his tale of early post-revolutionary Mexico City with "crazy motor-cars" and "frightful little Ford omnibuses called camions."⁴⁵ On an excursion to the outskirts of the capital, his characters travel along a major thoroughfare, likely Avenida de los Insurgentes, where they watch the rushing of "incredibly dilapidated Ford omnibuses, crowded with blank dark natives in dirty cotton clothes and big straw hats."⁴⁶

Similarly, Stuart Chase, a US polymath of sorts known for his studies of technology, economics, and society, traveled to Mexico in the early 1930s, where he witnessed firsthand the spread of such machines. In his widely read *Mexico: A Study of Two Americas*, Chase commented on the proliferation of automobiles and noted the "wildness of *mestizo* drivers."[47] To Chase, these mestizos had little, if anything, to do with the "authentic" Mexico that lay in places like Tepoztlán, where "flowers [were] more important to Mexicans than [were] motor cars, radios, and bathtubs combined, to Americans."[48] Fearful of the impact of road construction on these "machine-less men," as he called them, Chase noted, "My foreboding is concerned with the American tourist. In a few months (this is June, 1931) he will be able to drive his Buick through to Mexico City. Clouds of Buicks, swarms of Dodges, shoals of Chevrolets—mark my words, they will come."[49]

Throughout the 1920s and 1930s countless other visitors wrote home about their run-ins with the Mexican motorist. In 1925, Russian futurist poet Vladimir Mayakovsky made note of the "alarming agglomeration of automobiles and buses" and the "savage and battle-hardened competitions between chauffeurs."[50] A technophile, Mayakovsky was nevertheless annoyed by the manifestation of modern technology in traditional Mexico. Photographer Edward Weston, similarly, complained about the effect of traffic on the pleasant climate of the city; and in 1926, John Dos Passos wrote of the "noise, poverty, and chaos" in the capital during his trip to the country.[51]

More commonly, visitors simply censored the machine age from their representations of the country. While Weston was clearly perturbed by the effect of traffic, his body of work shows few sign of industrialism's incursions in the country. Unlike his lover Tina Modotti, who captured the new face of the capital in images of telephone wires, massive petroleum tanks, and modernistic, reinforced cement constructions, Weston confined himself to nudes, portraits, and depictions of an idyllic Mexico. Similarly, artists and writers like Winold Reiss, Langston Hughes, Hugo Brehme, and countless others limited their work to the traditional, while culture brokers like Anita Brenner and Frances Toor made careers of silencing images of industrialism and technological modernization in their representations of the country.

Some observers took a more nuanced approach to the advent of the machine age in Mexico. Radical US journalist Carleton Beals, for example, argued, "Roads are inevitable, and in their wake automobiles, gasoline stations, tourist inns, orange-crush stands, jazz dance halls—all of these things will probably, within the next quarter of a century, descend upon Tepoztlán, and Taxco, and Chilpancingo." Yet he concluded his 1931 study, *Mexican Maze*, by observing, "History teaches that unless a people is exterminated the conquered always conquer the conqueror."[52] For Beals, it was inevitable that the machine age would, in some sense, be Mexicanized. Indeed, even as foreigners wrote of the cultural destructiveness of motorized vehicles in traditional Mexico, avant-garde artists and writers were turning their attention to the unexpected and inventive adaptations that were taking place on the streets of the nation's capital in response to the advent of automobility.

## Avant-Garde Automobile Enthusiasts

Throughout the 1920s and beyond, the city's young avant-garde, figures like Manuel Maples Arce, Germán List Arzubide, Salvador Novo, and Arqueles Vela, among many others, would not only depict the nascent culture of the automobile, but in some cases use it to develop new forms of representation.[53] These individuals, less tied to idealized visions of Mexicanness, took inspiration from a cityscape in the process of profound transformation. Significantly, their activities were often diametrically opposed to efforts to rescue and revive the "authentic" nation; and, as Elissa Rashkin argues, their depictions of technology—cars, steel bridges, telegraphs, and such—"provided an important counterpoint to the folkloric stereotypes of the era."[54]

Examples of the exaltation of the very vehicles that Chase, Lawrence, and others largely rejected abounded during the early post-revolutionary era. In his first book of poetry, *Esquina*, Germán List Arzubide playfully asserted that "the voyage to Mars will finally be made by bus."[55] And in "Camiones," the sixth act of the *Teatro Mexicano de Murciélago*, Luis Quintanilla offered viewers an "anthropomorphic rendition of buses that cough and fall dead in the middle of the street 'like any one of the dogs they've run over.'"[56]

Many avant-garde intellectuals took part in a form of urban ethnography, tracing the impact of the automobile on the "City of Palaces." Salvador Novo, dubbed the "poeta chófer" (chauffeur poet) by friend Carlos Pellicer, was no doubt the most consistent chronicler of the capital's emergent car culture. In his mid-1940s *Nueva grandeza mexicana*, a work that drew on his earlier fascination with the bus and taxi industry, Novo recalled the social and cultural impact of a new, mechanized cityscape born in the late 1910s. He found that with the proliferation of motorized vehicles, surviving life in the big city forced residents to develop new skill sets. The early cobrador, for example, learned to move "on foot up and down the rear running board when he was not calling out 'room for two,' or asking for 'ten and one' at the gas station, or reciting a heterodox and hybrid litany of celestial and mundane names of routes."[57] In constant contact with the public, the cobrador thus developed unmatched ambulatory and linguistic agility. Likewise, bus and taxi industry workers learned a new technological vocabulary and developed their own neologisms to describe the transformed urban environment. "Our highly expressive folk language," Novo observed, "owes much to the verbal ingenuity of this new caste of drivers and conductors." He noted how, for example, the word *lambiscón*, which had become widely used by the mid-twentieth century to refer to sycophants and toadies, had originally been coined to describe "the young conductors of the buses." The verb *ruletear*, likewise, was first used to describe the offering of "services 'on the wheel.'" And *mordida* (a bite or a bribe), a word that had become widespread by the 1940s, had its origins in the nickname for the first traffic cops: *mordelones*.[58]

An early forum for exploring the city's dynamic car culture was the bus and taxi workers' magazine, *El Chafirete*, a publication believed to have been written largely by Novo and other local literary figures.[59] Established in 1923 in order to defend the interests of any and all chauffeurs in the capital, the magazine published numerous poems and fictional stories about the culture of motorized transportation. Themes addressed in this eccentric periodical, among others, dealt with conflicts between chauffeurs and traffic police, the enticements of prostitutes and cabaret women, and the masculinity of taxi and bus drivers.

The magazine frequently toyed with the increasingly intimate relationship between Mexicans and their machines, and cartoons, poems, and prose presented readers with numerous anthropomorphized automobiles and "technomorphized" transportation workers. In the March 1923 edition, for example, a piece entitled "Mi coche triste" (My sad car) tells the tale of lost love between an old Ford and its chafirete. In the poem, the *fotingo*, one of many slang terms for a Ford, recalls its abandonment by its chauffeur, "El Tenorio de Mixcalco." The car laments, "Ungratefully you left me / in a ghetto garage / leaving me without a steering wheel / and without tires, for free / knowing that I loved you, both night and day / we worked together / and in the station you changed me / and then you transformed me / into a ridiculous bus."[60]

Instead of fixing their gaze on the countryside and indigenous cultures, writers and artists like Salvador Novo, Luis Quintanilla, Manuel Maples Arce, and Germán List Arzubide, among countless others, turned their attention to the curious adaptations occurring in cities, on streets, in garages, and aboard taxis and buses. Rather than reject Mexico City's growing car culture as a manifestation of a "soulless Americanism," a variety of avant-garde intellectuals embraced these technological artifacts as agents of an emergent Mexican modernity and set about demonstrating the diversity of ways in which residents of the capital were not simply being transformed into carbon copies of their counterparts in the industrialized North Atlantic, but were producing their own local variant of the machine age.

## Touring Cultures

The return of a nominal peace during the 1920s encouraged the expansion of not only motoring in Mexico City, but automobile touring into the capital's hinterland and beyond. The most active and influential advocate for this relatively new practice was arguably the magazine *El Automóvil en México*, a publication that also established the first post-revolutionary auto club, the Automóvil Club de México. By 1921, the magazine had already done a study of the Mexico City to Pachuca road to facilitate reconstruction, organized the first "traffic agent contest," held a banquet for President Obregón and the Secretariat of Communications and Public Works, organized a variety of auto excursions into rural areas, and hosted the first "International Exposition of Automobiles" in the capital's National Theatre.[61] In the years before the state developed a clear policy for road building, *El Automóvil en México* clamored for *más y mejores caminos* (more and better roads) and set about mapping the extant road network in the central and northern portions of the country.

Initially founded in 1907 by English auto enthusiast A. R. Hogg, the magazine had been taken over by Rafael Alducín (later director of the newspaper *Excélsior*) and then purchased by Gustavo Alaña in 1918. By 1923, *El Automóvil en México* claimed to have a few thousand subscribers, up from three only five years before, and by 1928 there were twenty-five thousand issues of the magazine circulating. By March 1923, *El Automóvil en México* had completed an extensive surveying effort, producing a total of twenty-three road maps for automobilists; and in October participants established the Asociación Automovilística Nacional (National Automobile Association) in order to continue fighting for "the rights of automobilists."[62]

The magazine was particularly committed to extending roadways and motor vehicle use into provincial areas. During the 1920s, virtually every issue of the magazine featured recollections of recent excursions into the countryside, discussions of rural road conditions, and ideas on places to visit. Articles and images appearing in *El Automóvil en México* promoted such tours by appealing to the wealthy and emergent middle classes' desire to see the nation's natural wonders, view the ways of traditional peasants and indigenous peoples, and visit the relics of the Colonial and pre-Columbian past. As these motorists ventured beyond the confines of the capital, they did much to remap and catalog the cultural geography of an increasingly esteemed rural world. Employing other modern technologies like Kodak cameras, which were often advertised in the magazine, these explorers became active participants in the larger post-revolutionary effort to expand the visibility of the rural nation.

During the 1920s covers of *El Automóvil en México* were often illustrated with representations of ostensibly rural people and folkways. The May 1925 issue, for example, displayed a girl with long braids, wearing a traditional skirt and *huipil* (indigenous blouse). Pulling back a curtain, the girl revealed an image of the Palacio de Bellas Artes (Palace of Fine Arts). Next to her, an eagle appeared beside a cactus, holding a serpent in its talons and a twig in its mouth.[63] Other images that commonly appeared in the magazine included the famous chapel in Cholula, Puebla, the Chichen Itza and Mitla ruins, photographs of *charros* (cowboys), and the snow-covered Popocatépetl volcano. Such images reproduced easily recognizable tropes of Mexicanness, which, while predating the Mexican Revolution, had nevertheless become increasingly common during the 1920s as part of the national cultural revival.

The magazine was also often filled with depictions and pseudo-anthropological discussions of the nation's various ethnic groups. In May of 1928, the cover of the magazine presented a young boy holding a small guitar and wearing a belt with a set of arrows, with a caption stating: "A type of Huichol indian that populates in great extensions the states of Jalisco and Nayarit. This race of indians, which is still purely preserved in the departments of Totatiche and Mezquitic, speaks the Huichola or Huichichil language." Similarly, in December 1927, the magazine published a series of pictures taken from along the Mexico City–Acapulco road, one of which featured a tourist next to an Afro-Mexican man and several children. The caption asserted: "If the natural beauties along with numerous other interesting aspects were not a sufficient motive to stop off on any spot along the road to Acapulco, we publish here these photographs to show that the aficionado in historical studies will find numerous reasons in which to put to test his erudition, since constantly he will have occasions in which to discover archaic humanities and lost cultures, ancient customs, primitive clothing, etc., etc."[64] And in another case, in March 1929, the magazine offered images of the Isthmus of Tehuantepec. A caption appearing next to pictures of a group of local women stated: "The traveler interested in typical spectacles and in local color will find great pleasure in the village curiosities and the customs of the local natives. In the region of the Isthmus de Tehuantepec, in addition to Spanish, Tehuantepecano, Chontal, Trique, and Suave are spoken."[65] Motoring, as these pictures and corresponding captions made clear, not only was pleasurable, but also offered the opportunity to learn about the provincial nation.

The work of *El Automóvil en México* and other automobile advocates coincided with and no doubt helped encourage a series of early, if limited, efforts by the state to build a network of roads suitable for motorists. Since at least 1917 national political leaders had discussed the importance of good roads to the overall social and economic goals of the Mexican Revolution and proposed a Ley de Caminos y Carreteras (Law of Roads and Highways), which would consider the opening of local, municipal, and national roadways to be in the public interest. By the early 1920s, a new road had been constructed between the archeological zone of Teotihuacán and Mexico City to allow farmers and artisans to travel to the city as well as to encourage tourists to visit the famed archeological site.[66] And a few years later, in 1925, President Plutarco Elías Calles established the Comisión Nacional de Caminos (National Commission of Roads), which soon began construction on highways connecting Mexico City to Puebla, Pachuca, and Guadalajara.[67]

Although the post-revolutionary state attempted to improve transportation largely in order to unify the nation for standard political and economic reasons, leaders consistently discussed how road building would do much to attract foreign tourists as well. After a visit by Charles Lindbergh to Mexico City in 1927 during his famous "good will" flight, the national government eagerly anticipated "an influx of automobile tourists" and made plans to supervise the country's major roadways.[68] By the end of the decade the government had created a diversity of commissions and agencies that aimed to ease entry requirements for foreign tourists and improve coordination between offices dealing with travel, transport, public health, customs, and immigration.[69] By the early 1940s even PEMEX, the national petroleum company, began to produce its own guides on Mexican cultural attractions.[70] Significantly, these state-sponsored initiatives employed many of the same promotional tactics that had been pioneered by *El Automóvil en México*.

During the 1920s, then, private and public interests increasingly converged in the promotion of leisurely motoring by citizens and foreigners alike, efforts that would ultimately help to produce an explosion in automobile tourism during the 1930s. Indeed, according to a study done by the Banco de México, by 1934, 58.2 percent of foreign tourists traveled to Mexico in an automobile, while in 1939 the number had swelled to 86.1 percent.[71] What had begun as a casual effort by Mexico City automobile advocates to promote the benefits of motoring excursions into the countryside had by the early 1930s become an explicit government policy and, increasingly, big business.

## Conclusion

The first three decades of the twentieth century witnessed a diversity of technological transformations in Mexico and, indeed, around the world. Of these many changes, the advent of the automobile was certainly one of the most striking, and it is no surprise that artists, writers, intellectuals, business interests, and government officials all found opportunities to muse over its impact on society. By the early 1930s—due to both the Mexican Revolution and global changes in motorized vehicle use and production—the automobile had been transformed from the classic conspicuous consumer item and example of uneven modernization that it embodied during the Porfirian period to a common means of everyday movement in Mexico City. Indeed, the nation's capital had begun to look increasingly like the fast-paced modern metropolis it would be known as during the second half of the twentieth century. Yet as automobiles proliferated, they struck many, particularly foreign observers, as contrary to the postrevolutionary goals of fostering a new national identity rooted in indigenous and rural ways. As emblems of the machine age, automobiles were often seen as antithetical to the folk cultures promoted by Mexican intellectuals, government leaders, and foreign visitors. But these concerns, as this chapter has suggested, coexisted with the growing practice of leisurely motoring in rural areas, a practice that, in fact, did much to promote the very traditional cultures that opponents feared were being displaced by the spread of the machine age. By the late 1930s, Mexico had become a popular destination for the automobile tourist; and this dual vision of the country, which presented the nation's capital as a modern cosmopolitan center and the rural world as a delightful sanctuary from industrialism, had been established. This dual narrative would remain firmly in place throughout much of the middle decades of the twentieth century and would come to constitute a central feature of modern Mexican identity.

## Notes

1. Salvador Novo, "El Joven," in *Toda la prosa* (Mexico City: Empresas Editoriales, 1964), 538. There is virtually no work on automobile use in Mexico written in English and only general work in Spanish. See, for example, Héctor Manuel Romero, *Historia del transporte en la Ciudad de México: De la trajinera al metro* (Mexico City: Secretaría General de Desarrollo Social, 1987); Fanny Del Río and Carlos Vargas, *El Autotransporte* (Mexico City: Secretaría de Comunicaciones y Transportes, 1988).

2. There is a vast amount of literature on the construction, or forging, of an authentically Mexican identity during the post-revolutionary era. See, for example, Mary K. Vaughan and Stephen E. Lewis, *The Eagle and the Virgin: Nation and Cultural Revolution in Mexico, 1920–1940* (Durham, NC: Duke University Press, 2006).

3. For a discussion of what Eric Zolov terms a "cosmopolitan-folklórico discourse," see his "Discovering a Land 'Mysterious and Obvious': The Renarrativizing of Postrevolutionary Mexico," in *Fragments of a Golden Age: The Politics of Culture in Mexico since 1940*, ed. Gilbert M. Joseph, Anne Rubenstein, and Eric Zolov (Durham, NC: Duke University Press, 2001), 234–72.

4. See, for example, Steven B. Bunker, "'Consumers of Good Taste': Market-

ing Modernity in Northern Mexico, 1890–1910," *Mexican Studies/Estudios Mexicanos* 13, no. 2 (Summer 1997): 227–69. See also William H. Beezley, *Judas at the Jockey Club and Other Episodes of Porfirian Mexico* (Lincoln: University of Nebraska Press, 1987).

    5. "An Automobile," *The Mexican Herald*, July 1, 1898; Nathaniel J. Mason, "The City of Mexico," *Overland Monthly and Out West Magazine*, November 1906; "Nueva Mesa del Automóvil Club," *El Tiempo*, April 4, 1909.

    6. "Cuatro distinguidos sportmen efectuan una penosa jira automovilística," *El Diario*, October 9, 1910; "Cinco jovenes mexicanos realizan una brillante expedición en automóvil," *El Diario*, October 31, 1910.

    7. *El Universal*, January 22, 1895; "El coche del diablo," *El Universal*, January 16, 1895; "First Motor Carriage Built in Mexico," *The Horseless Age*, August 1897.

    8. Del Río and Vargas, *El Autotransporte*, 32–33.

    9. "El 'Chalmers Detroit,'" *El Imparcial*, January 11, 1910.

    10. "Packard '30' México a Guadalajara," *El Imparcial*, March 23, 1910.

    11. *El Diario*, August 7, 1910.

    12. *El Diario*, June 20, 1910.

    13. El Aguila petroleum company used similar strategies in marketing its "Naftolina" gasoline to motorists. *El Diario*, April 27, 1910; *El Diario*, May 6, 1910.

    14. Department of Commerce and Labor, Bureau of Manufactures, *Monthly Consular and Trade Reports, March 1908, No. 330* (Washington, DC: Government Printing Office, 1908), 20.

    15. *Dun's Review*, International Edition, September 1911, 49.

    16. Beezley, *Judas at the Jockey Club*, 44.

    17. For a discussion of conflicts over the use of the street during these years, see Pablo Piccato, *City of Suspects: Crime in Mexico City, 1900–1931* (Durham: Duke University Press, 2001), 24–26.

    18. "México antiguo y la ciudad moderna," *El Universal*, February 2, 1900.

    19. *El Diario*, August 5, 1907.

    20. Salvador Quevedo y Zubieta, quoted in *El Diario*, June 3, 1907.

    21. *Diario del Hogar*, March 14, 1909.

    22. Modernismo was a Latin American literary movement that lasted from the 1880s to around the early 1920s. For an introduction to modernismo, see Bart L. Lewis, "Modernism," in *Mexican Literature: A History*, ed. David William Foster (Austin: University of Texas Press, 1994).

    23. Amado Nervo, "El automóvil de la muerte," in *Obras completas: Tomo II* (Madrid: Aguilar, 1951).

    24. Rubén Lozano Herrera, *Las veras y las burlas de José Juan Tablada* (Mexico City: Universidad Iberoamericana, 1995), 96.

    25. Eduardo Chirinos Arrieta, "Oliendo las flatulencias de carburo: Poética de

la desconfianza en un poema de José Juan Tablada," in *Nueve miradas sin dueño: Ensayos sobre la modernidad y sus representaciones en la poesía hispanoamericana y española* (Lima: Fondo Editorial PUCP, 2004), 37.

26. José Vasconcelos, *Ulises Criollo* (Ediciones Botas, 1935), 499; Pablo Piccato, "Urbanistas, Ambulants, and Mendigos: The Dispute for Urban Space in Mexico City, 1890–1930," in *Reconstructing Criminality in Latin America*, ed. Carlos Aguirre and Robert Buffington (Wilmington, DE: Scholarly Resources, 2000), 113–48.

27. James J. Fink, *The Automobile Age* (Cambridge: MIT Press, 1990), 37.

28. "Mexico, a Land of Used Automobiles," *Los Angeles Times*, June 28, 1918.

29. "Poderosos autos cruzarán los campos de batalla sembrando a su paso la desolación y la muerte," *El Independiente*, March 6, 1914; "Motor Cars Useful in Mexico," *Washington Post*, February 1, 1914; "Autos Aid to Huerta," *Washington Post*, May 31, 1914; "Thousands for Bearing," *Los Angeles Times*, January 30, 1916.

30. "Ridiculeces de los Rebeldes," *La Iberia*, January 20, 1911. This case predates the 1920 car bombing in Manhattan that Mike Davis suggests was the first in world history. See Mike Davis, *Buda's Wagon: A Brief History of the Car Bomb* (London: Verso, 2007).

31. "Zapata Yields to Madero," *Washington Post*, June 21, 1911.

32. *The Automobile*, December 16, 1915, n.p.; *Motor West*, December 1, 1919, 23.

33. "Nueve personas fueron muertas y otras varias heridas por un automóvil, en las carreras de ayer, en la Condesa," *El Democrata*, April 9, 1917; "Auto Racing Gains Favor in Mexico," *Washington Post*, September 16, 1917. For an example of such criticism, see Thomas Edward Gibbon, *Mexico under Carranza: A Lawyer's Indictment of the Crowning Infamy of Four Hundred Years of Misrule* (New York: Doubleday, 1919), 33.

34. P. Harvey Middleton, *Industrial Mexico: 1919 Facts and Figures* (New York: Dodd, Mead and Company, 1919), 134.

35. Ibid., 23.

36. Moisés T. de la Peña, *El servicio de autobuses en el Distrito Federal* (Mexico City: Departamento del Distrito Federal, 1943), 13.

37. Ibid., 14.

38. Ibid., 14–15.

39. Ibid., 16–17.

40. Ibid., 17.

41. Ibid.

42. Ibid., 18.

43. Ibid., 35; "Taxis Commandeered," *Los Angeles Times*, February 3, 1924.

44. Mauricio Tenorio, "The Cosmopolitan Mexican Summer, 1920–1949," *Latin American Research Review* 32, no. 3 (1997): 224–42; see also Helen Delpar, *The*

*Enormous Vogue of Things Mexican: Cultural Relations between the United States and Mexico, 1920–1935* (Tuscaloosa: University of Alabama Press, 1992).

45. D. H. Lawrence, *The Plumed Serpent* (London: Wordsworth Editions, 1995), 2.

46. Ibid., 23.

47. Stuart Chase, *Mexico: A Study of Two Americas* (New York: MacMillan, 1935), v–vi.

48. Ibid., 9.

49. Ibid., 187–88.

50. William Richardson, "Maiakovskii en México," *Historia Mexicana* 29, no. 4 (April–June 1980): 633.

51. Rubén Gallo, "John Dos Passos in Mexico," *Modernism/Modernity* 14, no. 2 (2007): 330; Edward Weston, *The Daybooks of Edward Weston*, vol. 1 (Millerton, NY: Aperture, 1973), 129–30, 134, 139.

52. Carleton Beals, *Mexican Maze* (Philadelphia: Lippincott, 1931), 356–57.

53. For most scholars, 1916 represents the birth of the Latin American avant-garde, the very year that mass automobile transportation emerged in Mexico City. Hugo J. Verani, "The Vanguardia and Its Implications," in *The Cambridge History of Latin American Literature*, vol. 2: *The Twentieth Century* (Cambridge: Cambridge University Press, 1996), 116.

54. Elissa J. Rashkin, *The Stridentism Movement in Mexico: The Avant-Garde and Cultural Change in the 1920s* (Lanham, MD: Lexington Books/Rowman and Littlefield Publishers, 2009), 28; see also Ricardo Pérez Montfort, *Expresiones populares y estereotipos culturales en México, Siglos XIX y XX: Diez Ensayos* (Mexico City: CIESAS, 2007); and Rubén Gallo, *Mexican Modernity: The Avant-Garde and the Technological Revolution* (Cambridge, MA: MIT Press, 2006).

55. Rashkin, *The Stridentism Movement in Mexico*, 53.

56. Ibid., 102.

57. Salvador Novo, *New Mexican Grandeur*, 5th ed. (Mexico City: Petróleos Mexicanos, 1967), 25.

58. Ibid.

59. Very little has been written on *El Chafirete*. One of the few studies of the magazine is Adriana González Mateos, "El fifí y su chofer: Control social, homosexualidad y clase en un periodico del México posrevolucionario," *Signos Literarios* 2 (July–December 2005), 103–25. Discussion of Novo's involvement in the newspaper can be found in Viviane Mahieux, "The Chronicler as Streetwalker: Salvador Novo and the Performance of Genre," *Hispanic Review* 76, no. 2 (2008): 155–77.

60. *El Chafirete*, March 15, 1923.

61. *El Automóvil en México*, September 1921, 20.

62. Ibid., October 1923, 9.

63. Ibid., May 1925.
64. Ibid., May 1928 and December 1927.
65. Ibid., March 1929, 19.
66. Manuel Gamio, *Introducción, síntesis y conclusiones de la obra La población del valle de Teotihuacán* (Mexico City: Dirección de Talleres Gráficos, 1922), 89.
67. Ovidio González Gómez, "Construcción de carreteras y ordenamiento del territorio," *Revista Mexicana de Sociología* 52, no. 3 (July–September 1990): 50, 59; Wendy Waters, "Re-mapping the Nation: Road Building as State Formation in Post-revolutionary Mexico, 1925–1940" (PhD diss., University of Arizona, 1999), 56.
68. Russell Owens, "Lindbergh Impetus Decides Mexicans to Use Air Lines," *New York Times*, December 22, 1927, 1.
69. Alex Saragoza, "The Selling of Mexico: Tourism and the State, 1929–1952," in *Fragments of a Golden Age*, ed. Joseph, Rubenstein, and Zolov, 101.
70. Dina Berger, *The Development of Mexico's Tourism Industry: Pyramids by Day, Martinis by Night* (New York: Palgrave Macmillan, 2006), 104.
71. Ibid., 45–70; Banco de México, *El turismo Norteamericano en México, 1934–1940* (Mexico City: Gráfica Panamericana, 1941), 42–45.

# 13
# Railroad Culture and Mobility in Twentieth-Century Mexico

*Guillermo Guajardo and Paolo Riguzzi*
*Translated by Viviane Gomez*

> I say that the government—all the governments, since the railroads were nationalized—is the enemy of the railroads because it has never seen in them a means of worthy, useful, and necessary transportation showing great promise, but rather a bothersome relic from Mexico's earliest struggles to convert itself into an economically autonomous country. They are glorious, but they are of no use. It's as if a country had inherited, as a national treasure, Sarah Bernhardt at eighty years of age, with just one leg and determined to play the role of Hamlet.
> —Jorge Ibargüengoitia, *Instrucciones para vivir en México*

The purpose of this chapter is to analyze the relationship that came about in Mexico during the twentieth century between the railroad, on the one hand, and the organizational cultures that gravitated around it, on the other, including corporate-managerial, worker-union, and administrative-governmental cultures, the last of which concerns the state's role as owner and provider of a public service. Departing from the definition of (economic) culture as "a collection of attitudes, beliefs, and values that have a bearing in the performance of individuals and organizations," we are interested in mapping the contact zones between railroad cultures, as well as the knowledge they shared, especially with regard to the National Railroad of Mexico or Ferrocarriles Nacionales de México (or simply "Ferronales"), the state-owned company that between 1908 and 1996, when it was privatized, was a practical monopoly that dominated railroad service in the country.[1] In fact, the only important railroad lines other than Ferronales during these years included the Southern Pacific Railroad of Mexico, a subsidiary of the Southern Pacific Company, the British Mexican Railway (Mexico City–Veracruz), and United Railways of Yucatan. Through an examination of Ferronales, it is possible to reveal the manner in which technology (means of transportation), infrastructure (the tracks), organization (the company), public property, and ideas interacted with each other during the twentieth century. Although our analysis concludes with the privatization of the railroads, we shall focus more attention on the period up to the end of the 1950s, which coincided with the life cycle of the steam locomotive and during which the essential cultural characteristics of the railroad took shape.

The chapter is divided into five parts, each corresponding to one stage in the development of the railroad system, and concludes with an examination of the liquidation of the state-owned company. The objective will be to reconstruct and evaluate the changes in Ferronales's organizational culture and to explain the impact these changes had on the culture of mobility in Mexico. Our argument is that railroad transportation as a public service experienced a substantial decline, which forged an early and lasting image of obsolescence and a reality of deterioration and loss of capacity, which encouraged the culture of mobility in the country to gradually but drastically separate itself from the railroad. What remained was in large part an "invented" tradition that immortalized the railroad as an artifact of the Mexican Revolution. No less important, one must not underestimate how lasting and consistent the association has been between railroad union leaders and corruption, enrichment, and political manipulation, which since 1948 came to be embodied in the term *charrismo*.[2]

## The Railroad to 1910: Tracks, Clocks, and Stations

The first railway concession in Mexico was granted in 1837, but a combination of various institutional, technological, and political problems delayed construction until 1873, when the first railway line uniting Mexico City with the port of Veracruz was finally inaugurated. Nevertheless, political instability delayed the impact of railroads until the extended authoritarian regime of General Porfirio Díaz (1876–1880, 1884–1911), during which a large number of trunk lines were constructed and the Mexican railroad system began to take shape.[3] In 1910, Mexican rail transport had grown to approximately nineteen thousand kilometers and included more than two thousand stations, which made it the second largest railway system in Latin America, after Argentina.

Despite its considerable expansion, the railroad remained rather weak in terms of density, reaching a maximum level of one kilometer of tracks per one hundred square kilometers. This weakness expressed itself in geographic coverage, and at the beginning of the 1930s, when Frank Tannenbaum studied 3,611 rural towns that contained 17 percent of the entire population, he confirmed that 93.1 percent of the towns had no access to the railroad, and 95.8 percent had no telegraphs.[4] Despite such limited coverage, the layout of railway lines in the years before the Mexican Revolution would remain little changed during much of the twentieth century. Until the 1940s the system benefited from no significant new construction.

By the first decade of the twentieth century the railroad cultures alluded to above had developed a much more stable relationship as compared to the initial period in the 1880s. On the one hand, Ferronales's corporate-managerial culture had inherited organizational features of US corporations located in Mexico. US railroad unionism partially influenced Mexican work culture—by way of organizations created by US employees—and began to be superimposed onto the mutualist organizations typical of the Mexican context.[5] With regard to the company's administrative culture, a general regulatory framework emerged via the Railroad Law of 1899 and the creation of an Audit Committee for Fares, as well as through the formation of Ferronales as a privately managed but publicly owned entity.[6]

What impact did the aforementioned measures have on the culture of mobility? Toward the end of the nineteenth century the railroad had lost its mesmerizing appeal, its novelty and sense of wonder, which had been captured by the young intellectual Toribio Esquivel Obregón in his description of a field trip his school took to the León, Guanajuato, railroad station in 1882. Popular sentiment, which fell somewhere between wonder and fear, was expressed in a *corrido* (narrative folk song) composed in honor of the railroad's arrival in Zacatecas (1884): "from far is heard the hum / of a foreign machine . . . very large and very fast / we see that it comes running / and throughout the whole world / we see that it sweeps by."[7] Meanwhile, the train, the tracks, and the station had become a familiar sight for much of the Mexican population, and it is thus no surprise that the first known sketch by artist Diego Rivera, when he was just four years old, was of a train he saw at the Marfil, Guanajuato, station. What did persist was the typically nineteenth-century idea of a parallel between the railroad and civilization, which associated the linearity of the tracks with civilization's advancement.[8]

In order to establish the railroad's relationship with the culture of mobility prior to the revolution, it is helpful to distinguish between its "infrastructure," on the one hand (including time, space, language, and written culture), and the extent and rate of mobility, on the other. With regard to the former, it is essential to take into account the manner in which the railroad system became instrumental in establishing and spreading a system of national time. This was not merely the result of a technical requirement, the movement of trains, but rather the obligation, as established by federal railroad regulations issued in 1883, to place a clock on every station's exterior.[9] At the same time, the railways inspired a great number of written pieces and images.[10] Although these publications provided information essentially for the middle classes, the numerous corridos about railroads, as well as the spread of *exvotos* (votive offerings) in shrines as demonstrations of gratitude for keeping one safe from accidents, suggest the significant impact of the railroad on popular culture.[11]

With regard to the most fundamental aspect of mobility, the growth of passenger traffic, between 1880 and 1910 average annual ridership expanded at the very high rate of 10.3 percent, thereby generating a considerable increase in individual mobility. While in Mexico in 1880 there was one railroad passenger for every twelve inhabitants, in 1900 there were 1.3 for every one inhabitant, and in 1910, one passenger for every one inhabitant.[12] Meanwhile, railroads allowed hundreds of thousands of people to travel to festivals, participate in religious processions, and visit carnivals.[13] Furthermore, indigenous people, when they had access to adequate transport, took advantage of it, as demonstrated by the fascinating case of Zapotecan women from the Isthmus of Tehuantepec who became travel agents and established new commercial and social networks thanks to the advent of the railroad.[14]

In addition to the movement of people, the railroad expanded a variety of services. The postal system represented one of the principal applications of the new transport system to a service, since the Mexican government required railroads to carry mail free of charge, thus giving rise to an enormous increase in postal traffic. While more than 5.6 million pieces of correspondence were delivered in 1880, equivalent to 0.6 pieces per capita, more than 188 million, or 12 pieces per capita, were delivered in 1911. Together with the new mobility of cargo and people, the mobility of ideas and words likewise played a significant role in shaping national integration and producing a new culture of long-distance communication.[15]

The cultural changes introduced through new ways of moving nevertheless faced a great obstacle in the form of illiteracy. The administration of Porfirio Díaz was largely unsuccessful with regard to public education, and in 1895, 82 percent of the population older than age ten was illiterate, a figure that dropped to 61 percent in 1930, but which was still high and similar to underdeveloped nations (the Federal District represented an exception, and in 1910 literacy stood at 40 percent). These numbers reveal that at the start of the twentieth century, the level of education in Mexico was close to that of czarist Russia and of the Balkans, where until 1914 education levels were little better than those of medieval times.[16]

The Ferronales workforce, so crucial to the Mexican economy, formed without government support, and until 1910 unions assumed the responsibility for nationalizing job posts. In 1890, a labor movement began to take shape, which pushed for Mexicans to be hired for all positions. Meanwhile, the Gran Liga de

Empleados de Ferrocarril (Great League of Railroad Workers) along with the Unión de Mecánicos (Mechanics' Union) established a technical school between 1904 and 1908.[17] Upon creation of Ferronales, the company's Mexican administrators endorsed the establishment in 1910 of a Department of Instruction as well as other technical schools. In the aftermath of a 1912 Ferronales strike by US employees, which resulted in their departure from the country after their demands were not met, admittance rates increased.[18] The departure of US workers was further solidified in 1914 as a result of high levels of violence and instability associated with the revolution as well as the US invasion of Veracruz. While 1,075 foreigners had occupied positions in Ferronales in 1909 out of a total of 26,106 employees (4.1 percent), by 1917 there were only 179 (0.5 percent) out of a total of 32,796 employees.[19]

### Did the Railroad Win the Revolution?

To evaluate the impact of the Mexican Revolution on the railroad in historical rather than mythological terms, it is necessary to outline the different stages that Mexico traversed during the 1910s. From this point of view, it is quite problematic to embrace the idea that "the railroads won the revolution," an idea that has been reproduced so often that it has become part of official history.[20] The railway lines were, in fact, both a target and a military instrument for the various factions, including reformists, reactionaries, revolutionaries, and counter-revolutionaries. The intensity of these attacks inspired a curious and naive form of compassionate discourse, which anthropomorphized the train and then lamented its subsequent victimization.[21] Capturing the conflict over the railways, in the celebrated novel by Martín Luis Guzmán, *Memorias de Pancho Villa,* the movement of trains between the government and the rebels fragments into a confrontation between a series of factions that include "military trains," "Villa's trains," "hospital trains," and "the new enemy's trains."[22]

The revolts and unrest of the 1910s led Ferronales to lose considerable organizational control. Job posts and hierarchies were altered, while various revolutionary factions put an end to the technical schools and paralyzed the Department of Instruction. In 1916, the president of Ferronales admitted that a considerable part of its personnel had joined as a result of the Mexican Revolution and were in large part unqualified and controlled by the military.[23]

In order to reach a solution to such challenges, the central government, headed by General Venustiano Carranza, sought to oppose the "derechos de carabina" (gun rights), or worker rights obtained via force and by personnel who had joined under the protection of military leaders. But conflicts persisted, and at the beginning of the 1920s the railway workers who had pushed for the nationalization of their posts still exhibited low levels of technical ability, and their merits remained largely political in nature.[24]

## Consolidation of a Culture of Low Speed and Productivity

Railroad workers, who represented the first organized sector of the labor movement and the most important one during the better part of the twentieth century, obtained great benefits amid the confusion and chaos caused by the revolution. As the company grew weaker, they grew stronger. Socialist Rafael Nieto, a protagonist in the revolution, described this phenomenon in the mid-1920s in the following terms: the railroads were "one of the strongest and coherent parts of the Mexican proletariat," even as Ferronales was suffering from an "excess of personnel and a partial demoralization, deterioration of equipment, unorganized finances, military interference, and bureaucratic influence."[25] With the end of the armed conflict, the railroad union encountered favorable conditions.[26] Between 1912 and the mid-1920s Ferronales's workforce increased from 31,003 to more than 43,000, eventually reaching 47,000 employees. At the same time, the company's pay scale showed severe distortions, as evidenced by comparison with the average salaries of first-class railway workers in the United States. Among conductors, dispatchers, and engineers, the salaries at Ferronales were superior to their US counterparts by a range of 12 to 30 percent.[27]

Although the revolution has often been viewed with regard to its favorable impact on labor, it also played a part in the formation of a worker-union culture that distanced itself from formal systems of technical education as well as systems of merit. Faced with a lack of personnel formally trained in technical schools, post-revolutionary governments sought to design and institute programs aimed at reinvigorating railroad productivity.

To deal with the deficit of technical capacity within the company, the secretary of public education pushed for the creation of the School for Railway Workers in 1922 in order to produce workers in three years who would occupy various positions, from stoker to engineer. The school, located in Mexico City, began to be constructed in early 1922, but by the end of that same year, construction had been halted. Work was picked up again in 1923, but rather than serve as the railroad school, the building would house the Escuela de Artes y Oficios (School of Arts and Crafts). Meanwhile, the 1923 de la Huerta rebellion attracted a large number of railroad workers and thus obligated the government to establish a pact with this group of laborers in order to pacify them and consolidate their political support, thus placing political criteria above technical and economic efficacy.

In 1923, the Department of Railroad Instruction, which had been created during the movement to nationalize personnel in 1909 and 1910, was eliminated. The company's director shut down the department in May due to his judgment that it had failed to obtain satisfactory results, as well as due to dismal attendance.[28] Although the government and the railroad unions sought to create spaces for technical education, the merit system that existed prior to 1911 was becoming progressively marginalized as a work culture of low productivity based in unsystematic or frequently politicized methods of training was further institutionalized.

Yet productivity improved among only a very reduced number of laborers, and not among personnel who had been hired by the company during the armed conflict—personnel who continued to represent a powerful political base.[29] An important aspect of organizational deterioration and low productivity lay in union practices imposed on Ferronales in the contract of 1930, which included (a) an exclusion clause, which not only entailed forced unionization, but also granted the union the right to terminate the employment of disloyal workers; (b) the favoring of union workers for executive positions within the company; and (c) seniority as the almost exclusive criteria for advancement within the company. Together, these guidelines stifled the corporation's autonomy and limited in a decisive manner the merit system. Such requirements were present in no other labor legislation of the time.[30]

Meanwhile, another element present in the international railroad industry disappeared for Mexican railway workers: the professional affiliations of executives, technicians, and civil servants.[31] What took the place of the original organizational culture were elements that could not generate a new and stable culture because they emphasized political-governmental loyalty, on the one hand, and union solidarity, on the other.

Together, these elements had a significant impact on the culture of mobility, expressed, primarily, in the very evident decrease (by nearly half) in train speed, as compared to the period prior to the revolution. By 1930, average cargo train speeds were 22.6 kilometers per hour. In 1940, they increased to 23.4 kilometers per hour and then dropped to 18.0 kilometers per hour by 1950. And the trains' efficacy, measured in terms of the percentage of on-time arrivals, fell from 79 percent in 1934 to about 50 percent a decade later.[32] In reality, technical indicators obscure the actual proportions of the delays, which novels and testimonials describe. During the two or three days of waiting in the Culiacán station captured by Martín Luis Guzmán in *El Águila y la serpiente*, "the trains ... ran more irregularly than they would have if they moved by sail or were dependent on the weather." Commenting on the night that he spent on a stalled train in Monterrey in 1924, Alfonso Reyes wrote, "The next day, amazed at how well I had slept, I was surprised to still see outside my window the Silla mountain range!"[33] Meanwhile, due to the fact that development of modern roadways and automotive transport in Mexican was still in its initial stages, railroads continued to benefit from a large and captive public whose only option was to complain or simply accept an unstable service.[34]

## Nationalistic Expansion and the Rise of Diesel

Beginning in the 1930s, as the economy modernized, the railroad industry lost ground in the context of erratic policies and uncertain experiments. The expropriation of Ferronales in 1937 under President Lázaro Cárdenas was essentially a symbolic act that ended the presence of foreign shareholders, who for years had had virtually no influence in the company's affairs. Expropriation opened the way for the catastrophic Workers' Administration (1938 to 1940), in which the union exercised the role of general manager of the company and labor administrator, thereby increasing the rate of deterioration and bringing about a wave of accidents. As a result, after 1941 the corporation took a step backward and Ferronales became a decentralized public corporation that finally managed in 1948 to regularize its legal situation by way of new legislation. Nevertheless, it was no match for the powerful union, which remained firmly in control: about 90 percent of all executive positions were filled by union members, who, moreover, occupied three out of the six posts on the board of directors.[35]

At the start of the Second World War, the railroad became an essential means of transport for materials needed by the US military, as well as for hundreds of thousands of Mexican laborers who replaced US workers. In 1942, when Mexico entered the Second World War, a bilateral commission dubbed the United States Railway Mission in Mexico was created in an effort to rehabilitate the railways.[36] Mexico's strategic alliance with the United States and the introduction of diesel locomotives changed the state of affairs; and in 1946, with the endorsement of the United States Railway Mission, Ferronales opened two Escuelas Técnico Ferrocarrileras (Railroad Technical Schools) run by the Department of Instruction, one in the capital and the other in Aguascalientes.[37] In attempting to meet goals similar to those proposed in previous decades, a contract with the Union of Railroad Workers of the Mexican Republic (STFRM) was signed in December 1947 in order that personnel would be trained in the running and maintenance of the new diesel locomotives. In turn, the general management of Ferronales proposed to the National Polytechnic Institute that they join forces to open a railroad technical school, which would adjust course material to fit with the official technical education curriculum.[38] In 1957, the Department of Instruction was replaced with the Instituto de Capacitación Ferrocarrilera (Institute of Railroad Training); and in that same decade, as part of the shift to diesel, the Escuela Diesel (Diesel School) opened.

Yet given the dominant work culture in the company, those trained in the technical schools continued to be viewed with suspicion. Furthermore, the peculiarities in the collective contract encouraged hiring practices that favored personnel who could verify that a family member had also worked on the railroad, thus substantially reducing the impact of technical schools.

The changes that took place in 1950 significantly influenced the distinct railroad cultures whose imbalance and fragmentation were quite evident, despite nationalization. Ferronales's corporate-managerial culture had been transformed into an agency that provided transportation services but did not significantly increase the distances it covered even as its services deteriorated. Meanwhile, the company's work culture had been transformed into a nepotistic form of mutualism with limited technical ability, strongly nationalistic and xenophobic but, paradoxically, devoid of a sense of public service. In terms of mobility, the railway system went into decline as it faced competition from an expanding network of roadways and auto transportation. It is significant that these irregularities existed not because of a lack of resources, but rather *despite* huge resources, which were often cannibalized, so much so that until the 1950s the railroads were the main destination for public spending on transportation and communications.[39]

It is in this context that the political activity and mobilization of railroad workers in 1958 and 1959 should be understood, marked by an internal power struggle in its first year and by a salary dispute with the state in the second. In particular, the 1959 Ferronales strike, which ended in defeat for the labor movement and the firing of nine thousand employees, demonstrated the existence of a union culture burdened by a nationalistic ideology and an emotive combativeness. Meanwhile, the conflict demonstrated the absolute lack of a strategic plan among union leaders with regard to the economic, technical, and national reality of the railroads. There was a constant demand to increase salaries, as labor leaders argued that Ferronales's revenue had increased, without taking into account cost structures or even understanding that the corporation was in the red. Even the leader of the strike, Demetrio Vallejo, was unaware of the company's financial data, which he needed in order to effectively negotiate salaries, a fact he discussed during an interview in 1958: "We don't exactly know the company's financial situation, because it's never been made public."[40] This claim was made despite the fact that official reports on Ferronales, like yearly presidential reports, were publicly circulated.

The strike also revealed a diversity of clashes over the introduction of diesel, which since 1950 had brought about a variety of changes and rendered much of the old steam-engine equipment, workshop infrastructure, and job posts obsolete. Further aggravating the situation, retirement packages were suspended, thus leading the average age of personnel to reach forty during an era in which the average life expectancy of Mexican men was forty-seven years.[41] In this sense, the 1959 strike accelerated the elimination of the steam engine and forced railroad shops to modernize, even as union structures slowed down the process since control of the number of posts was linked not only to political power but also to the financial viability of STFRM, which received from Ferronales the union dues paid by the workers.

## Deterioration of a Public Service, Privatization, and Preservation of Memory

The deterioration of railroad services, which continued into the 1960s, '70s, and '80s, was described by Jorge Ibargüengoitia in the following manner: "The government rehabilitates the railroad every once in a while. It invests money in them not with the hope of actually fixing them, but rather in preventing them from dying altogether," it being "an impossible and unpleasant task . . . to preserve the life of something that is dying without going to the extreme of seeing it completely recuperated."[42]

In 1970, railroads carried 22.8 percent of cargo (in tons) and 8.0 percent of all passengers. By 1985, figures had dropped to 17.9 percent and 1.4 percent, respectively. Nevertheless, railroad personnel kept increasing at a faster rate than traffic, such that after an increase of almost 15 percent between 1968 and 1970, the total number of employees reached 92,338. The final attempt to establish a solid foundation rooted in technical efficiency, a modern business culture, and administrative improvement took place between 1970 and 1973 under the leadership of Víctor Manuel Villaseñor. An industrial organizer with a Marxist background, Villaseñor was, however, the subject of a very violent attack by the union, headed by Luis Gómez Z., which included slanderous accusations ("lackey of the imperialist Yankee"), intentional traffic slowdowns, and, in all likelihood, acts of sabotage. The October 1972 tragedy at Saltillo—the worst accident in the railroad history of Mexico—announced the end of the experiment in rationalization.[43]

Nationalization of the railroad system had produced such questionable results that in 1996 the administration of Ernesto Zedillo initiated the privatization of Ferronales, dividing it into different regional systems that would be sold to franchisees via public auction, although STFRM would maintain its collective contract with the new companies. The decision did not generate disputes or significant controversy in a society that no longer viewed railroads as a noteworthy public service. Even the union accepted the government's approach, as demonstrated by the reaction of its secretary general, who, when the maintenance shops were privatized, claimed in 1995, "Thanks to privatization, the future will be bright." The most important issue for the unions remained control of job posts: "In foreign companies there is machinery and cleanliness. Workers receive uniforms and cafeterías. The change is notable. The negative aspect of privatization is that they want to bring in trusted hands, and we aren't going to allow it."[44]

Privatization also meant the total elimination of regular passenger service, thereby ending entirely the traditional bond between the railway station and the population. Today, in many cities the cargo train is a nocturnal, occasional phenomenon that is generally perceived as obsolete and bothersome since operation often blocks automobile traffic. For others the railroad is the object of looting, and rails are often removed along abandoned lines or those with very little traffic. In other cases, trains are held up as they pass through urban areas, a fact that has led companies to assemble small armies of security guards.[45]

Finally, the ample and ossifying railway infrastructure has become part of the patrimony of museums. This initially took place a year after the nationalization of the railroad with the objective of spreading knowledge and preserving the memory of the material culture of the railroad, including old stations, machinery, equipment, and even tools. In 1988, the National Museum of Mexican Railroads opened in the city of Puebla, located on what was once the site of two old stations that had been completely abandoned and had been declared historical landmarks by the National Institute of Anthropology and History. Currently, the museum belongs to the National Council of Culture and Arts and has since acquired the status of National Center for the Preservation of the Cultural Patrimony of Railroads.[46]

## Conclusion

Throughout the twentieth century, railroads experienced a global decline in the economic and social, and thus cultural, life of nations. From this point of view, Mexico is no exception. Yet two aspects highlight the distinctiveness of the Mexican experience. The first is the intensity with which railroad technology went into decline. As has been shown, by the 1980s the railroad counted for little more than 1 percent of passenger traffic. The second has to do with a disequilibrium that came to characterize the relationship between the railroad's governmental, work, and business cultures, which began to develop early in the twentieth century. As the company increasingly came to be viewed as political loot, a space for clients, and a base for political support, its function as a provider of a public service declined in importance. Meanwhile, the work culture within the company increasingly rejected the importance of technical training as well as the merit system and instead opted for a system based in seniority and resistance to technical change. Ultimately, the railroad as a major feature within Mexico's culture of mobility virtually disappeared and subjected Mexican passengers to prolonged punishment.

## Notes

1. Michael Porter, "Attitudes, Values, Beliefs and the Microeconomics of Prosperity," in *Culture Matters: How Values Shape Human Progress*, ed. Lawrence E. Harrison and Samuel P. Huntington (New York: Basic Books, 2000), 14.

2. Kevin Middlebrook, *The Paradox of Revolution: Labor, the State, and Authoritarianism in Mexico* (Baltimore, MD: John Hopkins Press, 1995), 138–45.

3. Paolo Riguzzi, "Los caminos del atraso: Tecnología, instituciones e inversión en los ferrocarriles mexicanos, 1850–1900," in *Ferrocarriles y vida económica en México (1850–1950): Del surgimiento tardío al decaimiento precoz*, ed. Sandra Kuntz and Paolo Riguzzi (Mexico City: Colegio Mexiquense–Ferrocarriles Nacionales–UAM, 1996).

4. James Wilkie, *La Revolución mexicana: Gasto federal y cambio social* (Mexico City: Fondo de Cultura Económica, 1987), 249–50, cuadro IX-3.

5. Sandra Kuntz and Paolo Riguzzi, "La gran empresa de cabeza: Ferrocarriles Nacionales de México, 1908–1937," in *Empresas y grupos empresariales en América Latina, España y Portugal*, ed. Mario Cerutti and Javier Vidal (Monterrey: Universidad Autónoma de Nuevo León, 2003), 115–21; Lorena Parlee, "The Impact of United States Railroad Unions on Organized Labor and Government Policy in Mexico (1880–1911)," *Hispanic American Historical Review* 64, no. 3 (1984): 461–75.

6. Arturo Grunstein, "Surgimiento de los Ferrocarriles Nacionales de México (1900–1913): ¿Era inevitable la consolidación monopólica?," in *Historia de las grandes empresas en México, 1850–1930*, ed. Carlos Marichal and Mario Cerutti (Mexico City: Fondo de Cultura Económica, 1997), 78–86.

7. "The monster was resting, with the regular breathing of a perfectly healthy giant, immobile it extended its reach with the majesty of a lion at rest. We approach it in reverent silence, so surprised, we are speechless." Toribio Esquivel Obregón, "La llegada del ferrocarril," in *Recordatorios públicos y privados* (Mexico City: Universidad Iberoamericana, 1994), 276–77.

8. See Ian Carter, *Railways in Britain: The Epitome of Modernity* (Manchester: Manchester University Press, 2001), 7–19.

9. It is not clear whether clocks were really installed on the facades of the nation's two thousand plus stations. See Sergio Ortiz Hernán, "De estaciones, trenes y paisajes," in *De las Estaciones* (Mexico City: Secretaría de Comunicaciones y Transportes, Ferrocarriles Nacionales de México, y Museo Nacional de los Ferrocarriles Mexicanos, 1995), 40.

10. Regarding railroad photographer William H. Jackson, see Tania Gámez de León, "William Henry Jackson en México: Forjador de imágenes de una nación (1880–1907)," *Contratexto* 4, no. 5 (2007): 73–92, http://www.ulima.edu.pe/revistas/contratexto/pdf/04.pdf (accessed May 31, 2010).

11. Guadalupe Zárate Miguel, "El Ferrocarril y la modernidad en la cultura popular: El caso de los exvotos pintados del santuario de la Virgen de los Dolores de Soriano," in *Memoria del IV Encuentro de investigadores del ferrocarril* (Mexico City: Conaculta, 2000), 9–16.

12. Secretaría de Comunicaciones y Obras públicas, *Reseña de los ferrocarriles de jurisdicción federal* (Mexico City: Tipografía de la Dirección de Telégrafos Federales, 1911).

13. See Teresa Van Hoy, *A Social History of Mexico's Railroads: Peons, Prisoners, and Priests* (New York: Rowman and Littlefield, 2008), chapter 5.

14. Marcela Coronado Malagón, "Las viajeras zapotecas del Istmo de Tehuantepec y el ferrocarril," *Cuadernos del Sur* 14, no. 27 (2009): 60–67. We thank Daniela Traffano for bringing this source to our attention.

15. Alicia Gojman de Backal, *Historia del correo en México* (Mexico City: M. A. Porrúa, 2000); *Anuario Estadístico de la República Mexicana* (Mexico City: Secretaría de Fomento, 1900–1910).

16. Ivan Berend, "La indivisibilidad de los factores sociales y económicos del crecimiento económico: Un estudio metodológico," in *Historia económica: Nuevos enfoques y nuevos problemas; Comunicaciones al Séptimo Congreso Internacional de Historia Económica*, ed. Jerzy Topolski et al. (Barcelona: Editorial Crítica, 1981), 46–47.

17. Servando Alzati, *Historia de la mexicanización de los Ferrocarriles Nacionales de México* (Mexico City: Beatriz de Silva, 1946), 41, 54–57.

18. Ibid., 131–32, 175–80.

19. Ferrocarriles Nacionales de México, *Informes anuales* from June 30, 1909, to June 30, 1925.

20. Gilberto d'Estrabau, *El ferrocarril* (Mexico City: Secretaría de Comunicaciones y Transportes, 1988).

21. Victor Mata Temoltzin and Antonio Casanueva Fernández, *La economía mexicana y los ferrocarriles (1910–1920)* (Puebla: Gobierno del Estado de Puebla, 1999), 71–88; Lawrence Douglas Taylor, "Dinamiteros en acción: La conjura con-

tra el Ferrocarril de Chihuahua en 1912," *Nuestro Siglo* (2002): 87–93; Guillermo Guajardo, "Tecnología y campesinos en la Revolución Mexicana," *Mexican Studies/ Estudios Mexicanos* 15, no. 2 (1999): 291–322.

22. Jorge Ruffinelli, "Trenes revolucionarios: La mitología del tren en el imaginario de la Revolución," *Revista Mexicana de Sociología* 51, no. 2 (1989): 300. Guzmán reference is based on the the Ruffinelli work.

23. Kuntz and Riguzzi, "La gran empresa de cabeza," 122–23; Ferrocarriles Nacionales de México, *Octavo informe anual correspondiente al año social que terminó el 30 de junio de 1916* (Mexico City: American Book and Printing, 1916), 17.

24. Guillermo Guajardo, "Escuelas técnicas y derechos de carabina: Los problemas de la calificación y productividad de la mano de obra ferrocarrilera en México, 1890–1926," *Historias: Revista de la Dirección de Estudios Históricos del Instituto Nacional de Antropología e Historia, México* 37 (October 1996–March 1997): 91–106.

25. Barry Carr, *El movimiento obrero y la política en México* (Mexico City: Editorial Era, 1981), 143–44; Rafael Nieto, *Polémica laborista* (Rome: Failli, 1926), 303, 314.

26. In this manner, of the four general managers of Ferronales who held those positions between 1917 and 1925, two came from the most powerful and best organized group of railroad workers, the train dispatchers, and earned this post by virtue of being promoted over the Workers Movement. The other two had direct ties with military leaders or regional support with the president. Kuntz and Riguzzi, "La gran empresa de cabeza," 124–25.

27. Ferrocarriles Nacionales de México, *Los salarios y la empresa* (Mexico City: Editorial Cultura, 1931), 200–202. This is the publication of a study realized by a team headed by Jesús Silva Herzog, with data from 1926 to 1928.

28. Ibid.

29. See Ingrid Ebergenyi, *Primera aproximación al estudio del sindicalismo ferrocarrilero en México, 1917–1936*, vol. 49 (Mexico City: INAH, 1986); Arturo Grunstein, "Perspectivas gerenciales sobre el problema laboral de los Ferrocarriles Nacionales de México en la posrevolución 1920–1935," *TST: Revista de Historia de los Transportes, Servicios y Telecomunicaciones* 14 (2008): 42–89.

30. *Contrato de 1930 entre Ferrocarriles Nacionales y la Alianza Ferrocarrilera Mexicana* (Mexico City: Imprenta de E. Limón, 1930).

31. In 1931, the first board meeting of railroad executives since 1913 took place. This practice was discontinued in 1937. Kuntz and Riguzzi, "La gran empresa de cabeza," 136–37.

32. Administración de los Ferrocarriles Nacionales, *Datos generales sobre hechos registrados de 1930 a 1946* (Mexico City: Departamento de Estadística, 1947).

33. Guzmán and Reyes, cited in Secretaría de Comunicaciones y Transportes–

Ferrocarriles Nacionales de México, *Caminos de hierro* (Mexico City: Secretaría de Comunicaciones y Transportes–Ferrocarriles Nacionales de México, 1996), 95, 113.

34. By 1935, the extensions of oil-fueled and oil-coated routes were 1,713 and 4,313 kilometers, respectively. Estados Unidos Mexicanos, *Anuario estadístico: 1939* (Mexico City: Dirección General de Estadística, 1940), 273.

35. Sandra Kuntz and Paolo Riguzzi, "El triunfo de la política sobre la técnica: Ferrocarriles, Estado y economía en el México revolucionario, 1910–1950," in *Ferrocarriles y vida económica en México (1850–1950): Del surgimiento tardío al decaimiento precoz*, ed. Sandra Kuntz and Paolo Riguzzi (Mexico City: Colegio Mexiquense–Ferrocarriles Nacionales–UAM, 1996), 317–18; Gustavo López Pardo, *La administración obrera de los Ferrocarriles Nacionales de México* (Mexico City: UNAM-El Caballito, 1997), 140–50.

36. Guillermo Guajardo, "La tecnología de los Estados Unidos y la 'Americanización' de los ferrocarriles estatales de México y Chile, ca. 1880–1950," *TST: Revista de Historia de los Transportes, Servicios y Telecomunicaciones* 9 (2005): 110–30; Sergio Ortiz Hernán, *Los ferrocarriles de México: Una visión social y económica* (Mexico City: Ferrocarriles Nacionales de México, 1987–1988), 2:243; Guillermo Guajardo, "Aprendizajes de innovación y negocios en el petróleo y los ferrocarriles de México, 1952–1992," in *Innovación y Empresa: Estudios históricos de México, España y América Latina*, ed. Guillermo Guajardo (Mexico City: CEIICH/UNAM–Fundación Gas Natural, 2008), 203–24.

37. *Ferronales* 17, no. 2 (1946): 6–9; *Ferronales* 17, no. 8 (1946): 8.

38. Max Calvillo Velasco and Lourdes Ramírez Palacios, *Setenta años de historia del Instituto Politécnico Nacional*, vol. 1 (Mexico City: Dirección de Publicaciones del Instituto Politécnico Nacional, 2006), 289.

39. During the period spanning from 1939 to 1950, gross investment in transportation and communication allocated 50.9 percent for highway transportation, compared to 37.9 for the railroad.

40. Demetrio Vallejo, quoted in Antonio Alonso, *El movimiento ferrocarrilero en México, 1958/1959* (Mexico City: Ediciones Era, 1986), 127.

41. José Vasconcelos, "Cómo se forma un ferrocarrilero," *Ferronales* 37, no. 6 (1960): 16–17.

42. Jorge Ibargüengoitia, *Instrucciones para vivir en México* (Mexico City: Joaquín Mortiz, 1990), 233, regarding the article "¿La última curva?" published in *Excélsior* (a newspaper) in 1972.

43. Víctor Manuel Villaseñor, *Memorias de un hombre de izquierda*, vol. 2 (Mexico City: Grijalbo, 1976), 371–451. The accident in Saltillo, which provoked the derailment of a train carrying passengers on a pilgrimage, caused more than two hundred deaths and one thousand injuries.

44. Declaration of Víctor Flores Morales in the magazine *Proceso* (Mexico), October 30, 1995, 26.

45. The importance of the new railroad system is miniscule relative to that of auto transportation, which by the year 2000 employed 3.5 million people; 395,000 vehicles moved 56 percent of the country's cargo and another 452,000 vehicles carried 98 percent of all passengers. Secretaría de Comunicaciones y Transportes, *Anuario estadístico* (Mexico City: 2000, 2008), n.p.

46. For an essay on the museum's activities, see Samantha Alvarez, "Mexico's National Railways Museum," *Journal of Transport History* 26, no. 1 (2005): 112–14.

# 14
# From the Primordial Cave to Postmodern Velocity
## The Mexico City Subway

*Juan Villoro*
*Translated by Lorna Scott Fox*

To fly into Mexico City by night feels like descending into a glittering, chaotic galaxy. The city looks like a map of the stars. And yet this incandescent tide that fills up the valley cannot stop growing. Its nature is one of relentless expansion. How much further can it spread? All the arrows point downward. Despite spikes of vertical development in the business suburb of Santa Fe and along Paseo de la Reforma, the chief engineering works of modern Mexico City have been subterranean: the metro and the drainage system. The subsoil is our final frontier.

Underground constructions carry a powerful symbolic charge. In Mexico City, the concealed infrastructure of buildings always involves an accidental archeology: foundations, telephonic cables, waste pipes, all have to work around the buried Aztec city. But building downward not only implies a possible encounter with some pyramid steps, but also means coming up against a system of ideas. In his essay "Mitos mesoamericanos" (Pre-Hispanic myths), Enrique Florescano writes: "The dominant concept of Mesoamerican creation myths is the notion that the center of the earth enclosed a cave, where essential foodstuffs were stored and life was regenerated."[1] Beneath the earth lie our dead and our origins. It's not by chance that the most important indigenous legends (the tales of Quetzalcoatl or of the prodigious twins of the Popol Vuh) recount journeys to the underworld.

According to the story, the first inhabitants of the valley where the Mexican capital now stands came from a place called Chicomostoc (Place of Seven Caves). While Christian tradition gave us the custom of visiting the Seven Houses, our myth of origin refers to the same number of subterranean chambers. The tunnel builders of the twenty-first century are its heirs.

Today, underground caverns serve to confront a decisive urban challenge: that of transport. Big-city life is determined by movement. As Paul Virilio has suggested, whereas the modern city was defined by construction, that is, the dominion of space, the postmodern city is defined by journeys—in which time supersedes space. The governing principle is no longer the ordering of matter, but the control of velocity.[2]

The Aztec city of Tenochtitlan was founded on an island in Lake Texcoco. It grew along a network of canals that astounded the Spaniards with the vision of another Venice. Continued growth dried out the lake. Later still, dust and industrial fumes poisoned the air. The story of our landscape is a tale of two abolishments: of the water and of the sky.

Most cities flourish on the edge of a lake, a river, the sea. But this one destroyed the ecosystem that been its greatest asset. This was partly because of its constricted location, encircled by mountains. In order to preserve a healthy balance with the environment, it had to stay small. Rem Koolhaas has considered the "size," as in clothing, of urban centers: Mexico City, D.F. (Federal District) was an *S* by nature that swelled into an *XL*.[3] What folly produced such a mismatch? The main causes were the unequal development of the country and the habit of political concentration. Centralism was practiced by the Aztec empire, the colonial administration, and post-independence governments alike.

Apart from the sociological reasons for the capital's macrocephaly, it seems internally driven by a dynamics of expansion. In *Invisible Cities*, Italo Calvino suggests that every city obeys its own secret logic. The novel posits a conversation between the Tartar emperor Kublai Khan and the traveler Marco Polo, in which the Venetian describes imaginary cities, fantastical constructs to indulge the other's curiosity; but his inventiveness is subject to certain restrictions.

[Marco Polo spoke,] "From the number of imaginable cities we must exclude those whose elements are assembled without a connecting thread, an inner rule, a perspective, a discourse. With cities, it is as with dreams: everything imaginable can be dreamed but even the most unexpected dream is a rebus that conceals a desire or, its reverse, a fear. Cities, like dreams, are made of desires and fears, even if the thread of their discourse is secret, their rules are absurd, their perspectives deceitful, and everything conceals something else."

"I have neither desires nor fears," the Khan declared, "and my dreams are composed either by my mind or by chance."

[Marco Polo continued,] "Cities also believe they are the work of the mind or of chance, but neither the one nor the other suffices to hold up their walls. You take delight not in a city's seven or seventy wonders, but in the answer it gives to a question of yours."[4]

What is the thread that connects the discourse of Mexico City? To what question is it the answer? Its overriding impulse is always to advance, to replace the country with the city. Eco-extermination has proceeded ruthlessly: first the water, then the air, and now the strata deep beneath us. The earth was our last horizon. In its subjugation the city muddled its own name: Mexico City, Federal District, encroached upon the State of Mexico. This conurbation is known provisionally as *ciudad de México* (the city of Mexico). Until it becomes an official toponymy, the word *city* is not supposed to be capitalized (no one takes much notice, of course). At the time of writing, the Federal District's Assembly of Representatives was deliberating over the choice of an official name for the capital.

Following Marco Polo's line of thought, we may say that our capital offers an answer to two essential questions. How to keep expanding? And, how to get around in this expanding territory?

## The Challenge of Directions

The cartography of Mexico City has always combined two discourses: the consignment of objective data and narratives of use and usage. Successive city maps record a territory that is growing so fast that no one plan ever manages to show it all. In addition, since reports are often imprecise or conjectural, mapmakers have resorted to local descriptions to make sense of the space. Some built environments (Paris, Manhattan, the older parts of Puebla) extend themselves as an adventure of order. Others surrender, as Carlos Monsiváis has put it, to the "rituals of chaos."[5]

Within the sprawling immensity of the Federal District, GPS (Global Positioning System) navigation is to the driver what the sirens' song was to Ulysses: an alluring temptation that leads to madness. What happens when you activate the satellite navigation? The first problem is our street names, plentiful but indecisive. I live on one of the 178 streets named after the revolutionary general Venustiano Carranza. It's hard for the digital pilot to work it out when so many names are repeated so often. The rationality of GPS is also challenged by our lack of topographic hierarchy. In Paris, no modest side street would ever call itself the Champs Élysées. In the Federal District, however, any old alleyway may boast the name Reforma.

But the decisive factor is something else again. Basically, there is no point in knowing where Avenida de las Bombas leads, if you don't know how many street markets or marathons or cycling festivals are liable to block it on any given day. What's more, many streets have been removed from public access altogether, converted into gated communities. While this is admittedly a reaction to the rampant lack of security, it nonetheless breaches the majority's right to free and unimpeded circulation. This being the case, the capital's GPS service might usefully provide a timetable of street closures and local markets, a calendar of the many festive events that complicate our lives, and a list of the streets that have become permanently closed to traffic.

The advent of satellite navigation has had the effect of highlighting the irrationality of our approach to space. What functional principle determines our journeys? We have been combining a variety of cartographic methods for some time. In her fascinating study *Los letrados interpretan la ciudad* (The city interpreted by scholars), Marcela Dávalos looks at the representation of urban space during the eighteenth century. By that time the central grids of the Spanish city had been set down according to Renaissance cartographic conventions, but the Indian barrios were proliferating without planning or names. In an attempt to gain an overall picture of the city's development, the priest and scientist Antonio de Alzate drew up reports in which the planimetry of the center alternated with oral testimonies collected on the outskirts. Two different orders of knowledge thus informed his research: one part of the city was expressed in terms of geography, the other in terms of how people lived. Some indications were vague and relied on the interpretation of the learned Alzate: "when you reach the alley they call Vinegar Way," "opposite San Lázaro right by the gutter before the guards' bridge," "next to the Palacio pulque tavern," "the new house nearby the gallows."[6] These are some of the directions Dávalos found in the notarial archives.

In the Indian neighborhoods memory operated collectively to reproduce a kind of territorial knowledge, alien to the literate culture. In the absence of legal deeds, proprietorship was formalized in ceremonial rituals of stone throwing and tree carving. Concepts of territoriality and appropriation hinged upon usage.

It thus so happened that one part of the city was understood in terms of orderly topographic expansion, the other through spatial markers dictated by custom. These variables have persisted into the present, but with a shift of emphasis: toponymy remains a specialized, almost esoteric discipline, while the descriptive, customary appropriation of space is gaining ground irrespective of race, culture, or income. Whether in the corporate foothills of Santa Fe, around the campus of the Ciudad Universitaria, or in the featureless immensity of Chalco, it is rare to be given a proper address since everyone knows "the Oxxo store next to the diner." This goes a long way toward explaining the obsessive recurrence of patriotic street names: what counts is the narrative that distinguishes between their locations.

The word we now use in place of barrio—*colonia*—may well reflect an unconscious wish to preserve the efficient confusion of the viceroyalty. Nomenclature continues to be an abstraction dreamed up by cartographers, and space is still something one colonizes, dominating it by usage and practice.

Satellite navigation in the capital is for those who, like Father Alzate, improve on geography with lived experience. For those who know, it's not so much about pinpointing Plaza de la Conchita on a map, more about knowing that the Tamales Fair is held there on February 2.

Amid such geographic imprecision, the subway represents a weird symbolic shortcut. Travelers cannot get lost down there because they are, strictly speaking, always already lost. Underground there's no sense of the broader coordinates or the precise direction of travel. No "metronaut" uses categories like south or northwest. He or she knows the destination is a matter of connecting two stations, that of departure and that of arrival. That's it. Which makes this the single transport network in the capital that doesn't need to be narrated to be understood. On other transports—buses, collective taxis, minibuses, tramcars—the landscape is an accretion of landmarks, but these seldom consist of statues or civic monuments. Passengers are oriented by more everyday features, made familiar by use: shops, drugstores, beauty salons.

In Mexico City it is not only hard to find a place, but hard to grasp the directions that purport to help you find it. They tend to be expressed like this: "See that tree in front of the supermarket? Go right there, toward the tailor's shop." Your informer seems unable to put himself or herself in the position of a stranger to the area, who must assimilate it bit by bit; rather, he or she assumes that you know the layout and have simply forgotten the exact spot where you're headed. And what accounts for this is the urban culture studied by Father Alzate, in which maps were supplemented by narrative.

Hence, to an extent, there is the repetitive mania for naming streets after heroes. If no one cares about the name, who cares that the system is arbitrary? According to the *Guía Roji*, our most reliable street atlas, there are 252 Hidalgo Streets and 86 Hidalgo Closes, complemented by 336 Miguel Hidalgo Streets and 51 Miguel Hidalgo Closes. Lest the Father of the Nation feel short-changed, there are also 16 Privada Hidalgos and 9 Prolongación Hidalgos. To confound the hermeneutists we even have, in colonia Pueblo Santa Cruz Acalpixca, an apparently inexpugnable track that is labeled "Privada y Cerrada y Callejón Hidalgo."

John Berger has written of the old tactic whereby townspeople altered the street names in order to disorient invaders. The toponymy of Mexico City, Federal District, seems custom-made for disorienting its own inhabitants or reminding them that a street name is only a suggestion that can be ignored.

The most telling thing about the trusty *Guía Roji* is that taxi drivers don't use it. Instead, they defer to the passenger: "You tell me the way." Here is, once again, a narrative appropriation of space. The passengers are expected to describe where they are going. If they don't know the way, the taxi driver will stop to ask from time to time, not seeking road signs, which are rare, so much as usages embodied in stores, schools, or factories that will steer the driver in the right direction.

The subway altered that relationship to space. It's no accident that people ride in silence, with a sense of reverent awe toward what cannot be seen. From the psychological point of view, it's the city that is moving. The traveler is shut inside a metal box; the surroundings are static. If traveling by bus is cumulative, an addition, traveling by subway is a subtraction. The experience ranges from hypnosis to trance, via mere stupor. The territory is delocalized by this movement. Since its inauguration in 1969 the Collective Transport System has been building the story of an alternative city that both precedes and succeeds the teeming sprawl above; a sub-city where the mythical cave of the beginning unites with the postmodern god of speed.

The cartography of the subway is like a weave, the lines enmeshed like colored yarns. One thing's for certain, the only GPS that works in Mexico City is under the ground. Only there is usage fixed.

## Utopia below the Earth

A place ruled by the twin imperatives of growth and motion, whose final frontier is the underworld—the future leading into the past—will find no better site of definition than its subway.

In his essay "U-Bahn als U-Topie" (The subway as utopia), the Russian-German philosopher Boris Groys reflects upon the Moscow metro, which played the same role in the Soviet imaginary as the Collective Transport System does for Mexico City. Incapable of producing an egalitarian utopia, the Stalinist regime unleashed a massive industry of simulation. One of the most effective of such compensatory devices was the Moscow metro: a revolutionary dawn by electric light, below the earth. Alfonso Reyes translated *utopia* as "*no hay tal lugar*" (there's no such place). Designers of utopian spaces, which are nonexistent by definition, are forced to seek compromise solutions, middle grounds, heterotopias. "The appropriate strategy for constructing utopias," Groys writes, "consists in finding an uninhabited, preferably uninhabitable site in the midst of a populated area."[7] The projects of the early twentieth-century Russian "Disurbanists" failed due to their extreme unreality. Their answer to the urban jungle was an impossible outdoors lifestyle, with mobile houses and swimming pools that kept pace with the swimmer.

By contrast, the metro fulfilled many utopian requirements while remaining on, or in, solid ground. Its growth was limitless, it was completely dependent on a higher order, and it offered a controlled space, whose occupiers glimpsed fragments of reality while the whole remained obscured. "Though the subway forms part of metropolitan reality," says Groys, "it continues to exist in the realm of the fantastic; its totality can be conceived but never experienced."[8]

These features are common to all subway systems, but what set Moscow's apart was its capacity to defy time. In its tunnels, socialist-realist paintings and the austere futurism of the coaches coexisted with the palatial splendor of marbles and chandeliers, amid an architecture of no-place that mixes Islamic, Roman, and Renaissance forms.

Cinematic utopias usually conceive of time as linear, so that their settings are too pristine to be convincing. True utopia would lie outside time as we know it. This is the great merit of the movie *Brazil*, with its scruffy version of the future; the typewriters and clothes seem older than our own, which lends them an eerie plausibility: the future seems more credible when it's worn out.

By mixing periods and styles, the Moscow subway reinforced its utopian condition. And yet this was a utopia closer to Orwell's fantasy than to that of Thomas More, in that its efficient spatiotemporal otherness was repressive. To return to Groys: "The masses do not seem to enjoy the luxury afforded them by the metro. They are unwilling and unable to savor its art, to appreciate its elegant finishings, to decipher its ideological symbolism. They are deaf, blind and indifferent as they file through countless treasure chambers. The metro is not a serene paradise of contemplation, but an inferno of perpetual motion."[9]

The subways of Moscow and Mexico City are curiously similar. Both are packed with the symbols of failed revolutions and perform a clear compensatory function as "underground heavens." Perhaps it's no coincidence that both were inaugurated a year after dissident movements were crushed. In 1934, the Soviet Union's artists' and intellectuals' organizations were disbanded, and their members were forced to join a single state-controlled body; the metro opened in 1935. As for Mexico, the metro was the first major public work unveiled after the repression of the student movement and the Tlatelolco massacre of 1968.

But the most striking coincidence between the two systems is their manipulation of the past. The Mexican subway is remarkable for its signage. In 1969, the pictographs for the stations along line 1—color coded in a rhetorical "Mexican pink"—were presented as a codex of directions, to show that pre-Hispanic culture was alive and kicking while acknowledging that many passengers were illiterate.

The station icons were designed to be "typical," on varying levels of complexity. Balderas station is figured by a cannon, standing for the nearby fortress and alluding by extension to the Decena Trágica—the ten tragic days that sealed the fate of Francisco I. Madero, the visionary politician who launched the Mexican Revolution. The symbol for Chabacano (Apricot) station is simply an apricot.

This appropriation of the past is emphasized by the use of Aztec motifs on friezes and low reliefs, the presence of pre-Hispanic remains (including a pyramid at Pino Suárez), and the names of stations like Tacuba, Mixcoac, Tezozomoc, Mixihuca, Iztapalapa, Pantitlán, Tasqueña, Coyuca, Iztacalco, Aculco, and Moctezuma. At Panteones (the cemetery), an Aztec sculpture has a plaque to remind us that the earth is both "womb and tomb," enclosing the humid primeval cave and also Mictlán, the land of the dead.

A bastion of the informal economy, a venue for exhibitions, concerts, and book fairs, a site for suicides and births, the metro is a mobile city. Like the movie *Brazil* or the tunnels of Moscow, it deploys a scenario of temporal confusion. The trains are slick pieces of French technology, and certain stations are so futuristic that they have been used as sets for sci-fi dystopias. *Total Recall* was shot in Insurgentes and Chabacano (where the fake blood sprayed in one scene stayed on the ceiling for years, like a freak souvenir from the future). On the other hand, the cement friezes, station names, and pictographic signs recall the past. We're in some bewildering pre-Hispanic modernity.

But it is not the metro's architecture that makes the deepest impression; it is the travelers' blank faces, as if they had been bribed to be there. The Collective Transport System carries almost 5 million passengers a day (Pantitlán alone registers 360,000). Despite their numbers, they have been carefully selected. To descend the escalator is to cross a firm line of racial segregation. The underground city is populated by—pick your favorite slur—brownskins, Indians, Mexicans.

## The Virgin of the Subway

Such is the climate of violence in Mexico City that the most nervous people are those who have so far escaped a mugging—as if there were a compulsory crime quota for every citizen, a sort of tax on living in one of the most crowded, anarchic places on the planet. If you haven't found yourself at the wrong end of a knife yet, never mind: you're next up.

The scale of thievery ranges from the "instant kidnap," which consists in snatching a kid at the supermarket and returning the child in exchange for bagfuls of food, to the bank shootout with AK-47s, now one of the most sought-after items on the black market. Mexican Spanish has become enriched with sinister neologisms: to "milk" someone (*ordeñar*) is to march them to the ATM; to be "trunked" (*encajuelar*) is when the "client" is locked in the trunk of a car until he or she promises to hand over the Rembrandt or suffocates, whichever comes first; a "shard job" (*cristalazo*) is smashing a car window for purposes of reconnaissance.

During the mid-1990s, in response to the public outcry over the security situation, the last city mayor belonging to the Institutional Revolutionary Party (PRI) took some rather questionable measures. The Jaguar group (armed judicial police who thought every day was Rambo's birthday) was disbanded, that is, reincorporated into civilian life, which is to say, let loose in the hunting preserve known as the Federal District. In counterweight, a group of bicycle cops was created, known as *policletos*, who pedaled mildly around watching the robbers screech off in vehicles worthy of their names (Grand Cherokee, Intrepid, Phantom). It was hard not to sympathize with the bicycle cops' unequal struggle against crime and quite impossible to believe they served any purpose than to be pitied for their helplessness.

In a context of rising crime and diminishing police presence, only the Virgin could offer a way out. The apocalypse was mitigated by her presence. Mexico City felt like a warm-up for the Last Judgment until Saturday, May 31, 1997, when Eduardo González went down into Hidalgo station, as he did every day, to sell gum. In one of the passages he noticed a damp stain and stopped to examine the pinkish flags more closely. What he saw resembled a hallucination brought on by vitamin deficiency or the blissful delirium of the glue sniffer; there, under the surface, bathed in mineral translucence, lay the Virgin of Guadalupe.

González alerted the people of colonia Guerrero, which has a thriving community life, to his discovery. The rumor spread like wildfire: down in Hidalgo station the patroness of Mexico had come back for her children! The location could hardly be more symbolic. The subway is the city of moles, the chosen ones whose fate it is to burrow through this underworld. If a fair-skinned person appears, it's usually a post-hippie European with a backpack bought in Katmandu, on his way to the Frida Kahlo house. Except for these rare tourists, the Collective Transport System belongs exclusively to brown-skinned people stigmatized by poverty, who look so ill that it's a surprise when not everybody gets off at Hospital General or Centro Médico.

Not only did the Virgin manifest herself in the metro, she elected to do so at Hidalgo station—highly propitious for instant mythologies as the struggle for independence began when Father Miguel Hidalgo raised the standard with the image of the mutinous Guadalupe. Mexico was born under the wing of the dark-skinned virgin whose name now graces so many critical aspects of life, from Coyoacán's famous Cantina Guadalupana to staple goods such as Patrona safflower oil.

"All Mexicans are Guadalupans," the saying goes. Popular faith dwarfs organized religion. In 1996, the abbot of the Basilica of Guadalupe, Guillermo Schulenburg, revived the old dispute started by Fray Servando Teresa de Mier, casting doubt on the historical truth of the Virgin's appearance to the Indian Juan Diego on the hill of Tepeyac. This statement caused a furor: the custodian of the Guadalupan cult was an apostate! He was contradicted with a passion that proved the myth to be not only untouchable, but stronger than the Catholic Church. Every December 12, close on nine million pilgrims stream in from all over the country to pay homage to their patroness.

The debate about the Virgin flared up again in 1997, a year after Schulenburg's remarks, when her image appeared in Hidalgo station. There were plenty of prelates with Thomist leanings who explained that to revere a shape created by random mopping was tantamount to a hoax. Regardless, the flagstone was soon turned into an altar by the faithful, who flocked to see it in such numbers that lines 2 and 3, which intersect at this station, were brought to a standstill. The corridors filled up with candles and floral offerings until the Catholic Church finally gave way and built a shrine outside the station.

What heavenly message did the Virgin bring when she materialized beneath the shabby shoes of the subway crowds? One of her devotees told *Laberinto* magazine: "She came to tell us that Mexicans have to stop killing each other, because killing is a sin and even more sinful between brothers."[10] When bullets are flying, we reach for emergency talismans. This is the capital of fear, where the removal of a police chief is always good news, and the appointment of his successor a calamity.

Mexico City is safeguarded by nothing but a flagstone virgin. Those of us who made the pilgrimage can at least vouch for the striking resemblance of that form to the original image at Tepeyac, if not for its miraculous nature. How could such precise contours have emerged from a mere leakage? But it's a minor wonder compared to those reverent throngs, those candles burning in the night like the last stand of a vanquished city.

## The City Is the Heaven of the Subway

Navigating the bowels of the earth, the metro is a newness that travels the past whence all things came. It evokes the ancestors, what has been, and what is to come—a symbol outside of time. The garish orange paint and metal frame do not prevent it from occupying a perpetual beyond. Under the ground lies the no-place where time swallows its tail. However speedy the journeys, the space traversed bows to the law of what has no beginning or end, in the collapse of time such that a pyramid, a crucifix, and an electric vehicle exist simultaneously.

Octavio Paz sought to convey this ceaseless becoming in his writing. The structure of "Sunstone" reflects the revolution of the Aztec cycle of years; "Wind from All Compass Points" begins with these words: "The present is motionless." Paz's play *Rappaccini's Daughter* ends with this assurance: "What happened is happening still."[11] The subway circulates like a metaphor of the refutation of successive times, the confirmation of a unitary, stationary time.

On the Day of the Dead in 2008, I went to view the offerings in the Zócalo. That November 2, the Plaza de la Constitución was presided by a great statue of Mictlantecuhtli, god of the underworld. As is usual in festivals of the dead, most of the themes were pre-Hispanic. A huge triangular Mayan arch, adorned with the proboscis of the rain god Chaac, loomed on one side. There was little evidence of colonial Mexico. Significantly, the one installation that referenced the present day was a subway car filled with skeletons, whose destination was obvious: the graveyard at Panteones.

It's no surprise to find the subway present at the feast of the dead next to Mictlantecuhtli, an offering that underlined how the god of death does not represent finality: the underworld is a place of perpetual new beginnings, where the present, as Paz saw it, is motionless.

The coach of the offering was life size and open to visits. A long queue was waiting to enter this transport of skeletons. "Metronauts" who had gotten to Zócalo station on packed subway trains were enjoying the long wait to board a different carriage. Beginning and prophecy: the carriage of bones offered an intimate, realistic contact with pre-Hispanic cosmogony. Those who stepped on were the ordinary users, the common people who recognized in the skeletons their own future mirror, or their X-rays.

To the affluent sectors of society, the metro is what you take when in Paris. Down below, the hoi polloi are carried at postmodern speeds. It would be paranoid, and in a sense too generous, to suppose that this negative utopia—the operative control of history, the rotation of myths at sixty miles per hour—was deliberately planned as such. All the same, the Collective Transport System implements a transportation apartheid.

So why does no one pull the emergency handle? Does resignation to an adverse habitat create new modes of conformity? Does the displacement of tragedy, hoping the catastrophe has passed, leave people with no energy to complain about anything else? Might our post-apocalyptic ecosystem turn us all into docile passengers?

What's certain is that two axes of Mexican life intersect underground: the rhetorical importance of the past and functional racism. The subway as a high-speed model of injustice is not likely to upset most capital dwellers, inured as they are to the sight of small Otomí girls proffering chewing gum at traffic lights.

The indifference above trumps the mineral resignation below, at least until the temporal crossovers provoke a crossover of place: a Chiapas station in the Federal District?

Landscapes, signs, the notion of postmodernity, all are moving through the expanding city. Jorge Luis Borges believed that the future is different from the present and that that is all we can know about it. Where is the underground tempo headed? Where is it taking these drowsy, robotic, speechless hordes? Perhaps the lesson of the tunnels is to make us imagine the surface and assign a different value to the streets—hidden proof that the city is the heaven of the subway.

"Let's take heaven by storm!" cries an unrepentant utopian.

The mobs advance by the subway cars' fake daylight. Outside, virtual and powerful, the city awaits.

# Notes

1. Enrique Florescano, "Mitos mesoamericanos: Hacia un enfoque histórico," *Vuelta* 207 (February 1995): 25–35.

2. Paul Virilio, *L'horizon négatif* (Paris: Editions Galilée, 1984).

3. Rem Koolhaas and Bruce Mau, *S, M, L, XL: Small, Medium, Large, Extra Large* (New York: Monacelli Press, 1995).

4. Italo Calvino, *Invisible Cities*, trans. William Weaver (London: Vintage, 1997), 43–44.

5. Carlos Monsiváis, *Los rituales del caos* (Mexico City: Era, 1996).

6. Marcela Dávalos, *Los letrados interpretan la ciudad: Los barrios de indios en el umbral de la independencia* (Mexico City: INAH, 2010), 31.

7. Boris Groys, "U-Bahn als U-Topie," *Kursbuch* (Rowohlt, Berlin) 112 (June 1993): 1–9.

8. Ibid., 112.

9. Ibid.

10. *Laberinto* magazine is out of print.

11. Octavio Paz, *La hija de Rappaccini* (Mexico City: Era, 1990), 55.

# V
# ART, LITERATURE, AND ARCHITECTURE

# 15
# *Estridentismo*'s Technologies
## Modernity's "Efficient Agents" in Post-revolutionary Mexico

*Lynda Klich*

When the poet and law student Manuel Maples Arce pasted his broadsheet manifesto *Actual No. 1* onto building walls in the center of Mexico City in December 1921, he laid the foundations for *estridentismo* (Stridentism), a group of vanguard writers and artists who participated vocally in the cultural debates that took place during the first years of the post-revolutionary period. In the manifesto, exhorting Mexicans to praise "all the beauty of this century" and to "proclaim the aristocracy of gasoline," Maples Arce staked a position in favor of modernity and its technologies that distinguished the *estridentistas* from other intellectuals of the period.[1] Demanding an acknowledgment of present-day Mexico, as emphasized by the title of the founding manifesto, Maples Arce established a guideline for integrity that the movement consistently sought to follow in its cultural practices. For the estridentistas, recognition of modernity entailed the courage to see Mexico as it really was, rather than whitewash its faults and shortcomings through a filter of ancient grandeur or timeless rural culture. Maples Arce would articulate frequently the demand to remain true to the present in his typically bombastic public proclamations throughout the course of the movement, which remained active in Mexico City through 1924 before moving to Jalapa, Veracruz, until its demise in 1927.[2]

The artists and writers who entered Maples Arce's circle followed his lead and included technological imagery in their images and texts, most notably those they created for estridentista publications. Maples Arce, however, championed stylistic diversity (as long as it came with innovation) and simultaneously promoted works by the movement's visual artists that depicted subject matter more typically associated with the post-revolutionary period, including primitivizing masks and woodcuts of Mexican types. Artists closely associated with the group, including Ramón Alva de la Canal, Jean Charlot, and Fermín Revueltas, even painted some of the earliest murals for José Vasconcelos, the minister of public education. For Maples Arce, the novel formal languages of such works, despite their historical or rural subjects, sufficed to categorize them as estridentista.

Nonetheless, technological imagery remains the most emblematic of estridentismo and has rightly led to the characterization of the movement as antagonistic toward other projects that privileged ancient Mexican cultures and folk art production, as emblematized in SEP initiatives such as the Noche Mexicana and the Exhibition of Popular Arts, the Best Maugard Drawing Method, and the Escuelas de Pintura al Aire Libre (Open Air Painting Schools, or EPALs).[3] Simultaneously, however, the emphasis on modernity in estridentismo has led to it being seen as unrealistic and distanced from contemporary cultural concerns regarding post-revolutionary recovery and nation building.[4]

Yet, it is precisely these images of technological imagery that allow us to clarify and understand estridentismo's important, if undervalued, oppositional posture toward official discourse. An interrogation of the precise types of technology represented in the movement's images and the manner in which artists crafted such images (often in a primitivizing aesthetic) reveals that estridentismo strategically employed its vision of modern Mexico as a deliberate challenge to official constructions of the country. Countering representations of an "authentic" rural Mexico, the group's writers and artists insisted on infrastructural expansion as a post-revolutionary objective and, more importantly, made it seem within reach.

The characterization of estridentista imagery being distant from reality perhaps best applies to the images of architecture and city spaces created by Ramón Alva de la Canal and others for *Horizonte*, the estridentista journal produced in Jalapa from April 1926 to April–May 1927 under the auspices of Veracruz governor General Heriberto Jara, for whom Maples Arce served as deputy governor after receiving his law degree in 1925. Some of these images, such as Alva de la Canal's famous *Radio Station for Estridentópolis*, surely depict structures unlike anything seen in Mexico at the time. Their subjects, however, were closely related to what the estridentistas considered the very real achievements of the Jara administration, which undertook an extensive modernizing program in Jalapa and throughout the state of Veracruz.[5] As such, these works remained true to the movement's credo to represent the present, albeit in an admittedly exaggerated manner that was meant to convey symbolically their pride in the Jara regime, in which they played an integral role, notably as propagandists.

Indeed, because of their official sponsorship and ties to the government, the works from the Jalapa period must be separated from those created by the artists during the Mexico City period (something rarely done in the scholarship on estridentismo).[6] During the movement's initial years, however, estridentista urban imagery focused not on a modern metropolis, but on a city in flux and on the possibility of technology's arrival in Mexico City and its diffusion to the provinces. Such imagery, therefore, at once succeeded in depicting the movement's critique of post-revolutionary regimes that failed to implement infrastructural development in the capital and provinces and visualized the optimism of the immediate post-revolutionary period. Estridentista images of urban transformation, rather than point to a distant and impossible future, represented the challenges of modernization faced by actual, present-day Mexico, as demanded by estridentismo's leader from the movement's earliest days.

During the years it was active in Mexico City, estridentismo assiduously cultivated its oppositional stance and aimed to attract audiences further afield, issuing additional manifestos, launching a journal named *Irradiador* with confrontational editorials, and making provocative pronouncements in *El Universal Ilustrado*, a Sunday newspaper supplement that supported its cause. For example, the second estridentista manifesto, launched by Maples Arce and the writer Germán List Arzubide in Puebla on New Year's Day in 1923, declared a "profound disdain for the ideological rancidolatry of some functionaries" and claimed that only estridentismo would protect the nation's youth from the "quacks officially named to head official university departments."[7] A profile on the movement's founder in *Universal Ilustrado* was aptly titled "Maples Arce arremete contra todo el mundo" (Maples Arce attacks everyone).[8] The bombastic proclamations and antagonistic strategies, of course, point to the European avant-garde movements that inspired Maples Arce, such as F. T. Marinetti's Italian futurism and Guillermo de Torre's Spanish *ultraísmo*. Yet, when Maples Arce first had the opportunity to define estridentismo after its first year of activity, he invoked Germany's Novembergruppe and the Russian Suprematists as more relevant models for his post-revolutionary project. All three, he argued, represented "spiritual agitations" corresponding to political upheavals with the intent to reform.[9]

These same confrontational texts allowed Maples Arce to define modernity from the estridentista perspective and to establish it as a critique of projects, official and otherwise, that sought to eliminate suggestions of the present age from the national consciousness by constructing the fiction of a unified indigenous/rural nation. In determining its vision of modern Mexico, the group touted subjects such as electricity, iron and steel, asphalt, scaffolding, petroleum, motorcycles, automobiles, radiophones, horns, dynamos, gears, cables, and antennas, in lieu of what Maples Arce called the "romantic pottery shards" so esteemed by others, but that he considered "less original than a trash can."[10] The estridentistas also invoked aspects of popular culture, such as jazz bands, the cinema, and its international screen stars Mary Pickford and Charlie Chaplin. And they acclaimed consumer products such as Moctezuma beer, Buen Tono cigarettes, and Bayer aspirin, as well as shop window displays and signs, or "talking storefronts [that] bellow their piercing colors at the top of their lungs from one side of the street to the other."[11] Estridentista modernity thus included not only mechanical and technological marvels, but also the commonplace and non-Mexican, and it was decidedly noisy and disorderly. The movement welcomed, rather than shied away from, the messy assault of contemporary life. Yet, because such urban cacophony represented their lived experience, they considered their version of modernity to be unquestionably Mexican. From the beginning, the estridentistas not only embraced international culture on all levels as integral to their project, but also rejected conservative Mexican culture as inauthentic for the current era. The group, moreover, acknowledged that cynics—who thought elevators "an intellectual trick" and got headaches from simply crossing a busy city street—were not quite ready for its modern, paradoxical vision of Mexico.[12] Indeed, Maples Arce repeatedly described *retardataire* poets, critics, and even some governmental officials as afraid of such modern developments.[13]

Notably, therefore, Maples Arce did not shy away from the instability that accompanied reconstruction, differentiating his movement from concurrent projects in Mexico that are traditionally perceived as more nationalistic. Rather, he embraced the incompleteness and fragmentation of modernity, such as items "found at the side of the road [like] stupendous posters and geometrical signs [and] Haynes spare parts, tire rims, batteries and generators," along with its dirtiness, such as "the blue smoke of exhaust pipes."[14] Nor did he envision Mexico's redemption in terms of its pre-Hispanic past or what he perceived as the "mephitic efflorescence of our nationalist environment" but instead focused his interest on its present-day citizen.[15]

The beginning of Álvaro Obregón's presidency in 1920 ushered in a period of optimism in the capital; investment in the city increased as did its population; and many who had fled the country, including expatriates who had been in political exile, returned. Urban population growth, which expanded during the final decade of Porfirio Díaz's dictatorship, picked up at an even greater pace during the post-revolutionary period. Mexico City consequently found itself needing to confront housing and transportation shortages.[16] The city's technological modernization indisputably had begun during the Porfiriato as well.[17] Under Díaz, however, infrastructural modernization occurred separately from political and civil reform. During the post-revolution, the two were inextricably linked. Modernization of the city took on greater significance, as it represented recovery from the violent and destructive decade of the Mexican Revolution. Population growth, for example, necessitated the restoration and expansion of the city's trolley lines. But since rebel attacks on the trolley system's energy sources continually disrupted services during the revolution and continued into the 1920s, their smooth functioning was much more than convenient: it signified the restoration of order in Mexico.

From the start, Maples Arce and the estridentistas grasped the significance of modernity for Mexico's recovery. If the modernized settings of their poems and images laid bare the need for infrastructural development in Mexico, however, the post-revolutionary government faced enormous financial and practical barriers in the actual restoration of the nation. Venustiano Carranza, Obregón, and their successor Plutarco Elías Calles planned and began to implement the paving of streets and creation of highways, the expansion of electric and telephone networks, and the installation of trolley and bus systems. Such reconstruction signified not only the physical recovery but also the regeneration of the national outlook. Prohibitive economic and social conditions, however, sometimes led to delays in post-revolutionary modernization, especially during the Obregón administration.[18]

The estridentista perception of modernity is exemplified in *Andamios exteriores* (Exterior scaffolding, fig. 15.1), a watercolor created by Fermín Revueltas in 1923, shortly after Maples Arce published his first volume of estridentista poetry, *Andamios interiores: Poemas radiográficos* (Interior scaffolding: Radiographic poems).[19] More than just an inside joke, however, Revueltas's relocation of the scaffolds from the poet's mind to the street emphatically characterizes the city as incomplete and in process. The network of electrical and telephone or telegraph wires and enormous metallic storage tanks pushed to the foreground certainly create a dizzying, modernized space. Yet Revueltas filled it not with unrealistic skyscrapers but with the low-lying, nearly windowless, humble brick or adobe structures that characterized Mexico City architecture of the time much more so than did tall, steel-framed buildings, the first of which would not be built until several years later.[20] Moreover, his strategic placement of workers, adorned in simple worker's clothing and typical sombreros, on the scaffolds, suggests that they are modern citizens and active participants in national reform. The warm yellows and browns of the workers' zone of the image distinguish it from the surrounding cool, metallic grays and blues of the urban scene. Revueltas's formal language, locking the network of wires, scaffolds, and workers into the picture plane, encapsulates estridentismo's vision of modernity as an ongoing process. It also highlights the country's need for infrastructure development and the integration of the indigenous/mestizo working class into modernity as significant national concerns.

Estridentista artists repeatedly invoked the country's electrical and commu-

15.1. Fermín Revueltas, *Andamios exteriores* (Exterior scaffolding), watercolor on paper, 10¾ × 13⅜ inches (27.3 × 34.0 cm), 1923. Museo Nacional de Arte, CONACULTA, INBA, Mexico City. Gift of Blanca Vermeersch de Maples Arce. With permission, Collection Ing. Silvestre Revueltas.

nication systems in their works, often constructing their images in ways that emphasize the importance of extending such technology beyond the nation's capital. Another work by Revueltas, *Paisaje con líneas de alta tensión* (Landscape with high-tension wires, fig. 15.2), created in 1924, for example, embeds electrification directly into an otherwise empty landscape by depicting a series of pylons linked together by taut wires in front of a mountain. Since the volcano-dominated landscapes of José María Velasco represented Mexico at various World's Fairs at the end of the nineteenth century, artists filled their works with the country's peaks as signs of burgeoning nationalism.[21] Rather than naturalistic depictions of, for example, the famous Popocatépetl and Iztaccíhuatl, Revueltas heightens the sense of nationalism through a palette typical of folk art, such as earthy reds, blues, and greens, colors that Francisco Reyes Palma has called "national."[22] The employment of the colors underneath the electrical towers, moreover, recalls the Mexican flag—red and green flanking a strip of white—significant in a country where the main source of electricity, the Compañía de

15.2. Fermín Revueltas, *Paisaje con líneas de alta tensión* (Landscape with high-tension wires), watercolor on paper, 5¾ × 8 9/16 inches (14.7 × 21.8 cm), 1924. Museo Nacional de Arte, CONACULTA, INBA, Mexico City. Gift of Blanca Vermeersch de Maples Arce. With permission, Collection Ing. Silvestre Revueltas.

Luz y Fuerza, was foreign-owned, as were many industries essential to the country's reconstruction.[23]

In the estridentista view, such technological developments represented Mexico in its actuality, as indicated by contemporaneous photos taken by the artist.[24] And Revueltas's careful construction of his watercolor, like the colors he used, meaningfully situates technology within the landscape. The electrical wires come into the landscape from an imagined, extended space outside of the picture; they are thus part of a larger network. The artist then unites them with the mountain by using the same thick black outline for both. Moreover, the irregular, hand-drawn quality of the lines of the electric towers, as opposed to a precise, streamlined aesthetic, creates a harmony and naturalness in *Paisaje con líneas de alta tensión*. The image does not present what is typically regarded as urban imagery, but rather evokes the electrification of the Mexican landscape outside of Mexico City. It suggests the expansion of modernity outward from the capital, another contentious issue during the period, since, despite the widespread destruction of transportation and communication connections to the provinces, political pressures had led to a focus on improving the infrastructure of the capital.[25]

Estridentista artists repeatedly employed pictorial strategies to place emphasis on the development of the country's infrastructure needed to reach the parts of Mexico where the indigenous and mestizo populace (who were repeatedly invoked in official imagery as exemplars of Mexican authenticity) actually lived. Tina Modotti's famous photographs of telephone and telegraph wires, which belong to a group of her works generally considered to have been inspired by the movement, astutely addressed the issue of extending the country's communication networks.[26] Modotti's careful compositions both exalt the subject's symbolic value as a technological development and allude to its practical application as a method of communication. In one version, shot from below, the towers' network of horizontal beams forms a focal point and shifts the viewer's gaze upward. The verticality of the composition and the detachment of the poles from the land itself present the towers as icons of modernity. At the same time, the numerous wires that extend into the viewer's foreground space and the repetition of towers that recede seemingly endlessly into the unseen horizon remind us that what we see is only a small portion of a vast communications network, with much potential for expansion.

A cover for an estridentista publication by Alva de la Canal similarly emphasizes the movement's great faith in the modernizing effects of telecommunications. His red and black woodcut for Salvador Gallardo's *El Pentagrama Eléctrico* (The electric pentagram, fig. 15.3) also employs the motif of telegraph or telephone poles.[27] Like Modotti, Alva de la Canal utilizes repeated wires as the forceful central compositional line of the image. They originate, however, from the cross-like pole that Alva de la Canal tucks slightly off center from the bottom of the image. He propels the wires out of the upper left-hand corner of the page—like orthogonal lines that turn traditional one-point perspective askew. Alva de la Canal's wires, like Revueltas's, extend across a specifically Mexican landscape, identified by the customary mountains. The peaks, severely geometrized into sharp pyramidal forms, point in various directions amidst undulating lines that can be read either as clouds or as sound waves powerfully spreading communication throughout the land. A gigantic, spiky solar form hovers over the scene, packing the already energetic pictorial space to the point of overflowing. Like the works by Revueltas and Modotti, Alva de la Canal's cover suggests the possibility of expanding modernity to the provinces.

Although Maples Arce often employed the term "telegraphic" or variants of it, the telegraph itself was a nineteenth-century invention; in his works he was instead extolling the wireless telegraph or radio.[28] Nonetheless, the traditional telegraph still held significance in the 1920s in Mexico. During the revolution, the country's already insufficient telegraph system suffered significant damage; its recuperation proved vital to national reconstruction during the post-

15.3. Ramón Alva de la Canal, cover for Salvador Gallardo's *El Pentagrama Eléctrico* (Puebla: Ediciones Germán List Arzubide, 1925). With permission of Jorge Ramón Alva de la Canal.

revolutionary period.[29] Additionally, in the 1920s telephone lines began to replace the telegraph as Mexico's dominant method of communication, often using the same infrastructure. President Calles duly celebrated the initiation of long-distance lines in Mexico with a highly publicized first call to Emilio Portes Gil, governor of the state of Tamaulipas, in 1927.[30] In actuality, however, the planning of Mexico's phone system proved problematic. It consisted of two networks, owned by different companies (Ericsson and Mexicana), and it was impossible to call between networks, even within the capital itself, leading to many difficulties of communication. To combat the problem, the estridentistas (like other Mexicans) had phones on both systems, as indicated by their letterhead. Telephone technology existed for the bourgeois urban Mexican, but as the pages of popular publications like *El Universal Ilustrado*, *Revista de Revista*, and *Zig-Zag* attest, such novelties remained in reach only for those in the most central locations in the post-revolutionary period. One article, for example, noted that for the adventurous, even a move to one of the growing capital's new *colonias* (neighborhoods) could bring notable risks, "even losing the rapidity of communications" provided by the telephone companies. Notably, the article praises the Ericsson phone company's quick attention to the colonias that are home to the "most distinguished families."[31] If for middle-class residents of Mexico City such a delay could merely cause inconveniences, the implications for those in the outskirts or in poorer neighborhoods surely were greater, as they remained cut off from the advantages of this modern technology that was rapidly transforming the daily life of the better-off.

Maples Arce had envisioned a partnership between art and communication technology from the earliest days of estridentismo. "From now on," he wrote in *Actual*, "our unbeatable literature must insist on venerating the telephone and the perfumed dialogue that comes together haphazardly on its conductive wires."[32] Indeed, the publication of estridentista poetry and art in books and journals testifies to the importance the group placed on communication, although the obtuse nature of their formal language cannot be denied. One of the movement's key vehicles for self-promotion was its journal, *Irradiador*, published in three issues during the fall of 1923. Under the direction of Maples Arce and Revueltas, *Irradiador* included visual art, graphic design, poetry, theoretical articles, and editorials by a diverse group of contributors. It therefore meant to accomplish the objective proclaimed by its title, which translates to *radiator*—a transmitting device—simultaneously evoking technological modernity and the journal's mission to disseminate its ideas. Despite the journal's streamlined aesthetic, with the exception of a Buen Tono cigarette advertisement and a café scene (both by Revueltas) and a 1922 photograph of the Armco Steel factory by Edward Weston, who spent several years in Mexico during the 1920s, the art contained in *Irradiador* focused on Mexicanist imagery in vanguard styles, such as Jean Charlot's woodcuts of Mexican types and a Diego Rivera mural of indigenous workers. The journal did contain promotions for the Ward steamship line and the radio station owned by *El Universal*, which were more in line with the estridentista desire to promote modernity. Seemingly a minor detail, these ads nonetheless reveal that technologies of transportation and communication, in particular, appealed to the writers and artists as they conceived their approach to modernity during this period.

Maples Arce's *Urbe: Super-poema bolchevique en 5 cantos* (Metropolis: Bolshevik super-poem in 5 cantos), illustrated by Charlot, endures as the emblematic example of the estridentistas' promulgation of modernization as a feasible agent of social reform and their critique of those who did not harness its power. *Urbe*'s five cantos, which take place over the course of one day, convey the shift from the poet's optimism about political revolution to ambivalence because of his disillusion with its ultimate corruption, infighting, and ineffectiveness.[33] In his text, Maples Arce embodied this shift of position in the transformation of a confident city newly thriving and bustling into a wasteland marked by desolation and ruin, though, as Elissa Rashkin has pointed out, a landscape not altogether devoid of hope.[34]

Maples Arce not only filled his Mexican *Urbe* with the "mechanical rhythms" of cables and motors, but also populated it with "rivers of blue shirts," the "triumphal hurrahs of *obregonismo*," and striking masses.³⁵ The volume's cover (a sharply geometrized cityscape) and the word *Bolshevik* in the subtitle straightaway solidify the link between metropolis and social revolution. Charlot's five interior woodcuts, however, somewhat incongruously distance the reader from the hustle and bustle of the urban space, and in them the worker remains conspicuously absent (figs. 15.4 and 15.5). Instead, the artist favored austere depictions of universal icons of modernity, such as skyscrapers, an airplane, a train speeding over a viaduct, and ocean liners. Rather than convey the ruin and despair evident in the poem's ending, Charlot renders these modern emblems in monumentalizing fashion, suggesting permanence and stability. His pictorial style conflicts with the poem's disillusion with revolutionary ideals, embodied in the text by a decaying city, with blame placed at the feet of the bourgeoisie and politicians.³⁶ Charlot's icons of modernity instead remain more attuned to the yearning for lasting ideals and sense of hope that underlies Maples Arce's verses.

The apparent contradiction between the poem's depiction of a politicized urban experience in Mexico and the images' concentration on universalizing symbols of modernity reveals the complexities inherent in the estridentistas' approach to visual art, modernity, and the Mexican Revolution. Taken together, Charlot's five deliberately iconic representations convey an optimistic spirit as well as a sense of endurance. For example, compared to Italian futurism—the most frequent source mentioned for estridentismo's focus on modernity—Charlot's images have no sense whatsoever of the dynamism and sensorial suggestions favored by artists such as Umberto Boccioni and Giacomo Balla. Rather, Charlot, like Maples Arce in the text, searches for stability within modernity. As Reyes Palma has suggested, Charlot's *Urbe* woodcuts attempt through their scale and solidity to recreate the monumentalizing ambitions of muralism, by now the most pertinent artistic reference for estridentismo and for Mexicans in general.³⁷ But, they relocate the country's stabilizing force from its timeless rural traditions to the technological advances of the present day.

The images collectively engage issues of modernism and modernity within the context of the international avant-garde by depicting elements recognized worldwide as icons of twentieth-century modernity, including skyscrapers, perhaps the only such representations from this period of estridentismo. They do the same within post-revolutionary Mexico as well, by portraying elements that conjure economic and infrastructural stability. The artist's choice of and approach to the woodcut medium assert these dual contexts and the tensions between them. From a formal standpoint, Charlot renders his subjects with a solid linearity, instead of with the expressionistic splintering that prevailed in the modern Mexican woodcut. Nonetheless, although he flirts with the tenseness and tautness of the Italian futurist Antonio Sant'Elia's architectural drawings,

15.4. Jean Charlot, *Poet on Airplane*, woodcut for Manuel Maples Arce's *Urbe*, 1924. © The Jean Charlot Estate LLC. With permission.

15.5. Jean Charlot, *Ocean Liner*, woodcut for Manuel Maples Arce's *Urbe*, 1924. © The Jean Charlot Estate LLC. With permission.

the smooth, straight line of Fernand Léger's contemporary machine aesthetic, and the geometrized aesthetic of Kazimir Malevich's Suprematist compositions, Charlot ultimately favors a rougher, almost raw edge with rudimentary detailing. His juxtaposition of urban imagery with such a primitivizing, artisanal style suggests a formal approach that situates *Urbe* within the specific cultural environment of post-revolutionary Mexico. Even though the estridentistas embraced international trends that emanated from European centers, these images, especially, indicate a conscious differentiation of the Mexican experience of modernity.

With the exception of the illustration of twin towering skyscrapers (below which we see a scurrying crowd formed by the artist's hurried dots and dashes) that illustrates the first canto, Charlot's buildings share space with the mobile modern icons that connect Maples Arce's metropolis with the rest of the world. A giant, abstracted airplane fills the sky above a tilted street; in another, a great transatlantic liner dwarfs the cityscape. A powerful train steams across a tall viaduct, perhaps referencing the colonial-era aqueduct at Los Remedios, photographed by Weston, Charlot's close friend. Two razor-sharp steamships cut through the ocean, their forms deliberately angled to suggest the strong force that propels them; above them, two bright searchlights slice into and overwhelm the darkness, reinforcing technology's ability to conquer nature's obstacles. Like the telephone systems reported on in *Universal Ilustrado*, these powerful symbols of modernity graced the pages of Mexico City's popular magazines, as in an article that reported on the arrival to the Veracruz port of the powerful, new, "flaming" Toledo vessel of the Hamburg-Amerika Line in all its "gallantry and solidity," detailing its technical innovations and up-to-date appointments.[38]

Charlot gives his modern icons little descriptive detail but forms their generalized shapes from spare geometric forms that, taken apart, would not seem out of place in one of Malevich's abstractions. Nonetheless, the artist conveys the power of these modern machines with dramatic angles and monumentality. Charlot perhaps did not intend to materialize modernity as much as to embody the new perspectives wrought by technological development. Accordingly, he includes small and enigmatic figures within most of the woodcuts, including a poet (identified by his lyre) who as the plane's only passenger leaves behind a grounded Pegasus, the mythological symbol of intellectual flight and poetic inspiration, in the lower-left-hand corner in favor of a decidedly more modern symbol of creativity (fig. 15.4). In another woodcut, a small waving female form at the viaduct's base emphasizes its impossibly tall height, while two figures seem to have fallen from the enormous transatlantic ship (fig. 15.5). The juxtaposition of these tiny, wistful figures with outsized modern machines also tempers the optimistic spirit of modernity, making a melancholic nod to the poem's cataclysmic, if hopeful, ending and mirroring the ambivalence Maples Arce's poem conveys about revolution.

That Charlot (perhaps with input from Maples Arce) chose such universal icons of modernity coincides with the poet's concerted attempt to create an estridentista masterpiece, on par with those of his avant-garde counterparts throughout Europe and South America. Like other contemporary projects, notably Le Corbusier's *L'Esprit Nouveau*, the monumental depictions of these technological marvels in *Urbe* exalt the achievements of modernity.[39] But more than simply paying homage to technological modernity itself, the volume speaks of its potential to arrive in Mexico. Charlot depicts elements that Maples Arce considered instruments of modernity's diffusion. Objects such as the plane, train, bridge, and ocean liner all suggest interconnection, transference, and communication, much like telephone and telegraph wires. He renders visible what will be necessary to accomplish the arrival and spread of modernity in Mexico, and also elements that, for the most part, were already present in Mexican life.

In the woodcuts, Charlot monumentalized not the workers that take a forefront in Maples Arce's poem, or even its overtly political tenor, but the elements that the poet alludes to in creating the setting. *Urbe* is a burgeoning and transforming city, which Rashkin has characterized as "a kaleidoscopic swirl of ships, trains, telephone wires, streetcars, shop windows," though these elements are not emphasized in the text.[40] Charlot's images, then, serve to bring the poem's urban setting into heightened relief. Taken in conjunction with the text, they assert the estridentistas' positive approach to and fearless acceptance of technological innovation and expressly link it with the revolution's success, an ideal materialized by the majestic skyscrapers in the first woodcut (which is not illustrated in this chapter). Because they were capable of eliminating temporal and spatial distances between Mexico and the world (and even within the Mexican nation itself), the modern icons depicted by Charlot were the types that could transform the country and provide it with the sense of stability sought after since the foundation of the post-revolutionary state. In *Urbe*, Charlot communicates modernity as the lynchpin of post-revolutionary reform by employing a monumentalizing and spare visual language. Nonetheless, the tension between the sleek technological modernity of these icons and the inherent primitivism of the woodcut medium—a signature for the movement's publications—palpably visualizes the movement's paradoxical goals of achieving internationalism within a nationalist context.

Estridentista texts and images also represented the important technological element of the factory smokestack. From *Actual*'s invocation of their rising smoke as emblematic of twentieth-century beauty to the "sexual fever of the factories" that permeates *Urbe*, Maples Arce repeatedly called on industrial

15.6. Jean Charlot, cover woodcut for Germán List Arzubide's *Esquina*, 1923. © The Jean Charlot Estate LLC. With permission.

imagery.[41] Weston's *Steel: Armco, Middletown, Ohio* photograph, on *Irradiador*'s third cover, encapsulates the majesty of the factory's rising towers. In turn, three puffing smokestacks in Charlot's black and red woodcut cover for List Arzubide's poetry volume *Esquina* (Corner) energize a disjunctive city view that echoes the episodic disconnectedness of List Arzubide's poetic language (fig. 15.6).[42] The red flags of a workers' demonstration that wave nearby link labor and industry and politicize the urban space of the image, much like the poet does in his text, which he peppers with mentions of raised arms, protests, and negative characterizations of conservatives. The factory does not have as great a presence in estridentista visual imagery, perhaps because, unlike transportation and communication networks, Mexico's industrial infrastructure was not devastated during the revolution.[43] The appearance of smokestacks and labor iconography on the cover of a book by List Arzubide, however, takes on special resonance. As James Oles has noted recently, many of Mexico's most important factories were located in provincial zones, such as Orizaba, Querétaro, and Monterrey, thus making the factory smokestacks less identifiable with the capital in particular, and more representative of the expansion of modernization throughout Mexico in general.[44] A native of Puebla, List Arzubide not only was the most politically committed writer in the movement, but also agitated to keep the movement alive, following Maples Arce's departure from Mexico City in early 1925, and to spread its ideals to the provinces. (Estridentista manifestos were issued by satellite groups in Zacatecas in July 1925 and in Ciudad Victoria, Tamaulipas, in January 1926, prior to the group's relocation to Jalapa.) This coupling of industry and worker, in line with the estridentistas' vision for post-revolutionary reform, is infrequently remarked upon but was fundamental to the group's viewpoint. Maples Arce first linked "the industrialist diet of the great palpitating cities" with "the blue shirts of the workers, explosive in this emotional and moving hour," in *Actual*.[45] And, whether overtly stated (as in Revueltas's *Andamios exteriors*) or implicit in relationship between image and text (as in *Urbe*), revolution, labor, and modernity remained core to the movement's project throughout its trajectory, persisting into the Jalapa period.

From its beginnings, estridentismo embraced modernity's plurality and incompleteness. Envisioning the achievement of a modern Mexico as a transformative process, Maples Arce and his group welcomed images that described the city under construction. Rather than shy away from the fragmentation and uncertainty inherent in the process of modernity, Maples Arce and his cohorts called on precisely these elements to represent a progressive spirit. The group's writers and artists viewed modernity's disruptions as an integral and necessary aspect of Mexico's post-revolutionary rebuilding. Estridentismo sought not merely to conjure icons of modernity, but rather envisioned itself as a participant in the diffusion of modernity itself. Hence, the estridentistas' concentration on elements of transference and communication was part and parcel of the aim to spread cultural renovation through books, magazines, and public interventions. Such a platform made post-revolutionary ideals, such as extended literacy and infrastructural development, appear attainable. As such, the works with the subject of modernity that were created during the first period of the movement conveyed two ideas integral to its philosophy: a commitment to representing current-day Mexico and the concomitant faith that modernity could, in turn, disseminate the revolution and its ideals.

Their images tended not toward a futuristic metropolis, but, rather, concentrated on, as Maples Arce later noted, the "efficient agents" of technology—"the railroad, electricity, the automobile, the airplane, and editorial diffusion"—which would help "widen its radius toward the countryside."[46] The estridentistas therefore chose not simply to depict technological icons, but to represent specifically those elements capable of enacting and disseminating some of the goals of the revolution. As the works discussed above reveal, the visual images of modernity during the movement's first years, therefore, focused not on a realized or idealized urban space, but instead on one in a process of construction or awaiting the arrival of modernity. Technology as a subject held importance as a representation true to Mexico of the present, rather than as a representation of a future model city. In other words, the group's members sought to embody the process of modernization in their works, rather than to present it as a fait accompli. They brought to life, then, what estridentista writer Arqueles Vela called "the ignition of a society transforming itself."[47]

## Notes

I thank Araceli Tinajero and J. Brian Freeman for the invitation to contribute to this volume and for their editorial guidance. I am indebted to Anna Indych-López for her keen critical comments on this chapter and am also grateful to Edward J. Sullivan, Robert S. Lubar, and Ellen Adams, who have offered me crucial input for my work on *estridentismo*.

1. Manuel Maples Arce, *Actual No. 1: Hoja de vanguardia. Comprimido estridentista de Manuel Maples Arce*, broadsheet issued in Mexico City, December 1921. The manifesto is reproduced in Juan Manuel Bonet, *El ultraísmo y las artes plásticas*, exhibition catalog (Valencia: IVAM, Centre Julio González, 1996), 166–67. All quotations in this chapter are taken from my translation of Maples Arce's *Actual No. 1*; see Lynda Klich, "Revolution and Utopia: *Estridentismo* and the Visual Arts (1921–1927)" (PhD diss., Institute of Fine Arts, New York University, 2008), 508.

2. The foundational text on *estridentismo* is Luis Mario Schneider, *El estridentismo o una literatura de la estrategia* (Mexico City: INBA, 1970).

3. Rick A. López, "The Noche Mexicana and the Exhibition of Popular Arts: Two Ways of Exalting Indianness," in *The Eagle and the Virgin: Nation and Cultural Revolution in Mexico, 1920–1940*, ed. Mary Kay Vaughan and Stephen E. Lewis (Durham: Duke University Press, 2006), 23–42; Karen Cordero Reiman, "The Best Maugard Drawing Method: A Common Ground for Modern Mexicanist Aesthetics," *Journal of Decorative and Propaganda Arts*, no. 26, Mexico Theme Issue (2010): 45–79. The relationship between *estridentismo* and the EPALs is admittedly complex, as some artists affiliated with the movement also taught at EPALs.

4. Long known primarily for its literary side, *estridentismo* has taken a more visible place in Mexican art historical narrative since the landmark exhibition Modernidad y modernización en el arte mexicano, held at the Museo Nacional de Arte (MUNAL) in 1991.

5. Lynda Klich, "Estridentópolis: Achieving a Post-revolutionary Utopia in Jalapa," *Journal of Decorative and Propaganda Arts*, no. 26, Mexico Theme Issue (2010), 102–27.

6. A notable exception is Francisco Reyes Palma. See his "Arte funcional y vanguardia (1921–52)," in *Modernidad y modernización en el arte mexicano, 1920–1960* (Mexico City: INBA, MUNAL, 1991), 85; and "Otras modernidades, otros modernismos," in *Hacia otra historia del arte en México: La fabricación del arte nacional*

*a debate (1920–1950),* ed. Esther Acevedo (Mexico City: Curare/CONACULTA, 2002), 28.

7. Klich, "Revolution and Utopia," Appendix B.

8. Ortega, "Zig-Zag en la república de las letras: Maples Arce arremete contra todo el mundo," *El Universal Ilustrado* 7, no. 332 (September 20, 1923): 30–31.

9. Manuel Maples Arce, "El movimiento estridentista en 1922," *El Universal Ilustrado* 7, no. 294 (December 28, 1922): 25.

10. Manuel Maples Arce, cited in "Nuestras encuestas: ¿Qué piensa usted sobre el estado actual de la poesía en México?" *El Universal Ilustrado* 6, no. 290 (November 30, 1922): 50.

11. "Irradiación Inaugural," *Irradiador*, no. 1 ([September] 1923): n.p.

12. Ibid.

13. Maples Arce, cited in "Nuestras encuestas," 50.

14. Maples Arce, *Actual No. 1,* translated in Klich, "Revolution and Utopia," 508.

15. Ibid., 515.

16. Diane E. Davis, *Urban Leviathan: Mexico City in the Twentieth Century* (Philadelphia: Temple University Press, 1994).

17. Robert M. Buffington and William E. French, "The Culture of Modernity," in *The Oxford History of Mexico*, ed. Michael C. Meyer and William H. Beezley (Oxford: Oxford University Press, 2000), 397–432.

18. Jean Meyer, "La ciudad de México, ex de los palacios," in *La reconstrucción económica: Historia de la revolución mexicana, IV Periodo 1924–1928*, vol. 10, ed. Enrique Krauze (Mexico City, El Colegio de México, 1977); Stephen H. Haber, *Industry and Underdevelopment: The Industrialization of Mexico, 1890–1940* (Stanford: Stanford University Press, 1989); Alvaro Matute, *Las dificultades del nuevo estado: Historia de la revolución mexicana, 1917–1924*, vol. 7 (Mexico City: El Colegio de México, 1995); Thomas Benjamin, "Rebuilding the Nation," in *The Oxford History of Mexico*, ed. Michael C. Meyer and William H. Beezley (Oxford: Oxford University Press, 2000), 467–502.

19. Manuel Maples Arce, *Andamios interiores: Poemas radiográficos* (Mexico City: Editorial Cultura, 1922).

20. See Enrique X. de Anda Alanís, *La arquitectura de la revolución mexicana: Corrientes y estilos en la década de los veinte* (Mexico City: UNAM, 1990), 118–19 and fig. 31.

21. Esther Acevedo, "De lo nacional a lo arquetípico: La des-territorialización del paisaje (1900–1950)," in *Hacia otra historia del arte en México*, ed. Acevedo, 91–104.

22. Francisco Reyes Palma, "Otras modernidades, otros modernismos," in *Hacia otra historia del arte en México*, ed. Acevedo, 22.

23. Rubén Gallo, *Mexican Modernity: The Avant-Garde and the Technological Revo-*

*lution* (Cambridge, MA: MIT Press, 2005), 22–23; and de Anda Alanís, *La arquitectura de la revolución mexicana*, 42–43.

24. Carla Zurián, in *Fermín Revueltas: Constructor de espacios*, exhibition catalog (Mexico City: INBA/Museo Mural Diego Rivera/Editorial RM, 2002), 46–47.

25. Haber, *Industry and Underdevelopment*, 122–23; Davis, *Urban Leviathan*, 21–23.

26. Modotti's photograph became emblematic of the movement thanks to Germán List Arzubide's inclusion of it in his homage to the movement, written at the end of 1926; Germán List Arzubide, *El movimiento estridentista* (Jalapa: Ediciones de Horizonte, 1926), 55. The photograph was also reproduced in *Horizonte* 1, no. 2 (May 1926): 23.

27. Gallardo published his book in 1925, a transitional year for the movement. I discuss this work by Alva de la Canal in this chapter because it was published by List Arzubide in Puebla, not under the auspices of the Jalapa government.

28. Maples Arce, in his *Actual No. 1*, first employed the term "telegraphically urgent."

29. Matute, *Las dificultades del nuevo estado*, 206–9; Thomas Benjamin, "Rebuilding the Nation," in *The Oxford History of Mexico*, ed. Meyer and Beezley, 470; Haber, *Industry and Underdevelopment*, 132–34.

30. John W. F. Dulles, *Yesterday in Mexico: A Chronicle of the Revolution, 1919–1936* (Austin: University of Texas Press, 1961), 290.

31. "Los inventos modernos: Las nuevas estaciones de la Ericsson," *El Universal Ilustrado* 6, no. 301 (February 15, 1923): 48–49.

32. Manuel Maples Arce, *Actual No. 1*, translated in Klich, "Revolution and Utopia," 505.

33. Elissa J. Rashkin, *The Stridentist Movement in Mexico: The Avant-Garde and Cultural Change in the 1920s* (Lanham, MD: Lexington Books, 2009), 115.

34. Ibid., 119.

35. Manuel Maples Arce, *Urbe: Super-poema bolchevique en 5 cantos* (Mexico City: Andres Botas E. Hijo, 1924), n.p.

36. Schneider, *El estridentismo o una literatura de estrategia*, 98.

37. Reyes Palma, "Otras modernidades, otros modernismos," 34.

38. "El nuevo transatlántico 'Toledo' de la 'Hamburg Amerika Linie,'" *El Universal Ilustrado* 6, no. 302 (February 22, 1923): 56.

39. The estridentistas were avid readers of international avant-garde journals. See Klich, "Revolution and Utopia."

40. Rashkin, *The Stridentist Movement in Mexico*, 116.

41. Maples Arce, *Urbe*, n.p.

42. Germán List Arzubide, *Esquina: Poemas de Germán List Arzubide* (Mexico City: Librería Cicerón, 1923).

43. Haber, *Industry and Underdevelopment*, 133–34.

44. James Oles, "Industrial Landscapes in Modern Mexican Art," *Journal of Decorative and Propaganda Arts*, no. 26, Mexico Theme Issue (2010): 134.

45. Maples Arce, *Actual No. 1*, translated in Klich, "Revolution and Utopia," 508.

46. Manuel Maples Arce, "El espíritu nuevo," *Crisol: Revista de Crítica* 30 (June 1931): 416.

47. Arqueles Vela, *Evolución histórica de la literatura universal* (Mexico City: Ediciones Fuente Cultural, 1941), 409. Vela was specifically referring to Maples Arce's poetry.

# 16
# Technology, Labor, and Realism
## Diego Rivera's Secretaría de Educación Pública Murals

*Anna Indych-López*

The murals produced in the 1920s and 1930s by Diego Rivera and his compatriots, José Clemente Orozco and David Alfaro Siqueiros, among others, reflect the turbulent period of post-revolutionary reform in Mexico, when the country's elite attempted to unify the nation with an appropriate visual language. Perhaps most associated with the development of a pictorial *indigenismo*, Rivera called upon images of Mexico's rural populations—peasants, laborers, Indians, and mestizos, as well as the urban working classes—to construct an image of the nation. His most famous and ambitious murals, at the Secretaría de Educación Pública (SEP, Ministry of Public Education, 1923–28), center on representations of popular and folk festivities as well as regional traditions of labor. Typically, however, assessments of this mural cycle pay little heed to the images of technology—agrarian machines, aviation, radios—that form an integral part of his vision of post-revolutionary Mexico. This chapter elaborates how representations of modern machinery not only figure prominently in the settings of the artist's murals at the SEP, but also reveal a dialogue concerning the complex relations among technology, workers, politics, and realism.

Overt machine-age imagery is atypical in the work of the muralists in general. That we do not readily associate the iconography of the muralists with technology points to its ambivalent role in their broader critique of modernity. The artists' views and their work result from the uneven nature of technological development in the country as well as its primarily agrarian economy until the 1940s.

Lacking faith in the optimism and idealism of the machine age, Orozco, for example, resisted its concomitant standardization and regimentation and thereby, for the most part, avoided its direct representation. When technology does appear in the artist's murals—specifically, in *Epic of American Civilization* (1932–34) at Dartmouth College and in the portable mural, *Dive Bomber and Tank* (1940)—it emerges as a destructive, militaristic force, a foil for Orozco's broader critical project of universalistic humanism. Siqueiros, in turn, was the only one of the three for whom technology played a central role both in his theories on art, in which he advocated for a politicized public art that incorporated the latest technologies, and in his actual practice of art-making. Unlike the others, Siqueiros did not illustrate or depict technology (except in his 1939–40 mural for the Mexican Electricians' Syndicate); instead, he utilized mechanical means of reproduction to revolutionize the process of mural making in Mexico and abroad. Like Orozco, Siqueiros railed against the militaristic uses of technology in contemporary society, yet he exhibited exuberant faith in technological modernity as a means to provide an alternative model of muralism.

Of all the muralists, Rivera engaged most consistently and literally with the theme of the machine age. By 1932, when he was working in the United States, he made technology and industrial machinery one of the primary subjects of his production when he witnessed large-scale modernization firsthand. While the muralist's views on technology are conflicting and mutable over time and place, for Rivera technological progress, as part of a broader Marxist project, signified a means of liberation for the worker. He elaborated this view most consistently in several of his US commissions, yet an examination of Rivera's early mural cycles at the SEP reveals the ways in which these works established the foundation for the artist's life-long interest in machinery, technology, and science and their relation to both political and artistic revolution.

Rivera's SEP mural comprises 235 panels painted around the building's two different courtyards on three floors. On the ground floor, the Courtyard of Labor contains images depicting work and production characteristic of the different geographical areas of Mexico. The panels are oriented according to the cardinal directions so that a particular panel's theme corresponds to the part of the country that the wall faces. In total, the Courtyard of Labor panels depict aspects of physical labor with emphasis on native customs, rural culture, and the life of the working class and the peasants. Rivera centers attention on physical labor and uniformly exalts the worker. By ennobling laborers and revealing aspects of the oppression they faced, the artist imputes a Marxist interpretation onto Mexican national life. He focuses mostly on the agricultural and handicraft economies of Mexico's various geographic regions, yet the rise of industry is prominent as well. If the artist concentrates on physical labor in this cycle, technology and industry serve as the backdrops for the elaboration of his Marxist vision of Mexican culture. Here technology fits into his typical practice of playing opposites against one another within his murals. In addition to opposing physical to technical labor, the cycle contrasts rural and agricultural industry to more urban and industrial technology. Juxtaposing these different types of work, Rivera explores the very nature of modernity in Mexico and suggests the coexistence of different temporalities.

The north wall of the Courtyard of Labor privileges rural culture and premodern communal activities over industry: images of Tehuanas (both at leisure and at work) are interspersed with panels depicting artisanal labor (weavers, dyers), and plantation or rural labor (sugar mill workers). Here Rivera deliberately inserts labor into the region most identified with picturesque native culture by post-revolutionary intellectuals.

Murals on the east wall also depict preindustrial labor but focus more overtly on the incursion of industry into rural areas, specifically, two panels depicting a steel foundry in the northwest of the country. One panel depicts industrial laborers who gather around a crucible and push an iron rod into a furnace; the other panel shows two workers pouring molten metals into a crucible. The clothing choices made by the artist, denim pants and overalls and flat caps or fedoras, are associated with urbanity and differentiate these figures from peasant workers, typically shown in white clothing and sombreros. The foregrounding of industrial machinery here differentiates the foundry scenes from nearby panels, such as *Entry into the Mine* and *Leaving the Mine*, that exalt the worker as a political statement about the subjugation of rural laborers in western Mexican mining industries.

In the mining panels Rivera invokes the Marxist theme of workers' oppression. The pathos of the figures, amplified by the artist's appropriations of Christian iconography—the Crucifixion and the Way of the Cross—ennobles the workers and sanctifies their struggles. The muralist also monumentalizes the laborers, specifically, the figure of the miner who assumes the familiar pose of the crucifixion while an overseer performs an invasive body search. The steel industry panels, in turn, centralize and monumentalize the machinery of industry itself. They therefore introduce the artist's fascination with "mechanization as a source of progress and liberation of the worker," as Mari Carmen Ramírez has pointed out.[1] What Ramírez calls the "detached objectivity" of the steel industry scenes contrasts with the emotional engagement that Rivera instills in the mining panels.[2] The juxtaposition of these scenes, specifically, the unity of technological and social change, introduces the artist's Marxist ideas.

Rivera further emphasizes the importance of technology for his political views on the second floor of the Courtyard of Labor (1924), where he painted twenty-one grisaille frescoes that consistently link technology and scientific inquiry with social issues. One of the most direct representations of the machine age can be found here, in the panel *Electric Machine*, an abstracted image of an electrostatic generator (fig. 16.1). Replete with a rotating disc, electric sparks flying between two terminals, and a motor fan connected to two wire coils, it seemingly represents a quintessential icon of modernity. Rather than illustrate the most modern electrical equipment of the day, however, the artist depicts instead a Wimshurst machine, a device developed between 1880 and 1883 by British inventor James Wimshurst (1832–1903) for generating high voltages. Although such machines were frequently used to power X-ray tubes, by the time Rivera painted this panel in 1924, the Wimshurst machine would have been virtually obsolete.[3] It is plausible that only those with detailed knowledge of the history of electricity would be aware of the anachronism and that the mere image of a machine with electric sparks could be read as ultramodern in a country that only recently experienced the electrification of the landscape. That fiction aside, the artist's willful inclusion of an outmoded technological apparatus relates more to his particular vision at this moment of the educative aspects of science. Such electrostatic generators were still used as demonstration objects in physics lessons to show how electric charges accumulate and were illustrated in physics textbooks for educative purposes. The muralist's inclusion of the electric machine, with other panels focusing on scientific research, relates to the overarching need for the integration of science and technology with education in postrevolutionary society, precisely one of the SEP's broader mandates. In doing so, he self-consciously alludes to the overarching project of his patron and asserts the politics of muralism and the site of this specific commission.

Throughout the second floor, Rivera focused on such scientific and intellectual labor. Much of the work pictured here, moreover, specifically concentrates on technologies of vision involving looking, examining, and analyzing. As a result, the artist posits technology, scientific inquiry, industry, and mechanics as means to serve the interests of the laborer. Many of the panels can thus be linked to politicized issues of the post-revolutionary period. For example, in one panel, a land surveyor looks through the telescope of a leveling instrument. Land surveys took on a heightened significance in the post-revolutionary era, when land redistribution was one of the key platforms of reform. Rivera's panel surely makes reference to this social and economic problem, connecting the emotionally charged issue with scientific objectivity. Further, other panels include scientists in lab coats looking intently through microscopes in the midst of scientific research (fig. 16.2), and in one a surgeon with prominent goggles performs an operation. In the latter panel, the artist elides the human subject and focuses instead on an abstracted organ. The absence of a body can be explained, on the one hand, by the necessity of post-revolutionary muralists to efface corporeal violence for the sake of national healing.[4] The abstraction of

16.1. Diego Rivera, *Electric Machine*, 1924, fresco, north wall, Patio del Trabajo (Courtyard of Labor), second floor, Secretaría de Educación Pública, Mexico City. © 2012 Banco de México, "Fiduciario" en el Fideicomiso relativo a los Museos Diego Rivera y Frida Kahlo. Av. Cinco de Mayo, No. 2, Col. Centro, Del. Cuauhtémoc 06059, México, D.F. Photograph © Rafael Doniz.

the body or its reduction to internal organs, however, also relates to the muralist's evocation of medical technology as a broader historical agent of scientific inquiry that serves the collective good, rather than the individual's misfortune. Throughout his career Rivera explored the invention, dissemination, and uses of medical technology in the ancient and modern worlds.[5] For him, advancements in medicine, science, and scientific technology went hand in hand with the social condition; his technological utopianism envisioned medical progress and knowledge as a form of social justice and collective necessity. Although he was to solidify his theories about technology bonded to social change within the industrially advanced milieu of the United States, the roots of his thinking lay within the SEP.[6]

In this constellation of images the artist focuses notably on optical technologies, indicating that his scientific utopianism takes on an additional layer of meaning. More than just intellectual work, these repeated images of optical analysis—eyeglasses, microscopes, medical instruments, telescopes—become metaphors for the artist-constructor, making these images comments on not just social questions but aesthetic ones too. By connecting modern modes of vision with scientific objectivity and visuality, Rivera uses technology to make statements about the practice of realism. The artist's interest in technologies of visual inquiry links the notion of enhanced vision to the revolutionary artist's role in society: to

16.2. Diego Rivera, *The Investigation*, 1924, fresco, south wall, Patio del Trabajo (Courtyard of Labor), second floor, Secretaría de Educación Pública, Mexico City. © 2012 Banco de México, "Fiduciario" en el Fideicomiso relativo a los Museos Diego Rivera y Frida Kahlo. Av. Cinco de Mayo, No. 2, Col. Centro, Del. Cuauhtémoc 06059, México, D.F. Photograph © Schalkwijk/Art Resource, New York.

16.3. Diego Rivera, *X-Ray*, 1924, fresco, east wall, Patio del Trabajo (Courtyard of Labor), second floor, Secretaría de Educación Pública, Mexico City. © 2012 Banco de México, "Fiduciario" en el Fideicomiso relativo a los Museos Diego Rivera y Frida Kahlo. Av. Cinco de Mayo, No. 2, Col. Centro, Del. Cuauhtémoc 06059, México, D.F. Photograph © Rafael Doniz.

document and to expose the ills of state and society. Modern vision makes human and social perception and representation possible and enables the artist to expose critical truths about society.

Rivera's concern with modes of vision, optical technologies, and the nature of realism is exemplified by the long vertical panel *X-Ray* (fig. 16.3). The panel includes an abstraction of a 1920s X-ray tube, a lemon-shaped bulb with a conical spiral filament inside. Radiating outward are sharply angled sun-like rays that represent the invisible energy of the radiation waves. Beneath the tube the artist depicts paired images of an isolated hand. A naturalistic hand stretches upward, its palm facing the viewer; it is echoed by its transparent companion, an upside-down skeleton. The iconography of the hands refers to the historical first "medical" X-ray photograph by Wilhelm Röntgen in 1895 of his wife's hand.[7] The X-ray, of course, is also a form of imaging and represents a new way of seeing reality. By exposing and making visible the invisible, it is a metaphor for the artist's endeavor, especially one invested in social reform and socially engaged aesthetics.

As a technology of vision used for physical, bodily diagnosis, the X-ray also relates to the artist's insistence on figurative representation for his brand of modernist art. At the SEP he consolidated his signature modernist figuration, a blend of avant-garde, realist, and popular aesthetic strategies. In particular, the reference to the X-ray calls forth his Cubist period, during which he took great interest in the fourth dimension and new ways of seeing, before he ultimately rejected the aesthetic (though he continued to rely on its compositional principles). Rivera turned away from Cubism in order to create a more legible form of modernism that absorbed aspects of avant-garde abstraction yet remained firmly rooted in figuration and realism as a means by which to communicate with the masses. His interest in bodily technologies therefore serves as a metaphor for his adoption of figurative modernism as an innovative, politicized aesthetic.

In the large southern stairwell that connects all three patio levels, Rivera uses technology to make a statement about the modernization of Mexico's agrarian economy. Across the foreground of the large panel *Mechanization of the Country*, 1925, he places icons of post-revolutionary reform (fig. 16.4). At center sits an Indian woman in a field of corn; suggesting an ancient corn deity, she visually and ideologically roots the scene. Flanking her are a female figure who strikes down figures representing the hacienda, military, and the church with a lightning bolt, at left, and the revolutionary trinity of soldier, peasant, and urban worker, at right. Nestled into the background between these figures, a peasant farmer rides a state-of-the-art tractor, electrical towers and lines span the landscape, railroad cars bring goods to a factory, and an irrigation pipeline juts out

16.4. Diego Rivera, *Mechanization of the Country*, 1925, fresco, stairway, Secretaría de Educación Pública, Mexico City. © 2012 Banco de México, "Fiduciario" en el Fideicomiso relativo a los Museos Diego Rivera y Frida Kahlo. Av. Cinco de Mayo, No. 2, Col. Centro, Del. Cuauhtémoc 06059, México, D.F. Photograph © Schalkwijk/Art Resource, New York.

from a dam, while an airplane hovers in the sky above. This industrialized landscape contrasts with the premodern and pastoral symbolism invoked by the centralized image of the timeless corn goddess. And, coupled with the artist's signature symbols of evil and good at either side, industrial technology takes a place as an integral element in this call for reform. In Rivera's hands, modernity thus accompanies very specific political and cultural actions: the striking down of oppression in the form of the *hacendado* system and the church, the unification of the rural and urban proletariat, and the endurance of traditional Indian cultural beliefs.

As in the panels depicting technologies of vision, Rivera's scenes depicting modernization are accompanied by very specific and deliberate stylistic choices related to his brand of realism. In *Mechanization of the Country*, the markers of technology are jam-packed into a cramped space that is tilted up toward the viewer, a hilly landscape filled with awkward angles and multiple perspectives, all stylistic remnants of his Cubist period. The inclusion of the airplane, moreover, makes reference to a leitmotif in post–World War I European painting as well, specifically, the work of Robert Delaunay, who often included images of planes in his work.[8] This nod to French modernism and to icons of modernity joins his technological idealism and socialist interpretation of the Mexican future with the utopianism of the historical avant-gardes. Yet these references to Cubism coexist with emblems of Mexican revolutionary activity and a figurative modernism that is distinct, aesthetically and ideologically, from European painting. Indeed, local popular painting practices (*ex-votos, retablos*) and pre-Columbian/early Colonial visual culture (murals, codices) inform the abstraction present in all his works as much as Cubism. Nonetheless, he never fully forsook figure-ground relationships and painted in a predominately illusionistic style, insisting on a realist practice. As such, the muralist's particular conjunction of modernist, popular, and pre-Hispanic aesthetic sensibilities visualizes the social tensions between modernization and tradition.

*Mechanization of the Country* established a theme that Rivera revisited on the third floor in a cycle known as the Corrido of the Agrarian Revolution (1926). In this section, he translates the *corrido* into visual form. (Corridos are popular revolutionary songs used mostly by the illiterate and uneducated classes as a mode of storytelling to communicate current events and to narrate history.) Rivera runs a banner containing lyrics from several revolutionary songs across several mural panels. Beneath the banner in one panel, peasant laborers decorate a modern tractor with flower garlands for a festival or celebration. Significantly, the sprawling landscape in the background, plowed into perfect geometric rows as a result of the tractor in the foreground, includes two electrical towers that flank the peasant laborer sitting atop the farming machine, signaling the seamless integration of modernity into the rural landscape. Like the corrido itself, modernity fits naturally into Mexican life.

Similarly, the final panel of the series, *End of the Corrido* (also known as *All the World's Wealth Comes from the Land*), contains a landscape punctuated by icons of modernity and technology (fig. 16.5). In the foreground, however, the artist offers a new view of modernity integrated into the domestic life of the rural classes. A family of peasants fills the scene. A mother, holding a book, creates the strong central image; around her, children represent the young postrevolutionary generation and the future of the nation. At lower left, children study assiduously. Beside them, an older sister embarks on another productive task, sewing a white cloth on an electric machine, a jarring technological apparatus in a scene celebrating rural labor. Not only does interior domestic labor complement the agrarian scene, but also this image stands in contrast to the panel of weavers in the Courtyard of Labor who work on an elaborate premodern textile loom. The curvilinear forms of the sewing machine, as well as its dark black color, are mirrored by the modern radio with a large loudspeaker on the opposite side of the composition. In front of the receiver sits a child, looking intently toward the loudspeaker, as does his mother. The cabinet on which the child and the radio sit includes a real door that is part of the SEP's archi-

16.5. Diego Rivera, *End of the Corrido* (from the Corrido of the Agrarian Revolution), 1926, fresco, west wall, Patio de las Fiestas (Courtyard of Fiestas), third floor, Secretaría de Educación Pública, Mexico City. © 2012 Banco de México, "Fiduciario" en el Fideicomiso relativo a los Museos Diego Rivera y Frida Kahlo. Av. Cinco de Mayo, No. 2, Col. Centro, Del. Cuauhtémoc 06059, México, D.F. Photograph © Schalkwijk/Art Resource, New York.

tecture, which the muralist cleverly incorporates into his composition. To the right of the door he includes an illusionistic compartment filled with rifles and bandoliers that is both the literal and metaphorical support of the radio, echoing the revolutionary father on horseback who stands guard over the family in the middle ground.

Although it contains similar background technological imagery as other SEP murals (such as the modern tractor and airplane), this panel moves beyond industrial or medical imagery to incorporate communications technology into Rivera's broader revolutionary vision. The inclusion of the radio updates the oral tradition of the corrido and therefore modernizes folk culture. It also refers to the use of broadcast technology by the state to promulgate its revolutionary messages. In addition to using murals to educate, the SEP launched its own radio station in 1924, not long after the very first radio broadcast in Mexico City took place on May 8, 1923.[9] This government branch thus immediately seized upon the potential of the latest communication methods to collapse both geographic and ideological distances between the masses and the urban elites. Instead of a corrido, then, we might assume that the family listens attentively to a SEP broadcast. The radio's presence in this scene indicates not only the medium's role in shaping official discourse and the government's ability to disseminate new cultural programs through technology, but also Rivera's embrace of this state-sponsored modernization. His vision of cultural renewal, while utopian, is therefore not timeless nor merely a celebration of premodern Indian customs, as is typically assumed. Instead, it participates in both the avant-garde's engagement with modernity and post-revolutionary discourses of progress.

Similarly, in the Corrido of the Agrarian Revolution's counterpart on the third floor, the Corrido of the Proletarian Revolution, Rivera links the triumph of working-class revolution with the rise of industry and a technological utopianism. This series of ten panels, however, reveals a new phase of the artist's engagement with socialist politics. He completed it in 1928 after returning from an eight-month trip to the Soviet Union, which he took from October 1927 to May 1928 as part of a Mexican delegation honoring the tenth anniversary of the October Revolution. During his stay in the Soviet Union, he continued his long-standing connection with Russian intellectuals and artists, whom he had first befriended in Paris in the 1910s. More importantly, he engaged directly with socialist theories of art making and the task of creating a collective, revolutionary avant-garde art that integrated the proletariat. He arrived at a moment when Joseph Stalin was consolidating his power and campaigning against Leon Trotsky, and in the midst of a growing tension in the development of Soviet art. Rivera became a member of October, a group formed in 1928 by artists associated with productivism (including Alexander Rodchenko, Varvara Stepanova, El Lissitzky, and Gustav Klucis) who were committed to the construction of a new socialist society based on a highly developed industrial technology. The group was the last collective effort made by the avant-garde before an increasing authoritarianism began to dictate aesthetic policy against innovation, newness, and abstraction in favor of continuity with the past and more conservative naturalistic styles. Advocating for an art of mass production intended for collective use, the artists of October argued in their manifesto for a new "avant garde, revolutionary industrial proletariat [inspired by] methods of mechanical and laboratory scientific technology."[10] They continued, "The fundamental task of the proletarian artist is not to make an eclectic collection of old devices . . . but with their aid, and on new technological ground, to create new types and a new style of [art]."[11] They rejected the naturalistic realism of the AKhR (Association of Artists of the Revolution), the forerunner of Socialist Realism, which they felt promoted an "individualistic way of life; passively contemplative, [and] static," and the canonization of the past.[12] Significantly, however, they also renounced "abstract industrialism and unadulterated technicism," in other words, a view of technology and industry divorced from a social movement and from the actual situation of the country. Instead, they promulgated a "dynamic realism that reveals life in movement and in action."[13]

As a result of his associations with October, Rivera publicly denounced the aesthetics of AKhR and elaborated his views on the type of style that best corresponds to the proletarian revolution: "This will be a style which will create from painting an excellent, precise, clear and synthetic language, a style which will give the works of art . . . a deeply humane expression, a style which *reconciles art with our contemporary industrial life and our socialist economy*. Proletarian art must begin to speak a language comprehensible to all the proletarian masses of the world and powerful [enough] at the same time to penetrate by cultural means into the capitalist countries."[14] In essence, the Mexican artist's interactions with the Russian avant-garde confirmed his belief in a revolutionary, but legible, form of art. For him, that meant art that engaged a form of realism. While he resisted October's rejection of the past (by continuing to depict Mexico's ancient cultural heritage), his attention turned more directly toward incorporating contemporary industrial life into his vision for a new, if still realistic, form of proletariat art.[15] His stay abroad initiated his stance against "the old and bad academic painters" who were winning favor among the Russian masses. Rivera instead proposed a new form of public art, not as "inaccessible" or "hermetic" as the abstract art practiced by some revolutionary artists (the Constructivists), but still innovative and modern and not dictated by official precepts.[16] The muralist would balance the realism of the past with a new vision of the future.

When he returned to Mexico in 1928, Rivera came armed with a growing sense of how to link his aesthetics with his socialist consciousness and a renewed vision for a transformatory modernism depicting class conflict, which he would incorporate into the Corrido of the Proletarian Revolution panels. More than any other section, it is populated by images of factories, chimneys, smokestacks, industrial machinery, and the masses. With these new panels, the artist moved beyond the official discourse of *indigenismo* that dominates his earlier SEP murals to align the Mexican and Russian revolutions on an ideological basis. He left Russia with an unrealized mural commission for the Red Army Club, unused designs for the covers of Soviet magazines, and a sketchbook of forty-five small drawings of the tenth anniversary celebrations (now in the collection of the Museum of Modern Art, New York).[17] He thus returned to the SEP with a storehouse of revolutionary iconography, which Renato González Mello has called a "Soviet rhetoric of factory councils, assemblies, expropriations and workers."[18] Adding to this repertoire of Soviet-themed imagery is a preponderance of red flags and stars, hammers and sickles, and, notably, a heightened focus on industrial labor in general. Several panels (*He Who Works Eats*, *Cooperative*, and *United Front*) introduce a new type into his cast of characters: an urban factory worker turned revolutionary who wears blue coveralls with red stars and has decidedly "Slavic features," as Stanton Catlin has noted.[19] More combative and challenging than previous images of peasant and worker unity throughout the cycle, the Proletarian Revolution panels focus on direct action—the distribution of weapons, armed uprisings and protests, and even the assassination of a capitalist—creating a forceful political narrative that mobilizes the "dynamic realism" called for in the October manifesto. Rivera, typically, presents a highly idealized vision of revolutionary struggle (and a false notion of political achievement). But the panels also serve as warnings, both to government leaders (the patrons) and to the audiences of the SEP, that the threat of violence remains, should they not remain committed to fulfilling the goals of the Mexican Revolution.

The Corrido of the Proletariat marks Rivera's evolving views on technology, realism, and proletarian revolution. Furthermore, the murals represent the artist's adaptation of lessons learned abroad (international revolution and modernism) to the local context. Once he returned from Russia, he broadened his views of the Mexican conflict, moving from a solely national framework to the context of the international workers' struggle. Notably, throughout the Corrido of the Proletariat, he aligned the agrarian with the Soviet Revolution by incorporating "Land and Freedom," the motto of revolutionary leader Emiliano Zapata, onto a red communist flag containing the hammer and sickle. In these murals, workers of all races, not just Mexican *indios* or dark-skinned mestizos, unite in their call to arms. Rivera takes the racial discourse of indigenismo and makes it more international by merging it with a vision of global class conflict and the worldwide proletariat.

The backdrop for this politicized ideology is the factory and the industrial setting. Peasants and urban workers are still the main protagonists, but Rivera changes the rural setting to one dominated by industrial machinery. He fills the tops of the panels with smoking factory chimneys, drilling towers, and trains. Perhaps the most unexpected use of the factory setting is in *Our Daily Bread*, which includes an iconic Tehuana—the most frequent cipher for rural, indigenist Mexico—who stands over a "last supper," with figures of varied races and ages gathered around a table presided over by a dark-skinned urban worker whose blue shirt is adorned with a red star. The Tehuana, holding a cornucopia of fruits, represents local produce and nature. Beyond the commonplace imagery of the unity of peasant and worker, it is rare to see factories and large-scale industry invading the countryside. The artist thus couples the natural resources of Mexico with modern technology, an early version of a theme—the tensions between tradition and innovation that marked the nation—more fully explored in later murals in San Francisco and Detroit.

Rivera similarly foregrounded a technological society in the most well known panel of the series, *Distribution of Arms*. As in *The Cooperative* (fig. 16.6), the action takes place inside a factory; the focus is on armed insurrection and the eventual union of the rural peasantry (here relegated to the background corner) with the urban workers who cram the scene. The artist pointedly draws the viewer's attention to industrial technology in three instances. First, an industrial machine carefully delineated with details of small levers and moving parts sits prominently between portraits of Frida Kahlo, Julio Antonio Mella, and Tina Modotti, all depicted as communist revolutionaries. Human-scale, the machine seems to be the energy source for the revolutionary activity that is taking place in the mural. Second, the figure of a worker at center top points directly to a factory, in front of which peasants gather in protest. Finally, the same worker looks directly at another industrial machine at upper left. Around it, workers fill the interior of the factory, under a large ceiling window. The muralist employs a similar form of architecture for the interior factory setting of *The Cooperative*, where under the intricate latticework of the windowed roofing of the factory he inserts elaborate mechanical belt systems of two large industrial machines.

Rivera does not, however, merely depict technology or make it solely the setting for the scenes of the Corrido of the Proletarian Revolution panels. He also uses several visual and compositional devices to activate connections between figures and technology. In *Our Daily Bread*, the Tehuana's raised arms echo the stacked chimneys of the factory behind her. Her skirt is the same orange color as the rounded architectural form, in effect, turning her already-geometrized form into one of the factory's chimneys. Similarly, the infinitely repeating raised hands of the workers in *The Pledge* echo the sturdy uprights of the billowing smokestacks of the factory in the background, which prominently flies a red flag with a hammer and sickle. In *The Cooperative*, a modern telephone sits on the edge of a table in front of a factory worker who forcefully points to a list of demands placed in front of two industrialists. Placed near the worker's groin area, the tubular-shaped telephone becomes a mechanical phallus, an extension of the worker's body. Its strategic placement emphasizes the virility of the workers' movement in contrast to the slumping, weak forms of the factory owners. Finally, in *United Front*, a large orange spout projects from a concrete factory

16.6. Diego Rivera, *The Cooperative* (from the Corrido of the Proletarian Revolution), 1928, fresco, south wall, Patio de las Fiestas (Courtyard of Fiestas), third floor, Secretaría de Educación Pública, Mexico City. © 2012 Banco de México, "Fiduciario" en el Fideicomiso, relativo a los Museos Diego Rivera y Frida Kahlo. Av. Cinco de Mayo, No. 2, Col. Centro, Del. Cuauhtémoc 06059, México, DF. Photograph © Schalkwijk/Art Resource, New York.

that forms the backdrop for a scene of masses of armed peasants, soldiers, and workers uniting together. The artist positions the funnel so that it falls exactly in line with the tall, pointed sombrero of one of the anonymous peasants, making it appear as if the masses spill forth from the factory itself. With such deliberate compositional and iconographic devices, Rivera incorporates contemporary industrial life into his utopian vision of class harmony in Mexico. Yet, he also suggests a contradictory viewpoint by both humanizing technology and perhaps enacting a sense of technophilia that borders on the abstract industrialism critiqued in the October manifesto.

The Proletariat panels also demonstrate the artist's elaboration of a politicized realist modernism. Most of the SEP panels engage a plain visual language that, ostensibly, would be legible to the populace. Rivera's precise realism, accented with elements of avant-garde stylistic devices (stacking, multiple perspectives, flattened compositions), continued with the Corrido of the Proletarian Revolution. Nonetheless, the panels show traces of his attempts to modify his style in ways that perhaps relate to the concept of the dynamic realism so prized by the October group. In *Distribution of Arms*, for example, the bottom edge of the factory windows creates a powerful diagonal line that is picked up by other orthogonals throughout the composition (including the machine's long axle, the rifles, the bayonets in Kahlo's arms, and the small industrial machine). Together, these repeated diagonals animate the scene, seemingly mobilizing the viewer and the figures into action. The artist's use of force lines and strong diagonals is unusual in the Corrido scenes, which are mostly tall, narrow panels in which he emphasizes verticality and asserts rectilinear compositions. His use of architectonic and spatial elements to create active compositions suggests the development of a dynamic realism parallel to Soviet propaganda posters (which also featured technology and industrialism as subjects).[20]

Examining the presence and significance of technology at the SEP reveals that the murals painted here played an integral role in the development of Rivera's political and aesthetic ideal. In their utopianism, the Corrido of the Proletarian Revolution panels elide the social strife associated with official modernizing programs. Yet, the artist integrates technology into scenes that he conceivably intended as warnings for governmental leaders, while in earlier SEP cycles, technology seemed to engage official discourses seamlessly. Together, however, the SEP panels promulgate a vision of technology bonded to human agency. Workers and soldiers still dominate, and technology forms a backdrop for politicized social interactions.

An investigation of Rivera's imagery of technology also aids in understanding his continuing practice of realism. In reference to Rivera's mural in Detroit, scholar Rubén Gallo has censured the muralist for his use of techniques from the past (fresco) and use of traditional pictorial modes (muralism/naturalism/realism) to make statements about modernization, the future, and technology.[21] Rather than an ideologically faulted contradiction, however, the artist's working practice (and use of an ancient medium) was a deliberate paradox and mode of production that paralleled specific social conflicts between indigenous culture and modernization; making figurative paintings in a legible manner also served as a mechanism to address Mexico's lagging industrialization and the country's need to grapple with distinct cultural temporalities. Furthermore, by aligning the practice of realism with technologies of vision, the muralist asserted the revolutionary artist's political objective of social reform. Scientific objectivity, in

Rivera's hands, became a metaphor for artistic perception and the social imperatives of the realist's politicized practice. Lastly, his renewed embrace of technology in the context of the Soviet Union allowed him to experiment with slightly more innovative aesthetics that yet were still rooted in revolutionary activity. While his realist murals and frescoes did not literally employ a technological medium, they nonetheless enlisted and communicated a politicized idea of realism.

An analysis of Rivera's early views on technology reinforces the awareness of Mexican muralism as an avant-garde movement that dealt broadly with the issue of modernity and modernization. His SEP murals indicate that previous to his sojourn in the United States, he was already engaging directly with machine-age iconography. The exploration of the muralist's early views on technology augments an understanding of the complex relationship between art, political ideology, and modernity in his early murals to reveal the artist's fundamental Marxist belief in the ability of technology to liberate the worker and to modernize the agrarian nation for the benefit of the masses. An analysis of his images of technology also demonstrates the artist's attempts to reconcile the perceived timelessness of indigenous culture with modernization. While his views might be condemned today as hopelessly utopian or even deeply flawed, they need to be understood within the context of their making and as products of a developing and working practice of realism.

## Notes

I am grateful to Araceli Tinajero and J. Brian Freeman for inviting me in 2009 to contribute a chapter to this volume. Acknowledgments are due to Lynda Klich for bringing my attention to Rivera's depictions of technology at the SEP and for her insightful suggestions after reading several drafts of this chapter.

1. Mari Carmen Ramírez, "The Ideology and Politics of the Mexican Mural Movement: 1920–1925" (PhD diss., University of Chicago, 1989), 256–57.

2. Ibid.

3. "History of Electrostatic Generators," http://www.hp-gramatke.net/history/english/page4000.htm (accessed April 29, 2010).

4. Leonard Folgarait, *Mural Painting and Social Revolution in Mexico, 1920–1940: Art of the New Order* (Cambridge, MA: Cambridge University Press, 1998). It should be noted that two panels at SEP depict warfare and bodily injury: *In the Trench* and *The Wounded*.

5. Mural cycles that focus on medicine include those at La Secretaría de Salud (1929), Instituto Nacional de Cardiología (1943–44), El Cárcamo (1951), and Hospital de la Raza (1953). For an analysis of Rivera's approach to medicine and

technology, specifically as reflected in the Cardiología murals, see David Lomas, "Remedy or Poison: Rivera, Medicine, and Technology," *Oxford Art Journal* 30, no. 3 (October 2007): 454–83.

6. See Robert Linsley, "Utopia Will Not Be Televised: Rivera at Rockefeller Center," *Oxford Art Journal* 17, no. 2 (1994): 48–62.

7. Bettyann Holtzmann Kevles, *Naked to the Bone: Medical Imaging in the Twentieth Century* (Camden, NJ: Rutgers University Press, 1996), 19–22. The X-ray also summons images of the occult and religious symbols; for discussions of Rivera's esoteric symbolism, see Stanton Catlin, "Mural Census," in *Diego Rivera: A Retrospective*, ed. Cynthia Newman Helms, exh. cat. (Detroit: Detroit Institute of Arts, 1986), 245; and Renato González Mello, *La máquina de pintar: Rivera, Orozco, y la invención de un lenguaje* (Mexico City: UNAM, 2008), especially 47–72. Wilhelm Röntgen appears again in the mural cycle at the Instituto Nacional de Cardiología (1943–44).

8. Two famous examples of Robert Delaunay's work are *Sun, Tower, Airplane*, 1913, Albright-Knox Art Gallery, Buffalo, New York, and *Homage to Blériot*, 1914, Kunstmuseum, Basel, Switzerland.

9. Rubén Gallo, *Mexican Modernity: The Avant-Garde and the Technological Revolution* (Cambridge: MIT Press, 2005), 125–26.

10. The October group published their manifesto in *Sovremennaia arkhitektura* [Contemporary architecture] (Moscow), no. 3 (March 1928): 73–74. Translated in John Bowlt, *Russian Art of the Avant Garde: Theory and Criticism* (New York: Viking Press, 1976), 276. For more on Rivera in Moscow, see Leah Dickerman, "Leftist Circuits," in *Diego Rivera: Murals for the Museum of Modern Art*, ed. Leah Dickerman and Anna Indych-López (New York: MoMA, 2011), 14–20.

11. The October group published their manifesto in *Sovremennaia arkhitektura* [Contemporary architecture] (Moscow), no. 3 (March 1928): 73–74. Translated in John Bowlt, *Russian Art of the Avant Garde: Theory and Criticism* (New York: Viking Press, 1976), 276.

12. Bowlt, 276.

13. Ibid., 277.

14. Diego Rivera, "AKhRR i stil proletarskogo revoliutsionnogo iskusstva (otkrytoe pismo v redaktsiiu)," *Revoliutsiia i kultura*, no. 6 (1928): 43. Cited in William Richardson, "The Dilemmas of a Communist Artist: Diego Rivera in Moscow, 1927–1928," *Mexican Studies/Estudios Mexicanos* 3, no. 1 (Winter 1987): 61 (italics mine).

15. Rivera had already incorporated imagery of industrial production and technology into his mural at the Universidad Autónoma de Chapingo (1923–27); his images of industrial workers also appeared on the cover and in the pages of *Mexican Folkways* in 1926.

16. Diego Rivera, "Revolutionary Spirit in Modern Art," *Modern Quarterly* (Baltimore), no. 3 (Fall 1932), quoted in Diego Rivera, *Diego Rivera: Illustrious Words, 1922–1957*, vol. 2 (Mexico City: Editorial RM, 2008), 197.

17. Richardson, "The Dilemmas of a Communist Artist," 63–64; Bertram Wolfe, *The Fabulous Life of Diego Rivera* (New York: Stein and Day, 1963), 216–23.

18. Renato González Mello, "The Murals in the SEP," *Diego Rivera: The Complete Murals* (New York: Taschen, 2008), 32.

19. Catlin, "Mural Census," 245.

20. González Mello, "The Murals in the SEP," 32; Dickerman, "Leftist Circuits," 20.

21. Gallo, *Mexican Modernity*, 7, 16.

# 17
# Cyborg versus "Homo scribens"
## Mexican Literary Expressions in the Era of "Technoculture"

*Erja Vettenranta*

Throughout the twentieth century, authors in Mexico and the rest of the world have needed to adjust to the ever-increasing pace of modern life, where new modes of transportation and media technologies have been shifting the position of the writer and public intellectual within society. It might still be too early to determine whether the contemporary digital culture will produce its own chief narrative form in the same manner that the novel characterizes the print culture of the last five centuries.[1] However, we may currently observe more and more writers taking advantage of new online communication tools as well as incorporating the language and the landscape of cyborgs into their work. This chapter aims to offer a look at these developments in Mexican literary expression. The first part provides a rough overview of various twentieth-century narrative and poetic responses to the scientific and technological manifestations of modernity. The second examines works by three contemporary authors to give an idea of how digital media and other intelligent technologies are played out in literature. Although very different in their approaches, Cristina Rivera Garza's writing in the cyberspace, Naief Yehya's studies and short stories on "technoculture," and Carmen Boullosa's treatment of a virtual novel in *La novela perfecta* all highlight the importance of reading and writing as essentially human acts of meaningful dialogue that distinguishes us from our increasingly sophisticated machines.

## A Sketch of Mexican Literature and Technology in the Twentieth Century

Toward the turn of the twentieth century, the increasing industrial development in Mexico resulted in the growth of urban centers and a rapidly changing social and cultural environment. In this new reality, Realist and Naturalist novelists described and denounced various social and political ills, influencing later writers of the Mexican Revolution. At the same time, the Modernist poets launched their esthetic and moral reaction against the increasing pragmatism, secularization, and materialization of society.[2] They condemned, specifically, the low cultural level and the lack of artistic and intellectual sensibility of the bourgeoisie.[3]

As a rule, the Modernists regarded machines and commercial artifacts as representations of the deterioration of esthetic and spiritual values in modern society. As Rubén Gallo points out, the generation's attitude may be observed in an 1881 statement by Manuel Gutiérrez Nájera, where the Mexican poet laments the distractions that those with higher esthetic standards suffer due to "the asthmatic wheezing of locomotives, the irritating screeching of railways, and the whistling of factories."[4] A more ominous tone may be observed in Amado Nervo's tale "La locomotora" (The locomotive), where the march of technology seems to have surpassed human capacity to deal with it. Approached by a "powerful, violent [locomotive], crested with fire," a father crossing a wide yard of railway tracks with his son gets completely paralyzed in confusion over which pair of rails the train is running on.[5] At the end, father and son are saved by sheer luck as the train passes right next to them before continuing its journey toward the horizon, crowned by its "threatening crest."[6]

Of the novels written during the Mexican Revolution, Mariano Azuela's *The Underdogs* (1915) is the most well known and has been heralded as a precursor of the modern Hispanic American novel in general.[7] Modern narrative techniques, together with Azuela's treatment of conventional symbols of progress, engage the readers actively in their discovery of the pessimistic vision that the novel presents of the revolution and its goals of modernization and social justice.[8] In the first part, the rumored technological advantage of Pancho Villa—the legendary leader of the northern division of the revolution—makes Azuela's peasant fighters dream of the kinds of trains and airplanes (never-before-seen "metal bird[s]") that are said to be filled with arms and food in the North.[9] By the end of the second part, however, these promises of social and technological progress are completely contradicted by the train that takes the different victorious factions to meet in Aguascalientes. The dehumanization of the revolution is symbolized by the terrible crowding on the train, which the narrator describes as being "more packed than a car full of pigs."[10] In addition, the endless stories of looting by these filthy, drunk, and crude soldiers foreground the atrocities committed in the name of reform and modernization.

A completely different attitude toward technology developed during the 1920s as avant-garde artists all over the world embraced machines and new communications devices as symbols of modernity. In Mexico, *estridentista* poets (Manuel Maples Arce, Arqueles Vela, Luis Quintanilla, and Germán List Arzubide, among others) followed the lead of the Italian poet Filippo Tommaso Marinetti and his futuristic ideas to celebrate the dynamic character of the twentieth century. For instance, in the inaugural program of the first Mexican radio transmission in 1923, Manuel Maples Arce read a poem called "TSH," an acronym for *telefonía sin hilos* (wireless telephony).[11] He incorporated the new language of radio technology and visualized different technical puzzles of this "nuthouse of Hertz, Marconi, and Edison."[12] Even more radical in its assimilation is Luis Quintanilla's poem "IU IIIUUU IU," where the poet not only describes the marvels of the radio, but also tries to emulate the acoustic listening experience. Onomatopoetic and void of punctuation marks or clear syntax, the poem is viewed by Rubén Gallo as a kind of "smorgasbord of sound programming" that truly responds to Marinetti's call for a "telegraphic style."[13]

Another vanguard group of the 1920s was organized around the magazine *Contemporáneos* (Contemporaneous) and included writers like Gilberto Owen, Xavier Villaurrutia, Salvador Novo, Carlos Pellicer, and José Gorostiza; they called themselves Contemporáneos. Their esthetic approach was radically opposed to that of the estridentistas, although Gallo recounts a 1924 radio talk by Novo, where he explores the nature of this new medium as well as its possible future uses, such as a tongue-in-cheek suggestion to replace baby bottles and wet nurses with radios.[14] Novo's subdued enthusiasm is characteristic of the kind of "polished sophistication" with which the Contemporáneos approached modernity and their immediate reality.[15] Embracing a philosophical and cosmopolitan attitude, they turned away from strictly nationalistic perspectives in order to participate as Mexican poets in a wider universal culture.

Although he was not a member of the Contemporáneos group, Alfonso Reyes's work shares its cosmopolitan interest in international humanism. In "Capricho de América" (America's caprice), for instance, he emphasizes the creative imagination behind all human endeavors, such as the first European voyages to the New World: "America, before having been discovered by the mariners, was invented by humanists and poets."[16] The same focus on the human spirit of adventure and poetic fancy may be observed in Reyes's essays on the phenomenon of aviation. An essay on the seventeenth-century Spanish humanist Antonio de Fuente la Peña shows how, against all rational ideas of his time, this visionary had the courage to propose the theoretic possibility of flying. In his case, as in numerous others, reasons Reyes, "The poetry suggested the paradigm, fulfilled later by mechanics."[17]

Beginning with Lázaro Cárdenas's presidency in 1934, avant-guard literary innovation waned in favor of more nationalistic themes and traditional narrative techniques to reflect the new political direction of the country. One of the best second-generation novelists of the Mexican Revolution was Gregorio López y

Fuentes, whose *El indio* (1935) combines its revolutionary theme with the marginal position of the Mexican indigenous population in a kind of historical allegory. As in Azuela's *The Underdogs*, the novel emphasizes a sense of alienation from the revolutionary action that has failed to fulfill its promises even when a brand new road is built right around an indigenous village. Ironically, the road, a symbol of the progressive revolutionary project, only serves the needs of the wealthy while forcing the Indians to travel even further to pay their tributes.[18]

In later novels of the revolution, especially from the 1940s onward, and reflecting the government's attempts to increase Mexico's international profile, the modern narrative strategies familiar from French, English, and North American fiction started to be combined with otherwise national themes.[19] As John S. Brushwood affirms, Agustín Yáñez's novel *Al filo del agua* (1947) indicates the new direction by blending anecdote, *costumbrismo*, and social protest with calculated esthetic effects and a level of narrative interiorization not witnessed earlier.[20] José Revueltas and Juan Rulfo are other important writers of this period. In Rulfo's *Pedro Páramo* (1955), the gloomy and pessimistic world of *caciquismo* is given universal meaning by a complex narrative structure. The novel breaks away from chronological time and uses multiple narrative voices, blending the world of the living with the mythical world of the dead in a way that truly anticipates the move toward greater universalism and innovation in Latin American literature during the 1960s and 1970s.

This "new narrative," both in Mexico and in the rest of Hispanic America, emphasized the imaginative freedom to experiment with language, perception, and different levels of conscience to mirror simultaneous developments in linguistics, philosophy, psychology, and other sciences. Early examples of the fantastic are some of the short stories collected in Juan José Arreola's *Confabulario* (1952) and Carlos Fuentes's *Los días enmascarados* (1954). Showing influence of the Argentine Adolfo Bioy Casares's novel *The Invention of Morel* (1940), Arreola's science-fiction story "Baby H. P." announces the creation of a device that converts the incessant movements of children into useful electrical energy. Fuentes, for his part, critiques the materialism of contemporary society in "El que inventó la pólvora," where a vicious cycle of consumerism is provoked by everyday manmade objects that start to disintegrate mysteriously at an ever-increasing pace.

As Jerry Hoeg observes, the writers of the second half of the twentieth century tend to react against modernization projects led by science and technology because they are often seen as tools of those in power.[21] A somewhat late Magical Realist novel in Mexico, Laura Esquivel's *Like Water for Chocolate* (1989), juxtaposes the gringo doctor John Brown's scientific rationality with the protagonist Tita's intuitive science of cooking and her connection with the indigenous world through Nacha, the late family cook.[22] The opposition between reason and intuition, medicine cabinet and food pantry, may not seem so strange considering the long history of unfulfilled hopes and promises of modernization in post-revolutionary Mexico. After all, by the 1980s, dreams of progress through scientific and technological advancement seemed to be crumbling due to a severe economic downturn and the devastation caused by the 1985 earthquake in Mexico City.

In his novella *Battles in the Desert* (1981), José Emilio Pacheco paints the miraculous years of growth and optimism during the late 1940s and early 1950s from a cynical perspective devoid of all confidence in utopian or nationalistic promises of social liberation. The childhood memories of Carlos, the preadolescent narrator of the Pacheco novella, are filled with Cadillacs and Buicks in the streets of Mexico City, US entertainment on the airwaves and in the theaters, and toys, Wonder Bread, and Kraft cheese at his friend Jim's house. Now, in Carlos's view, all that remains of that city of innocent love and hope is demolition and a sense of disillusion: "That city came to an end. That country was finished. There is no memory of the Mexico of those years. And nobody cares. . . . Everything came to an end just like the records on the jukebox."[23] In a more ironic take on urban subcultures, José Agustín's *Ciudades desiertas* (1982) looks into the life of a Mexican couple against the backdrop of the uninspiring and alienating life of US cities.

In the 1990s, the writers of the Crack Movement—just as the more internationally organized McOndo writers—consciously broke away even further from the kind of Latin Americanism promoted by Magical Realist texts. Instead, they returned to the cosmopolitan ideals of their own tradition, producing novels like Jorge Volpi's *In Search of Klingsor* (1999), where it may be difficult to even guess at the author's nationality. This novel intertwines world history with science, politics, and fiction as two physicists, one American and one German, try to solve the identity of Klingsor, a scientist they believe was in charge of Nazi Germany's research for an atomic bomb. The ultimate cause of these contemporary authors tends to be literary, but, as Ignacio Padilla highlights, without being evasive of reality.[24]

The twenty-first-century Mexican authors, together with other Latin American voices, are thus moving increasingly toward what Julio Ortega calls "the new internationality."[25] One of the most booming literary scenes of the moment is taking place in the North of Mexico and, specifically, in Tijuana, where the new global and digital discourses are starkly contradicted by a wall that physically separates two historically and economically joined border regions. Writers such as Heriberto Yépez (*Al otro lado*, 2008), Javier Fernández Aceves (*Se-

*ñora Krupps*, 2010), and Fran Ilich (*Circa 94*, 2010) re-create in their fiction the hallucinating contemporary existence of "in-betweenness" in Mexico and the United States, in Spanish and in English, on paper and online. Similarly, the work of the three authors examined below departs from a pursuit of crossing frontiers both literally and literarily. Let us now examine how Cristina Rivera Garza, Naief Yehya, and Carmen Boullosa approach the shrinking boundary between human beings and their computers and other electronic devices.

## Cristina Rivera Garza: Reading and Writing Collectively with Digital Media

Cristina Rivera Garza is a novelist, poet, and essayist whose work in print as well as in cyberspace engages with problems of point of view, memory, shifting identities, and the relationship between fiction and reality. Her novel *Lo anterior* (2004) begins as an inquiry into the origin of a note found in the pocket of a moribund man, saying: "Love always occurs after, in retrospective. Love is always a reflection." What interests the author is not to tell an amorous anecdote, however, but to explore the way in which life experiences are transformed through reflection and language into meaning.[26] The novel's play on photographic focalizations from multiple perspectives makes a realist reading of the story impossible, highlighting the fact that for Rivera Garza writing is not about conveying messages, but about producing realities and meanings through time, and part of this whole process is an active reading of the other.[27]

The opportunity that technology offers for exploring this type of heteroglossia or dialogic language is what draws Rivera Garza to write in cyberspace.[28] She defends the use of both Twitter and weblog as a type of writing lab in real time: "What truly interests me is the production of short and interconnected texts that affect each other through principles of juxtaposition and montage in order to produce a narrative . . . very much in sync with how we live here and now."[29] Most importantly, electronic writing enables the constant interaction with and awareness of the other, which is a defining factor of Rivera Garza's philosophy of reading and writing as communal practices. A concrete example of Rivera Garza's literary take of Twitter is her monthly column "La Cámara Verde" (Green camera) in the poetry magazine *Periódico de poesía* of the UNAM (Universidad Nacional Autónoma de México). In this space, which the writer also reproduces on her personal blog, she reflects on the polyphonic and even subversive nature of tweeting: "The very passing of the [Timeline] forces us to intersperse, to zigzag, to make indirect notions and to have a dialogue."[30]

A more playful tone may be observed in her submission titled "Tuitrulfo:

Una Convocatoria," a call for papers to rewrite Juan Rulfo's short story "We're Very Poor" in a series of no more than twenty tweets of 140 characters each.[31] Rivera Garza humorously calls for any type of "appropriation, recycling, copy, transcription, hijacking [or] retouching" of Rulfo's story in order to diversify the kind of entries usually found in Mexican cultural and academic forums. The contest may certainly be read as a slightly sardonic commentary on the cultural value of a book like *Twitterature*, a tongue-in-cheek book by Alexander Aciman and Emmett Rensin, in which the two young Americans abridge the most beloved works of world literature into status updates of less than 2,800 characters.[32] Nevertheless, it also goes to demonstrate how technology may open up new possibilities for literary engagement by encouraging new creations based on a meaningful dialogue with another text.

A friend and contemporary of the Mexican Crack writers, Rivera Garza does not consider herself a member of this group, although she shares their interest in questioning existing forms of narrative and searching for alternative ones. In addition to Twitter, she carries out this quest on her blog *No hay tal lugar: U-tópicos contemporáneos* (There's no such place: Contemporary U-topics). As the title of the blog reveals, the medium intrigues Rivera Garza for the utopian freedom it offers her to experiment and to discover the processes of producing texts, or what the author calls the "reverse side" of writing with all its false beginnings and endings, different drafts and revisions, everything in inverse order with the most recent post appearing first.[33]

Therefore, what started as a *blognovela* was rebaptized as a *blogsívela* to highlight its fundamental difference from a novel. As Rivera Garza's play on words demonstrates, a novel does exactly the opposite of what its name in Spanish implies (*novela* in Spanish means "does not veil"): it veils (*sívela*), by masking and disguising its own processes, giving the impression of its own inevitability, of "this-is-how-it-was-going-to-be-from-the-beginning."[34] It is the blogsívela that makes all this apparent, ironically "show[ing] what the novel refuses to show: that the writing is always about to become something else."[35]

For Rivera Garza, technical innovations and devices are a tool for thinking about reality and our place in it through the act of writing. In her narrative, the principles of optics illustrate the relational character of our experience as human beings. On the other hand, her experiments with Twitter and her blog become instruments for understanding writing as a constant work in progress, as well as for a continuous reading of and dialogue with others. This way, Rivera Garza's texts demonstrate the nontraditional identity that Julio Ortega observes in current Latin American narrative: it embraces the new and the momentary as well as the networks of interaction where the human subject is, above all, someone that exchanges words.[36]

Literary Expressions in the Era of "Technoculture" / 309

## Naief Yehya: Selected Writings of a "Homo Cyborg"

Rather than use the Internet community to explore the dialogic aspect of humanity, Naief Yehya uses his printed texts to engage with different human phenomena in cyberspace. As the title of his blog, "Homo Cyborg—Naief Yehya," suggests, the writer adopts the posthuman assumption that our existence today is fundamentally characterized not only by our use of computers and electronics, but by the fact that these devices have actually become almost like extensions of our being. Yehya has written extensively on the increasing role of technology in society. In *Tecnocultura* (2008), for example, he explores the reach of cybernetics, electronics, and artificial intelligence and the way in which our communication and information technologies are actually beginning to shape and redesign us as human beings.[37]

Yehya's fictional treatment of the cyborg figure may be observed in several stories in his *Historias de mujeres malas* (2001), where he plays with the obsessions fed by scientific and high-tech solutions for perennial human problems. As futuristic as many of his scenarios might seem, they engage with emerging twenty-first-century advances in robotics, nanotechnologies, and computational neuroscience, while also participating in the ethical debates concerning the direction of these developments. In "La nueva dirección de María" (María's new address) and "El ladrón de los sueños" (Dream thief), the protagonists' lives are transformed by technologies that enable ever more accurate measurements of our being through our sensory, cognitive, and emotional systems. In "1969, año cero" (1969, year zero), two computer wizards face a reality where artificial intelligence has not only surpassed human thinking, but also develops consciousness and a parallel universe to replace the one inhabited by humans.

Yehya's treatment of these problems tends to question the power of science to crack the fundamental core of human experience and to solve our existential crises. In "La nueva dirección de María," the baffled and disillusioned narrator is left with a few lines of email correspondence from his beloved flesh-and-spirit wife after a scientist paradoxically pretends to transmigrate the essence of María de Mi Corazón (María of my heart) by storing into binary code each of her brain cells only. In "El ladrón de los sueños," a sleep researcher claims to have invented a machine that reads people's dreams, enabling him to reduce the secrets of the human mind into a single overarching paradigm. However, his scientific explanation of one of the most enigmatic human experiences provokes the fury of people unwilling to accept the exchanging of "the dreams of humanity for the nightmare of progress."[38]

The other consistent theme in Yehya's stories is the fascination with communications technologies and the effects of mass media on culture. In "1969, año cero," humans have gradually lost the control of their intelligent machines, which have hence proceeded to take charge of all major decision-making processes. The takeover by computers has been so successful thanks to various "strategies of *positive alienation*" to satisfy all the immediate desires of humans, immersed in an artificial reality saturated with news, tragedies, and images presented as entertainment.[39] The complete reversal of the order of things is highlighted at the end of the story when the narrator realizes, to his horror, how his own daily rituals are irrevocably repeating "like a program" written by a computer.[40]

While the growing sophistication of artificial intelligence is thus forcing people in the twenty-first-century to rethink what makes them truly human, the answer in Yehya's story is ironically provided by a computer. As one of its auto-generated posts reminds, humans have forgotten the fact that rather than originating from an essential ego, "each individual is the result of a net."[41] In other words, the humans of "1969, año zero" are at such a loss with the world and themselves because they have lost the network of true communication and dialogue with each other. Hypnotized by the virtual reality of media entertainment, they seem helplessly trapped in what Fredric Jameson has called the "perpetual present," where the only meaning is provided by the perception of breaks and differences in sensory data. However, as Jameson points out, this is actually a "way of getting rid of content" since "this kind of view does not pose the problem 'How do we relate to those things, how do we turn those things back into continuities or similarities?'"[42]

As may be observed both in his essays and fiction, the type of social and historical amnesia that Jameson describes is of great concern for Naief Yehya. While it makes no sense for this self-proclaimed cyborg to turn back the clocks on technology, he also makes it clear that "the printed word has one quality that the screen cannot offer: it gives us a respite to reflect, it allows us to take time to think and consider things" without the ever-pressing glare of television or computer monitors.[43] This way, the "Homo Cyborg" of Yehya's blog could be rebaptized as "Cyborg scribens," a writer who reflects—and makes his readers reflect—on the deeper essence of María de Mi Corazón, on the dreams and fears of his other characters, and on how we use technology to improve or to impair our fundamentally human qualities.

## Carmen Boullosa's *La novela perfecta*: Negotiating Meanings through Text

Like the texts analyzed above, Carmen Boullosa's work also addresses the fundamental importance of human language and the act of writing as a way to engage more meaningfully with reality. Her historical novels *Llanto: Novelas imposibles*

(1992), *Duerme* (1994), and *Cielos de la tierra* (1997) are cases par excellence of texts that strike a dialogue with the national past in order to negotiate the authority of official historiography and to offer new perspectives on how history informs the present.[44] In *La novela perfecta* (2006), Boullosa extends her revisionist look from Mexico to the increasingly transnational realities typical of New York City, where the novel is set. In addition, it comments on the possibilities of technology to translate human experiences and to replace writers and readers as interpreters of history.

The narration of Boullosa's novel consists of a sort of diary or a written venting by José Vértiz, a lazybones (*holgazán perezoso*) Mexican novelist living in Brooklyn, whose virtual, computer-generated "perfect novel" has just been destroyed in a disastrous chain of events. The novel has been a joint project with his next-door neighbor Paul Lederer, an inventor of a computerized system capable of capturing and transmitting the entire imagination of a writer without him or her having to go through the fastidious task of putting it into words. As the visionary neighbor claims, words are simply not efficient enough in truly conveying anything, because "phrases always have loopholes, they always have hollows and there is always some space, no matter how small, where the reader is able to squeeze things out of the author and go where he shouldn't."[45]

Vértiz's initial hesitation before Lederer's prophesies is quickly erased after observing his novel emerge from his mind exactly like he has imagined it: "charged with emotion, color, light, odor, presence, [and] sense of touch."[46] In all its sensory richness, the computer-generated reality of the "perfect novel" is completely irresistible to anyone who observes it and impassive to any other interpretation of the (his)story than the one envisioned by its author. The captivating power of this new novel bears a striking resemblance to the media torrent used by the computers in Yehya's story "1969, año cero" to keep people happily unaware of things going on around them. In Boullosa's novel, the chilling insinuations of such an apparatus become dreadfully clear when Vértiz jokingly comments on the commercial possibilities of a machine that produces such "impeccable, exemplary beings without any will of their own."[47]

While the shallow surface value of Vértiz and Lederer's perfect novel simply denies any interpretations, the reality of the city that the characters of Boullosa's novel inhabit symbolizes the kind of dynamic textual space where true meaning is created through contact, dialogue, and translation. Vértiz's walks through the cultural mosaic of his Brooklyn neighborhood offer endless opportunities for challenging and negotiating different cultural readings and preconceptions. As the cross-cultural critic Homi Bhabha asserts, it is the space of cultural difference and exchange, "the *in-between* space . . . that carries the burden of the meaning."[48] Just like engaging with a dynamic, transcultural city like New York, the process of writing and reading a novel also requires one to step beyond one's own subjectivity and story of origin.

Of course, none of this is possible through the perfect novel, where the reader is completely at the mercy of the writer's will. Vértiz himself starts to observe the strangely disturbing nature of his work as it "represent[s] something totally unreal, a piece of reality that [does] not obey the rules of the real world."[49] As mere sensory appearance, the novel turns as incomprehensible as Lederer's work studio, an empty shell of a brownstone, where the lack of all interior dividing structures seems to have erased its very essence and capacity to function as a house.[50] Likewise, Vértiz's characters—devoid of any meaningful point of contact with the reader or the real world—lose all identity and end up destroying each other in the apocalyptic final scene that highlights their dehumanization. It is the missing social-historical depth that makes these characters so despicable, and they could be seen to symbolize the kind of literature that Boullosa describes as "disposable, pure play of 'anecdotes,' without any value in the words."[51]

Ironically, although Vértiz doubts the literary value of his account of the events, his text acquires significance precisely because it is not perfect, because it has "loopholes" that connect it meaningfully to real histories. As for Naief Yehya's "Homo Cyborg," writing is a way for Vértiz to make sense of what has happened as well as to center his own self within reality. Moreover, as he lets his innermost thoughts and ideas "shoot into the whole wide world," he also subjects them to the scrutiny of his readers.[52] Therefore, as does Cristina Rivera Garza, Carmen Boullosa creates in her novel another intermediate space where the dialogue between the reader and the text truly completes its meaning. As the cultural crossroads of the city space, this literary space is dangerous because, in it, all previously held conceptions about reality are put to the test.[53]

## Conclusion

The work of writers like Cristina Rivera Garza, Naief Yehya, and Carmen Boullosa makes it obvious that we are not yet witnessing the end of the book or the novel in favor of an endlessly diverging virtual "hypertext," a development that Robert Coover anticipated back in 1992.[54] Instead, these authors use their writing to step away from the incessant media saturation to do what we humans still do better than our computers: engage in a meaningful and creative dialogue that increases our social, cultural, and historical consciousness. As one of Naief Yehya's stories highlights, "each individual is the result of a net"; in other words, we are all products of an ongoing negotiation of meaning.[55] This is the fundamentally human characteristic that Cristina Rivera Garza's narrative and virtual writing explores and what Carmen Boullosa's *La novela perfecta* foregrounds against culturally empty or unifying discourses.

Of course, the authors studied above only represent the tip of the iceberg when it comes to texts being produced in and about cyberspace, about computers, artificial intelligence, and the chaotic contemporary world of media overflow. It is, in fact, hard to think of a current writer who is not, in one way or another, exploring the impacts of technology and mass media on culture. Among the many contemporary Mexican narrators whose work deals with these themes are Juan José Rodríguez (*El gran invento del siglo XX*, 1998), Guillermo Fadanelli (*Lodo*, 2002), Juan Villoro (*El testigo*, 2004), Pablo Soler Frost (*Yerba americana*, 2008), Daniel Krauze (*Cuervos*, 2008), and Sergio González Rodríguez (*Infecciosa*, 2010), just to name a few. In science fiction, Bernardo (BEF) Fernández and Pepe Rojo are examples of writers who explore the relationship between humans and their machines. In addition, essayists such as Fausto Alzati and the aforementioned Fran Ilich have written on the reach of technology, cyberspace, advertising, and mass culture into the private and public lives of Mexicans. What characterizes the work of all these writers is the poetic imagination and human creativity that underlie and give coherence and meaning to the perennial human quests beyond the reach of scientific and technological progress.

## Notes

1. George P. Landow, *Hypertext 3.0* (Baltimore: Johns Hopkins University Press, 2006), 266.

2. Realism and Naturalism are terms used to describe novels and other prose texts, whereas the Latin American Modernists are mostly poets, although many of them also wrote short stories and chronicles as part of their labor as journalists.

3. José Olivio Jiménez and Carlos J. Morales, *La prosa modernista hispanoamericana: Introducción crítica y antología* (Madrid: Alianza, 1998), 30.

4. Manuel Gutiérrez Nájera, quoted in Rubén Gallo (and translated by Gallo), *Mexican Modernity: The Avant-Garde and the Technological Revolution* (Cambridge, MA: MIT Press, 2005), 4–5.

5. Amado Nervo, *Obras completas*, vol. 2 (Madrid: Aguilar, 1967), 597. All translations are mine unless otherwise indicated.

6. Nervo, *Obras completas*, 2:597.

7. Carlos Fuentes, *Valiente mundo nuevo: Épica, utopía y mito en la novela hispanoamericana* (Mexico City: Fondo de Cultura Económica, 1990), 193.

8. For a more detailed account of the narrative point of view in Azuela's novel, see Dick Gerdes, "Point of View in *Los de abajo*," *Hispania* 64, no. 4 (1981): 557–63.

9. Mariano Azuela, *The Underdogs*, trans. Sergio Waisman (New York: Penguin Books, 2008), 65.

10. Ibid., 112–13.

11. Manuel Maples Arce, quoted in Gallo, *Mexican Modernity*, 123.

12. Ibid., 126.
13. Gallo, *Mexican Modernity*, 165.
14. Ibid., 159.
15. John S. Brushwood, "Innovation in Mexican Fiction and Politics (1910–1934)," *Mexican Studies/Estudios Mexicanos* 5, no. 1 (1989): 83.
16. Alfonso Reyes, *Obras completas*, vol. 11 (Mexico City: Fondo de Cultura Económica, 1960), 75.
17. Ibid., 6:299.
18. Juan Bruce-Novoa, "La novela de la Revolución Mexicana: La topología final," *Hispania* 74, no. 1 (1991): 41.
19. Brushwood, "Innovation in Mexican Fiction," 87.
20. John S. Brushwood. *México en su novela* (Mexico City: FCE, 1987), 23.
21. Jerry Hoeg, "*La ciudad de los prodigios* de Eduardo Mendoza frente a una visión latinoamericana de ciencia, cultura y tecnología," *Revista Iberoamericana* 73, no. 221 (2007): 867.
22. Ibid., 866.
23. José Emilio Pacheco, *Battles in the Desert and Other Stories*, trans. Katherine Silver (New York: New Directions, 1987), 116–17.
24. See Ignacio Padilla, "McOndo y el Crack: Dos experiencias grupales," in *Palabra de América*, ed. Roberto Bolaño et al (Barcelona: Seix Barral, 2004), 147.
25. Julio Ortega, introduction to *Antología del cuento latinoamericano del siglo XXI: Las horas y las hordas* (Mexico City: Siglo Veintiuno, 1997), 12.
26. Cristina Rivera Garza, in an interview with Jorge Luis Herrera, "El amor es una reflexión, un volver atrás," *El Universo del Buhón* 60 (2005): 50.
27. Ibid., 48–49.
28. *Heteroglossia* is a term defined by the Russian linguist Mikhail Bakhtin to describe the diversity of voices and points of view that characterizes the novel as a genre. See Mikhail Bakhtin, "Discourse in the Novel," in *The Dialogic Imagination*, ed. M. M. Bakhtin and Michael Holquist (Austin: University of Texas Press, 1981), 259–422.
29. Cristina Rivera Garza, in Carlos Hugo González's article "Tecnología y literatura en tiempo real," *Milenio*, May 21, 2010.
30. See www.cristinariveragarza.blogspot.com (accessed March 10, 2011). Timeline, or "TL," on Twitter refers to the timeline that collects and visualizes a user's stream of tweets in real time.
31. See www.cristinariveragarza.blogspot.com (accessed March 19, 2011).
32. Alexander Aciman and Emmett Rensin, *Twitterature* (New York: Penguin, 2009).
33. Cristina Rivera Garza, "Blogsívela: Escribir a inicios del siglo XXI desde la blogósfera," in *Palabra de América*, ed. Roberto Bolaño et al., 177.

34. In Rivera Garza's column "La mano oblicua," *Milenio*, July 24, 2007, http://impreso.milenio.com/node/7093766 (accessed April 15, 2011).

35. Ibid.

36. Julio Ortega, "Prólogo," in *Antología del cuento latinoamericano del siglo XXI: Las horas y las hordas*, ed. Julio Ortega (Mexico City: Siglo XXI, 2001), 15.

37. Naief Yehya, *Tecnocultura* (Mexico City: Tusquets, 2008), 10.

38. Naief Yehya, *Historias de mujeres malas* (Mexico City: Plaza & Janés, 2001), 125.

39. Ibid., 15.

40. Ibid., 19.

41. Ibid.

42. Fredric Jameson and Ian Buchanan, *Jameson on Jameson: Conversations on Cultural Marxism* (Durham, NC: Duke University Press, 2007), 47.

43. Yehya, *Tecnocultura*, 116.

44. Around the 1992 centenary of Christopher Columbus's first voyage to America, there was a surge in artistic interest for the conquest and the colonial history both in Mexico and in the rest of Hispanic America. For a more detailed description of this phenomenon in Mexico, see the first chapter of Carrie Chorba's book *Mexico, from Mestizo to Multicultural: National Identity and Recent Representations of the Conquest* (Nashville: Vanderbilt University Press, 2007).

45. Carmen Boullosa, *La novela perfecta* (Mexico City: Alfaguara, 2006), 24.

46. Ibid., 42.

47. Ibid., 69.

48. Homi Bhabha, *The Location of Culture* (New York: Routledge, 1994), 38.

49. Boullosa, *La novela perfecta*, 122.

50. For a further description of this type of residential buildings that started to define new neighborhoods in New York City from the mid-nineteenth century onward, see Edwin G. Burrows and Mike Wallace, *Gotham: A History of New York City to 1898* (New York: Oxford University Press, 1999), 716–18.

51. Carmen Boullosa, "La destrucción en la escritura," *Inti: Revista de literatura hispánica* 42 (1995): 216.

52. Boullosa, *La novela perfecta*, 130.

53. Boullosa expresses this idea about the riskiness of all literature in her article "La destrucción en la escritura."

54. Robert Coover, "The End of Books," *New York Times*, June 21, 1992, http://www.nytimes.com/books/98/09/27/specials/coover-end.html?_r=2 (accessed April 16, 2011).

55. Yehya, *Historias de mujeres malas*, 19.

# 18
# Technology and the Architectural Culture of Mexico from the 1968 Olympic Games to the Onset of the New Millennium

*Edward R. Burian*

The complex relationship of technology and architectural culture in late twentieth-century Mexico can best be described in terms of inherent tensions related to Mexican building production, situated in the midst of shifting architectural paradigms. These inherent tensions are the result of intricate dichotomies of materiality and tectonics, handcraft and industrialization, and economics and labor, as well as the production of works that engage both national and local issues and aspirations to participate in contemporary global culture.[1] Shifting architectural conceptions of the era are reflections of broader architectural discourses, including a number of postmodern agendas in architecture. They include the residual effects of modern technical rationalism and structural determinism during the late 1960s, the postmodern resistance to modern globalization that characterized critical regionalism in the 1970s and 1980s, and the rediscovery of orthodox modernism (although without its social utopianism) as an unfinished project in the 1990s onward.[2] Due to the fact that architectural cultural in Mexico is typically discussed as merely an aesthetic practice, excluding its technological dimensions, a number of key innovative technological conceptions prior to 1968 will be briefly reviewed here and provide a context for the discussion of architecture after the 1968 Olympic Games.

## Background: Avant-Garde Composition, Materials, and Technology, 1792–1968

The desire to participate in the global avant-garde in Mexican architectural culture initially emerged during the Neoclassical era (1792–ca. 1867), a period that signaled the emergence of the modernist architectural concepts of rationalism, functionalism, modularization, reductive typologies, and static abstraction, which would play a large role in the manner in which technology was conceived in modern architecture.[3] Composed in the Neoclassical vocabulary, the Palacio Buenavista, designed by Manuel Tolsá (Mexico City, 1803), consciously utilized gray quarry stone to represent an avant-garde and globalized architectural vocabulary and construction technology, instead of the red volcanic stone *tezontle* that was associated with the outmoded architecture of the preceding Baroque era of Mexico City.[4]

The aspiration to embrace European and US culture only became stronger during the Porfiriato (1876–1911). The Banco Mercantile, designed by Alfred Giles (Monterrey, Nuevo León, 1901), reflected the cosmopolitan nature of technological exchanges from the United States after the arrival of the railroad in 1882, as well as a hybrid attitude toward the work methodology, technology, and building imagery.[5] Giles was an English-born and educated architect who immigrated to San Antonio, Texas, where he established a successful practice and started a branch office in Monterrey. While the general contractor was the American W. H. Hollingworth, the project was built by Mexican labor. The cladding of the façade of finely cut stone from Durango was crafted by highly skilled workers from San Luis Potosí over a bolted cast-iron structural frame that was fabricated in the United States.[6] The façade of the project reflected global US and European eclectic Beaux Arts architecture but was locally inflected with the Mexican national symbol of an eagle with a snake at the top of each parapet.

Post-revolutionary architecture (1920–1939) reflected the search for an appropriate representation for a post-revolutionary society in a number of vocabularies and also explored the emerging technology of the time, reinforced concrete. One of the most innovative projects of the period was the Kahlo/Rivera Studios, designed by Juan O'Gorman (Mexico City, 1929), which merged the globalized architectural principles and vocabulary of Le Corbusier with local and national concerns. A landscaped wall of cordon cactus at the edge of the site was used to establish outdoor rooms in the traditional Mexican manner. Unlike the contrasting, discrete enclosure systems of Le Corbusier's work of the period, the studios were tectonically conceived as a rational system of concrete frame and waffle slab, where walls were acting as infill between the concrete frame. Glazing consisted of steel, industrialized windows that formed a fine overall pattern. Traditional vernacular colors utilized in the exterior façades of the studios related the architecture to the culture of the place.

The interiors were quite distinct from Le Corbusier's interior surfaces, which were smooth and uninterrupted, with concealed wiring and plumbing and with industrial objects, cubist paintings, views to landscapes, and objects of a healthy

18.1. Palacio Buenavista, Manuel Tolsá, Mexico City, 1803. Edward R. Burian Collection of Mexico and the American Southwest.

life to project an image of modernity and physical well-being and a controlled sense of order. In the Kahlo/Rivera Studios the interiors were expressed as distinct, separate elements. Mechanical and electrical systems were exposed in the studios. All pipes were visible, and electrical power runs and connections were used as expressive compositional elements reminiscent of veins and arteries. Industrial components of the house, such as the shower, were proudly presented to aid in health and cleansing. Ironically, although machine technology was utilized as an expressive device, the building was produced by inexpensive hand labor. Thus the representation of industrialized systems of construction did not correspond to the realities of handcraft technology.

With an influx of income during World War II, Mexico's economy and cities rapidly grew (1940–1968); and a variety of presidential administrations pursued an agenda of industrialization and progress. This was reflected in one of the earliest modernist churches in Mexico, the Iglesia la Purísima, designed by Enrique de la Mora y Palomar with Félix Candela and built by Armando Ravizé Rodríguez (1940–1946). Although composed as a traditional Latin cross plan,

18.2. Iglesia la Purísima, Enrique de la Mora y Palomar with Félix Candela, Monterrey, Nuevo León, 1940–1946. Edward R. Burian Collection of Mexico and the American Southwest.

the church is tectonically expressed as a series of parabolic concrete ribs with a thin-shell concrete vault spanning between. The smaller, scaled volumes of perpendicular vaults intersect the main parabolic vault at its edges and create side altars. In contrast, the façade behind the altar is rationally composed with a curved strip of glazing between the structural vault and an infill wall of local stone masonry. A contrasting vertical bell tower made of local stone masonry featured an articulated belfry with horizontal slots. Here the economic realities of labor, handcraft, structural determinism, and structural efficiency created a religious symbol associated with modernity and progress. With abundant, inexpensive labor, handmade wooden forms were made to efficiently use concrete in compression, with the interior plastered smooth to suggest machine production. Widely acclaimed throughout Mexico as a symbol of modernity, the church became known to many in Mexico and around the world when it was featured on three of the regular postage stamps of Mexico issued between 1950 and 1976.[7]

With this context in mind, the changing agendas of architectural ideology and the approaches to technology from the Olympic Games of 1968 to the new millennium are revealed by examining pairs of specific works of architecture that offer insights into various agendas, approaches, and problems during the time period.

## Rational Representations of Modernity and Progress: 1968–1980

While many scholars posit that post-modernity began as a global phenomena in 1968, characterized by the questioning of power, authority, and the trajectory of modern culture, modernist notions of technical rationalism and structural determinism continued in Mexico during the late 1960s and into the early 1970s. The Aceros Planos Office Building, designed by Rodolfo Barragán Schwarz (Monterrey, Nuevo León, 1973–1975), was explicit in its rationally driven tectonic expression. The project was the result of an invited competition for the master plan for the corporate building complex for the nearby La Fundidora Monterrey steelworks, one of the "mother industries" of Monterrey and the largest in Latin America. This two-story office building, which served as the computer and information systems facility, housed a mega-computer and support office space.[8]

The project was an essay on the modernist ideas of articulated systems of nested enclosure and expanded on many of the ideas explored by Eero Saarinen in the John Deere World Headquarters (Moline, Illinois, 1963). According to the architect, "the project was designed and developed under the premise of an extensive use of the company's [steel] products" and utilized steel for the primary structure and the hanging perimeter sunscreen, glass for the exterior skin enclosure, and brick for the vertical circulation and service cores.

However, the project is more than merely the rational coding of systems at multiple scales as the architect also had an agenda to develop a harmonic system of proportions for the project. The relational and classical system of proportions of the façade is based on Francesco di Giorgio's *harmonia mundi*, or "proportional system."[9]

Student unrest occurred around the world in 1968, including the tragic massacre at Tlatelolco in Mexico City prior to the Olympic Games and the symbolic protests by US African American athletes during the awards ceremony. These were important events in the emergence of postmodern discourses. The 1968 Olympic Games gave Mexico an opportunity to present itself to the world as a developed country yet also embrace national folkloric culture. According to Lance Wyman, graphic designer for the Olympic Games:

> Graphic design contributed to the ambiance of the Mexican games and helped to make a meaningful visual impact for fewer pesos. . . . The Mexico 1968 logotype, based on traditional forms from the Mexican culture as well as being Sixties Op-art kinetic typography, set the tone for the en-

18.3. Aceros Planos Office Building, Rodolfo Barragán Schwarz, Monterrey, Nuevo León, 1973–1975. Edward R. Burian Collection of Mexico and the American Southwest.

tire graphics system . . . to create a parallel line typography that suggested imagery found in Mexican pre-Hispanic art and Mexican folk art. The logotype powerfully expressed a sense of place and culture and visually exclaimed the Games were in Mexico.[10]

Among the few new buildings constructed for the Olympic Games was the Palacio Deportivo, designed by Félix Candela and E. Castañeda Tamborell (Mexico City, 1966–1968). Built for the basketball competition, the arena was also a multipurpose arena, seating 22,370. Organized in section in three main levels, it provided facilities not only for athletes, but for judges, officials, and organizers, as well as services for radio, television, and the press. A mezzanine provided access to boxes and the middle- and upper-level arena seating.

Unlike most of Candela's well-known work, which explored thin-shell concrete construction, here the structural frame is a series of long-span steel trusses, spanning 380 feet, with tubular aluminum space frames as infill.[11] These loads were rationally transferred to an eccentric concrete frame at the perimeter.

The project was an example of globalized, rational, structural determinism, but with subtle references to traditional types. However, unlike most of his preceding thin-shell concrete work, which is primarily the result of analytic geometry and structural determinism, there is a determination here to engage local culture. Instead of merely covering the steel-frame structure in a single plane, the exterior skin is a multifaceted copper dome. Each panel is a hyperbolic paraboloid covered with waterproof copper-sheathed plywood that spans in between the tubular space frame. The form and reflectivity recall eighteenth-century Baroque *mudejar* architecture and the shimmering interiors of Churrigueresque *retablo* altar pieces. The building became one of the symbols of the Olympic Games broadcast on television around the world and was also featured on posters and postage stamps issued for the games.

## Resistance to Globalization: 1980–1992

The simultaneous resistance to globalization and the response to site, regional building types, and materials characterized critical regionalism. Furthermore, the desert has been viewed as an ideal setting for religious experience not only in Christianity but also in other religions.[12] The Monasterio Jesús María, designed by Antonio Attolini Lack (San Luis Potosí, 1981), is on an arid desert site with cholla cactus and mesquite trees. The stepped massing of the wall-dominant project reflects and has an empathetic, mimetic relationship to the imposing craggy mountains beyond. This inward-turning project reinterprets traditional wall-dominant colonial monasteries and yet also utilizes modernist notions of abstraction. The wall-dominant composition provides thermal, acoustical, and psychological protection.[13] The choice of materials is both limited and local, intended to create a more severe, austere, abstract, and inward-focused experience with plastered adobe walls, wood-frame roof, glass, and hand-woven rugs. Attolini Lack's work undoubtedly deserves wider critical attention in terms of the discourses of modern architecture in Mexico and the Americas.

The Christ Episcopal Church, designed by Carlos Mijares Bracho (Mexico City, 1992), is an exploration of expressive tectonic systems, materiality, and sensory experience. Here traditional Baroque masonry architecture and its system of arches, vaults, and domes are abstracted. The woven tectonics of the brick recall weaving and the traditional masonry technology of *cappuccino*, where brick is laid in an alternating, coursing pattern in the plane of the wythe and also spans between the two wythes to provide a structural tie. Wall openings are expressively made as articulated surrounds that reveal the unitary nature of the material and handcraft. Mijares's intimate relationship with the material and means of production is reflected in his statement, "I am in close relation to the brick masons in particular and the workers in general."[14]

18.4. Christ Episcopal Church, Carlos Mijares Bracho, Mexico City, 1992. Edward R. Burian Collection of Mexico and the American Southwest.

## Modernity as an Unfinished Project in a Cosmopolitan Culture, 1994–1999

The rediscovery of orthodox modernism as an unfinished project in the 1990s and onward was reflected in Monterrey, Nuevo León, in the embracing of modernity, progress, engineering technology, industrial production, and entrepreneurship.[15] The city evolved largely unburdened by appeals to a mythic pre-Columbian past or a Colonial past that was largely destroyed in the project of modernization along with its nineteenth- and early twentieth-century urban fabric.[16] Monterrey's proximity to the United States, only two hours away by car, also produced a multiethnic society with a notably strong Protestant culture.[17] These factors forged a culture of pragmatism, technology, and commerce. This strong sense of a commercial culture of applied technology would lead to the regional development of the extruded metal, glass, structural steel, and concrete industries.

18.5. Gimnasio Dojo, Agustín Landa, Monterrey, Nuevo León, 1997. Courtesy of Landa Arquitectos.

In this context, the Gimnasio Dojo, designed by Agustín Landa (Monterrey, Nuevo León, 1997), reflects the cosmopolitan nature of Monterrey with a program for Japanese martial arts in the industrial giant of northern Mexico. Sited on a bustling avenue, the project is focused inward to create a sense of quiet serenity. The plan organization is generally symmetrical, with support services, offices, and stairs on the perimeter that surround the central square interior void that serves as the heart of the building, recalling the spatial and formal organization of Louis Kahn.[18] Not only did Kahn's masonry and concrete vocabulary make sense in terms of economics, materials, and labor in the tectonic culture of Mexico and Latin America, but Kahn also brought a renewed concern for tectonic rigor, materiality, and light. Kahn's work abstracted and reinterpreted a tradition of classical composition, as well as a timeless, deliberately static architecture of idealized, platonic forms. Kahn also emphasized the dignity of the individual and the human spirit, yet also ambitiously attempted to represent the archetype of the "collective room," in which many could gather, in his architecture.

The serene abstraction and proportions of the Gimnasio Dojo also refer to the traditional Japanese temple, while the building's proportions are based on modules of the dimension of the *tatami* mat and the square.[19] The tectonic order of the building reflects the dominant, local industries of Monterrey: steel and concrete. A slender, bolted-steel frame supports a narrow-profile steel-frame roof. The exterior of the steel frame is infilled with cellular concrete panels, which form the modular exterior cladding that is strongly horizontal in proportion. The central indoor court is the most memorable space in the building and is illuminated by four symmetrical skylights that mark the central space and float atop the long-span steel-frame trusses with their light-weight cable tensile elements.

In contrast, the National Theatre School, designed by Enrique Norten and TEN Arquitectos (Mexico City, 1994), was part of the favored final project of outgoing president Salinas de Gotari (1988–1994), noted among other things for his agenda for Mexico of fully embracing global markets. Conceived in the then-soaring Mexican economy, the National Arts Center was to be designed and built in less than two years, before Salinas de Gotari left office at the end of his term, to bring together the national schools scattered around the city in dance, music, fine arts, and cinematography and to facilitate student and faculty collaboration. The public project was loosely master-planned by Ricardo Legorreta, who established parking, major circulation through the site, building footprints, and building heights and then invited noted architects to have an autonomous "free hand" in designing each building, not unlike a World's Fair. Given the hurried time schedule, the Department of Public Works acting as general contractor with three of its engineers serving as construction managers, up to twenty thousand workers on the site during construction, and the collapse of the Mexican economy in 1994, it is something of a miracle that the project was completed.

Located south of the historic center among a discontinuous urban fabric of several large film studios and the existing National School of Cinematography, and perpendicular to a loop freeway (*periférico*) system, the project creates a landmark discernable from the high-speed freeways. Norten's National Arts Center project takes a global, optimistic view of technology and primarily concerns itself with formal compositional concerns, materiality, and expressive, discrete tectonic systems. The program consists of three performance areas and service and support areas with administration, cafeteria, gym, set design shops, costume design labs, reading rooms, and a library. Each of the major programmatic elements is clad in a different material from globalized sources, including redwood slats and travertine, among others.

Significantly, Norten's project fully embraces globalization in terms of the use of materials with a bent, tubular, steel frame that was fabricated in Houston, Texas, and a corrugated steel cladding imported from Germany, which allowed the architect to be formally innovative. The primary, discrete, enclosing shell spans thirty-two meters and consists of twelve sixty-centimeter-diameter steel tubes that are joined at the ground and at eighteen meters high with a cable system to resist bending. This tube frame is stabilized with steel, rectangular tubes perpendicular to the bent, tubular, steel frame to resist lateral earthquake and wind loads. The covering membrane of the shell is formed by two layers of grooved and rolled steel decking with insulation inside.

This unitary shell expression also appears in Enrique Norten's Televisa studio project, which engages typical urban conditions on the edges of the city, where discontinuous fabric on a triangular urban site is framed by television transmission towers and a high-rise support office building. A shell symbolically protects the user from the environmental degradation of the city, while the discrete enclosure system of an elliptical shell seems to be capable of limitless extension, a fitting symbol for a company dedicated to global communications. However, in this project the environmental response is largely visual, formal, and compositional.

## Problems and Opportunities Related to Technology and Modern Architecture in Mexico

As discussed earlier, architecture in Mexico has engaged the international avant-garde since the late eighteenth century and the modernist agenda of technical rationalism, functionalism, and modularization since the mid-nineteenth century largely in terms of globalized technology. The successive developments of architecture until the end of the twentieth century can be viewed as basically variations within this larger utopian, scientific model, but with a continuous tension, in the case of Mexico, between globalized systems of technology and formal systems amid the constraints and opportunities of local realities.

Technology is never a neutral issue, especially in Mexico, and encompasses ethical and experiential issues related to underemployment, economics, ecology, and craft, as well as sensory experience and architectural meaning. The relationship of technology and architecture in contemporary Mexico presents both a series of problems and possibilities. Contrary to the glossy images of architectural books and periodicals, and not nearly enough discussed, is the fact that very little of what is currently built in Mexico is actually designed by architects. Probably close to 75 percent of buildings in the country are designed and self-built by their owners.[20] Thus, those public buildings and institutions designed by architects in Mexico have the opportunity to serve a didactic role for the public regarding possible futures, in terms of contemporary lifestyles *and* maintenance over time, in terms of appropriate technology *and* ecology, in terms of sensory experience *and* poetic resonance.

While many large cities in Mexico are clearly connected to global markets and electronic information technology, the painfully obvious issues of poverty, massive unemployment, and underemployment remain. This cannot continue to be ignored by architects in Mexico and elsewhere. In contrast to the National Theatre School discussed earlier in this chapter, large portions of the construction of works of architecture could be conceived at the initial stages of schematic design to be readily built by low-skilled, unemployed workers. Furthermore, the wise use of local materials, where possible, is absolutely necessary for the Mexican economy to grow; and goals of utilizing 90 percent of the material and labor from local sources seem attainable.

However, the exploration of architectural technologies and materials that recognize both local and global conditions not only reflects the contemporary reality of Mexico but also provides a multitude of opportunities. Global technology probably can be best integrated at the level of information technology, electrical control systems, elevators, and energy-saving, back-up HVAC systems.[21] Even modest self-built housing offers both pragmatic and lyrical lessons regarding hybrids, in that masonry or rubble is used for the walls and industrialized long-span steel deck is used for the roof.

Furthermore, the fact that almost every city in Mexico is teeming with buildings no more than forty years old, which are poorly maintained and in unfortunate disrepair, is a reflection that not enough emphasis or capital is placed on maintenance of works of architecture. Given this reality, the design of building components that can be easily replaced after they have become worn out or broken is absolutely essential in Mexico. For example, masonry's capacity to be replaced in standard, individual modules that are readily available has worked extremely well over the millennia.[22] Extending this argument, the design of buildings with components that can be easily replaced, such as curtain walls, electrical service, or HVAC systems, with the original components recycled into new products, offers advantages of updating the performance of the buildings.

There are numerous talented and hardworking individuals in architecture, engineering, urban ecology, development, construction, and landscape craft who might re-imagine an architecture that addresses the appropriate role of technology in Mexico. The design of public architecture and landscape and utility infrastructure can play a particularly crucial role given its potential to serve as both a didactic and a symbolic model, as well as to inspire and delight. The question remains if architects and other leaders in Mexico will be able to persuade their fellow citizens to causes greater than their own immediate, short-term self-interest, to ensure both economic development and ecological well-being. In this larger task, architects can begin to imagine the poetic possibilities of a sensory architecture that utilizes appropriate technology and engages contemporary lifestyles, ecological realities, and maintenance over time.

## Notes

1. "Tectonics" is a term deployed most notably by the architectural critic Kenneth Frampton; it focuses on architecture as systems of building construction, materials, craft, and labor. See Kenneth Frampton, *Studies in Tectonic Culture* (Cambridge, MA: MIT Press, 2001).

2. "Critical regionalism" is a term coined by Alexander Tzonis and Liane LeFaivre and expanded upon by Kenneth Frampton in several books and essays. It is essentially a resistive act against the processes of the globalization of culture. It is also critical in that the intention is that the user will be made aware of the contemporary issues of place in relationship to contemporary means of production. The emphasis is placed on design principles related to the particularities of place, for example, typologies of site, plan, and section organization, response to climate (including solar orientation and natural ventilation), cultural practices, local materials, and local labor practices. These principles do not resort to nostalgic technologies and generally utilize principles of modern architecture. The spatial and technical possibilities of modern architecture are viewed as opportunities to respond to local conditions; thus, at times, the more general term "modern regionalism" is sometimes used, although this emphasis is somewhat less ideological and critical.

Rem Koolhaas fundamentally changed the direction of the contemporary discourse of architecture and has profoundly influenced a generation of contemporary architects. His was a radical reaction at the height of neo-traditional postmodern architecture and urbanism. His strategies include a pragmatic acceptance and embracing of globalization, which he views as both liberating and exhilarating; an extension of the potential and language of orthodox modernist architecture as an unfinished project; extremely inventive spatial compositions and, especially, section ideas; a vacillation between modernist abstraction, modularity, industrialization, and the expressive aspects of modernism; an embrace of the liberating aspects of urban density and largeness; an interpretation of the program in terms of contemporary culture as a "narrative plot" of the building; and an interest in object buildings not limited by context. Architects he has brought back into the contemporary critical discourse include Mies van der Rohe, Ludwig Hilberseimer, Konstantin Melnikov, Walter Harrison, and Oscar Niemeyer, among others.

3. "Typology" is a classification of physical characteristics commonly found in works of architecture and cities, according to their association with different categories, such as formal building organization, compositional principle, or a characteristic pattern in a work of architecture or city that persists over time (for example, courtyards in architecture and plazas in cities, in Colonial Mexican architecture). For a further discussion of these issues, see my interview with the architectural

theorist Alberto Pérez-Gómez in Edward R. Burian, ed., *Modernity and the Architecture of Mexico* (University of Texas Press, Austin, 1997), 13–24.

4. Due to the intersection and residual spaces of the oval courtyard and curved front façade on the rectangular, orthogonal plan, some of the spaces recall Baroque architecture yet were executed in a Neoclassical vocabulary.

5. Giles designed eight major buildings within four blocks in downtown Monterrey, as well as eleven buildings in the state of Chihuahua for General Luis Terrazas, who literally owned the state. The revolution of 1910 curtailed Giles's practice in Mexico, but he continued to find work there until his death in 1920. Ironically, his buildings in Mexico have long been a source of great pride, and most have been maintained in a better state of preservation than their counterparts in the United States.

6. Email, May 24, 2010, from Juan Casas in Monterrey, Nuevo León, who is now restoring Giles's Banco Mercantile project.

7. La Purísima was featured on these regular stamp issues of Mexico: 1950–1952, 3 centavo blue; 1975, 3 peso orange, and 1975–1976, 3 peso orange, Scott 1097. James E. Kloetzel, ed., *Scott Standard Postage Stamp Catalogue*, vol. 4: *Countries of the World, J-M* (Sidney, OH: Scott Publishing Company, 2010), 967, 973.

8. An email transmission from his son, Rodolfo Barragán Delgado, on February 22, 2006, explained that this was the only building that was constructed from the entire master plan designed by Rodolfo Barragán Schwarz. The other two offices on the site were completed by Ricardo Guajardo and Eduardo Padilla. The project was developed in Barragán Schwarz's office in Mexico City when he was director of the School of Architecture of the Universidad Iberoamericana, when some of his students, including Aurelio Nuño and Alfonso Govela, worked for the firm. His early residential work parallels the well-documented *Arts and Architecture* Case Study House program of the 1960s in Southern California of Pierre Koenig, Rafael Soriano, Charles Eames, and Craig Ellwood, among others. Barragán Schwarz's works are examples of his tectonically rigorous and meticulously detailed steel architecture that grew out of the industrial culture of steel production in Monterrey and a concern for site, context, and sensory experience.

9. For more on Francesco di Giorgio's *harmonia mundi*, or proportional system, see Rudolf Wittkower, *Architectural Principles in the Age of Humanism*, part 4, "The Problem of Harmonic Proportion in Architecture" (New York: W. W. Norton & Company, 1971). The project was one of the great buildings produced during this time in Northern Mexico and deserves wider critical attention. Barragán Schwarz also designed the Umbrella House (circa 1958), now destroyed, and his own residence, the Casa Barragán (circa 1984), which explored traditional courtyards and ideas regarding flexibility. In a more monumental public scale, he designed the

poured-in-place concrete State Supreme Court Building on the Macroplaza (1985). In 2004, he received the Premio Gallo from the Universidad Iberoamericana, where he had formerly served as director of the school. His work undoubtedly deserves wider critical attention in terms of the discourses of modern architecture in Mexico, the region, and the Americas.

10. See "Olympic Games 1968 Mexico City: The Look of the 1968 Olympics Was Described by Games Designer Lance Wyman from New York," http://olympic-museum.de/design/lancewyman/wyman.htm (accessed May 20, 2010).

11. Félix Candela worked as an architect upon his arrival in Mexico, until 1949, when he started to engineer many concrete structures utilizing his well-known thin-shell design. Reinforced concrete is extremely efficient in a dome or shell-like shape, which reduces tensile forces in the concrete. In regard to shell design, he tended to rely on the geometric properties of the shell for analysis, instead of complex mathematical formulas. For further information, see Enrique X. de Anda Alanis, *Candela* (Cologne, Germany: Taschen, 2008).

12. These include the biblical wandering of the Israelites in the desert and Christ's contemplation in the desert for forty days and nights, as well as the development of Judaism and Islam.

13. See Louise Noelle Mereles, "Mexican Contemporary Architecture: An Approximation," *Process Architecture* 39 (July 1983): 87.

14. From an email discussion with Carlos Mijares Bracho on March 27, 2009.

15. For a discussion of modernity in relation to the architecture of Mexico, see Burian, *Modernity and the Architecture of Mexico*. Like Mexico City, Monterrey grew rapidly with the influx of capital due to the commercial demands for both raw materials and products of the Allied Powers during World War II. In a city whose image was determined in large part by engineers after World War II, hillsides around the city unapologetically and proudly sprouted television and radio transmission towers, while the banks of the Rio Santa Catarina became a route for electric transmission towers and a freeway. During the 1970s and 1990s the city continued its explosive growth, with the city rapidly doubling and tripling in size.

16. In 1882 the arrival of the railroads linked Monterrey not only to Mexico City, but also to the United States and, thus, by means of the extensive network of railroads and ports in the United States, to markets of Europe and the rest of the world. The railroads also spurred the birth of the "mother industries" of Monterrey: the *cervecera* (beer brewery) and a glass production facility for bottles, both located close by the railroad tracks on what was then the northern edge of the city, soon followed by a metal fabrication facility that initially produced bottle caps and soon evolved into La Fundidora, the largest steelworks in Latin America.

17. The pervasive influence of the United States in Monterrey is cleverly revealed in a humorous joke I heard in Monterrey in the fall of 2003. "What is the difference

between a young person in Monterrey and New York City?" "The young person in Monterrey is more polite and speaks better English."

18. Louis Kahn was one of the most important architects of the twentieth century; he combined a concern for modernist, rational, structural expression with forms and materials from the ancient world. The influence of Kahn in Latin America is an important subject that has never been properly assessed and is worthy of a major study. Certainly, he influenced not only Carlos Mijares, Agustín Landa, and others in Mexico, but also architects such as Rogelio Salmona in Colombia and many other significant practitioners around the world.

19. A *tatami* mat is a woven straw mat used as a floor covering in traditional Japanese architecture. They are also used for protective purposes in Japanese martial arts, such as judo. In Japan the size of a room is typically measured by the number of tatami mats that fit on the floor; however, the size of tatami differs between different regions.

20. For example, self-built housing for the working poor is economically viable, meets a huge need the formal market cannot provide, allows a degree of choice for each family, and can be built by an individual family slowly over time as funds become available.

21. HVAC is an abbreviation for heating, ventilating, and air conditioning system.

22. For the use of masonry in a heavy-wear school building by Antonio Méndez-Vigatá that is wearing extremely well, is energy efficient, and responds to its climate, see Edward R. Burian, "A Contemporary Response to the Conditions of Place: El Colegio Cervantes de Torreón, Coahuila in Torreon, Coah," *Texas Architect* 54, no. 1 (January–February 2004): 22–23, 59.

# Bibliography

Acevedo, Esther, ed. *Hacia otra historia del arte en México: La fabricación del arte nacional a debate (1920–1950)*. Mexico City: Curare/CONACULTA, 2002.
Administración de los Ferrocarriles Nacionales. *Datos generales sobre hechos registrados de 1930 a 1946*. Mexico City: Departamento de Estadística, 1947.
Agostoni, Claudia. "Las delicias de la limpieza: La higiene en la ciudad de México." In *Historia de la vida cotidiana en México, volume 4, Bienes y vivencias, el siglo XIX*, edited by Anne Staples, 563–97. Mexico City: Fondo de Cultura Económica–El Colegio de México, 2005.
———. "Las mensajeras de la salud: Enfermeras visitadoras en la ciudad de México durante la década de 1920." *Estudios de Historia Moderna y Contemporánea de México* 33 (2007): 89–120.
———. *Monuments of Progress: Modernization and Public Health in Mexico City, 1876–1910*. Calgary: University of Calgary Press, 2003.
———. "Popular Health Education and Propaganda in Times of Peace and War in Mexico City, 1890s–1920s." *American Journal of Public Health* 96 (2006): 54–55.
Aguayo, Sérgio. *1968: Los archivos de la violencia*. Mexico City: Grijalbo/Reforma, 1998.
Aguilar Camín, Héctor, and Lorenzo Meyer. *In the Shadow of the Mexican Revolution*. Translated by Luis Alberto Fierro. Austin: University of Texas Press, 1989.
Aguirre Anaya, María del Carmen. *El horizonte tecnológico de México bajo la mirada de Jesús Rivero Quijano*. Puebla: Benemérita Universidad Autónoma, Instituto de Ciencias Sociales y Humanidades, 1999.
*Agustín Jiménez: Memorias de la vanguardia*. Mexico City: RM and Museo de Arte Moderno, 2008.
Agustín, José. *Tragicomedia mexicana 1: La vida en México de 1940 a 1970*. Mexico City: Planeta, 1990.
Alonso, Ana María. *Thread of Blood: Colonialism, Revolution, and Gender on Mexico's Northern Frontier*. Tucson: University of Arizona Press, 1995.
Alonso, Antonio. *El movimiento ferrocarrilero en México, 1958/1959*. Mexico City: Ediciones Era, 1986.

Alva de la Selva, Alma Rosa. *Radio e Ideología*. Mexico City: El Caballito, 1982.
Alvarez, Samantha. "Mexico's National Railways Museum." *Journal of Transport History* 26, no. 1 (2007): 112–14.
Álvarez Amézquita, José, Miguel E. Bustamante, Antonio López Picazos, and Francisco Fernández del Castillo. *Historia de la salubridad y de la asistencia en México*. Vol. 2. Mexico City: Secretaría de Salubridad y Asistencia, 1960.
Alvarez Garín, Raúl. *La estela de Tlatelolco: Una reconstrucción histórica Movimiento estudiantil del 68*. Mexico City: Grijalbo, 1998.
Alzati, Servando. *Historia de la mexicanización de los Ferrocarriles Nacionales de México*. Mexico City: Beatriz de Silva, 1946.
*Anuario Estadístico de la República Mexicana*. Mexico City: Secretaría de Fomento, 1900–1910.
Aréchiga Córdoba, Ernesto. "Dictadura sanitaria, educación y propaganda higiénica en el México Revolucionario, 1917–1934." *DYNAMIS* 25 (2005): 117–43.
Atkinson, Robert D., Daniel K. Correa, and Julie A. Hedlund. *Explaining International Broadband Leadership*. Washington, DC: ITIF, 2008.
Azuela, Luz Fernanda, and José Luis Talancón. *Contracorriente: Historia de la energía nuclear en México, 1945–1995*. Mexico City: Plaza y Valdés, 1999.
Azuela, Mariano. *The Underdogs*. Translated by Sergio Waisman. New York: Penguin Books, 2008.
Baer, Miriam Delal. "Television and Political Control in Mexico." PhD diss., University of Michigan, 1991.
Bakhtin, Mikhail. "Discourse in the Novel." In *The Dialogic Imagination*, edited by M. M. Bakhtin and Michael Holquist, 259–422. Austin: University of Texas Press, 1981.
Ball, S. J. *The Cold War: An International History, 1947–1991*. London: Arnold, 1998.
Ballent, Anahí. "La publicidad de los ámbitos de la vida privada: Representaciones de la modernización del hogar en la prensa de los años cuarenta y cincuenta en México." *Alteridades* 6, no. 2 (1996): 53–74.
Banco de México. *El turismo Norteamericano en México, 1934–1940*. Mexico City: Gráfica Panamericana, 1941.
Barragán, Juan. *Juan F. Brittingham y la industria en México, 1859–1940*. Mexico City: Urbis Internacional, 1993.
Barthes, Roland. *Camera Lucida*. Translated by Richard Howard. New York: Hill and Wang, 1981.
Beals, Carleton. *Mexican Maze*. Philadelphia: Lippincott, 1931.
Beatty, Edward. "Approaches to Technology Transfer in History and the Case of Nineteenth-Century Mexico." *Comparative Technology Transfer and Society* 1, no. 2 (2003): 167–200.
———. "Bottles for Beer: Business Strategy and the Challenge of Technology Transfer in Mexico." *Business History Review* 83 (2009): 317–48.
———. *Institutions and Investment: The Political Basis of Industrialization in Mexico before 1911*. Stanford, CA: Stanford University Press, 2001.
———. "Patents and Technological Change in Late Industrialization: Nineteenth-Century Mexico in Comparative Perspective." *History of Technology* 24 (2002): 121–50.
Becerra Celis, Luis. "Mexico." In *International Television Almanac: Who, What, Where in Television*, edited by Charles S. Aaronson, 733. New York: Quigley Publications, 1961.
Becker, Carl. *The Heavenly City of the Eighteenth-Century Philosophers*. New Haven: Yale University Press, 1932.

Beezley, William H. *Judas at the Jockey Club and Other Episodes of Porfirian Mexico*. Lincoln: University of Nebraska Press, 1987.
Benjamin, Thomas. "Rebuilding the Nation." In *The Oxford History of Mexico*, edited by Michael C. Meyer and William H. Beezley, 438–70. Oxford: Oxford University Press, 2000.
Benjamin, Walter. "On Some Motifs in Baudelaire [1939]" In *Illuminations*, translated by Harry Zorn, 152–96. London: Pimlico, 1999.
Berend, Ivan. "La indivisibilidad de los factores sociales y económicos del crecimiento económico: Un estudio metodológico." In *Historia económica: Nuevos enfoques y nuevos problemas; Comunicaciones al Séptimo Congreso Internacional de Historia Económica*, edited by Jerzy Topolski et al., 37–48. Barcelona: Editorial Crítica, 1981.
Berger, Dina. *The Development of Mexico's Tourism Industry: Pyramids by Day, Martinis by Night*. New York: Palgrave Macmillan, 2006.
Berumen, Miguel Ángel. *1911: La batalla de Ciudad Juárez/II; Las imágenes*. Ciudad Juárez, Mexico: Cuadro por Cuadro, 2003.
Bhabha, Homi. *The Location of Culture*. New York: Routledge, 1994.
Bijker, Wiebe E., Thomas P. Hughes, and Trevor J. Pinch, eds. *The Social Construction of Technological Systems: New Directions in the Sociology and History of Technology*. Cambridge: MIT Press, 1987.
Birn, Anne-Emanuelle. *Marriage of Convenience: Rockefeller International Health and Revolutionary Mexico*. Rochester, NY: University of Rochester Press, 2006.
Blanco Labra, Victor. *Rockstalgia: Crónicas rocanroleras, años 50 y 60*. Mexico City: Diana, 2007.
Blum, Ann S. "Cleaning the Revolutionary Household: Domestic Servants and Public Welfare in Mexico City, 1900–1935." *Journal of Women's History* 15, no. 4 (2004): 67–90.
Bonfil Batalla, Guillermo. *México profundo: Una civilización negada*. Mexico City: Grijalbo, 1990.
Bordwell, David. *On the History of Film Style*. Cambridge, MA: Harvard University Press, 1997.
Bordwell, David, and Kristin Thompson. *Film Art: An Introduction*. 7th ed. New York: McGraw-Hill, 2004.
Boullosa, Carmen. "La destrucción en la escritura." *Inti: Revista de literatura hispánica* 42 (1995): 215–20.
———. *La novela perfecta*. Mexico, DF: Alfaguara, 2006.
Bowlt, John. *Russian Art of the Avant Garde: Theory and Criticism*. New York: Viking Press, 1976.
Bray, Francesca. *Technology and Gender: Fabrics of Power in Late Imperial China*. Berkeley: University of California Press, 1997.
Brehme, Hugo. *Fotograf-Fotógrafo*. Berlin: Verlag Willmuth Arenhövel, 2004.
———. *México: Una nación persistente*. Mexico City: Conaculta-INBA, 1995.
———. *México pintoresco*. 1923. Reprint, Mexico City: Porrúa-INAH, 1990.
———. *Pueblos y paisajes de México*. Mexico City: Porrúa-INAH, 1992.
Brown, Jonathan C. "Foreign and Native-Born Workers in Porfirian Mexico." *American Historical Review* 98, no. 3 (1993): 787–818.
Browne, Donald R. *Electronic Media and Indigenous Peoples: A Voice of Our Own?* Ames: Iowa State University Press, 1996.
Bruce-Novoa, Juan. "La novela de la Revolución Mexicana: La topología final." *Hispania* 74, no. 1 (1991): 36–44.

Brushwood, John S. "Innovation in Mexican Fiction and Politics (1910–1934)." *Mexican Studies/Estudios Mexicanos* 5, no. 1 (1989): 69–88.

———. *México en su novela*. Mexico City: Fondo de Cultura Económica, 1987.

Buffington, Robert M., and William E. French. "The Culture of Modernity." In *The Oxford History of Mexico*, edited by Michael C. Meyer and William H. Beezley, 397–432. Oxford: Oxford University Press, 2000.

Bunker, Steven B. "'Consumers of Good Taste': Marketing Modernity in Northern Mexico, 1890–1910." *Mexican Studies/Estudios Mexicanos* 13, no. 2 (Summer 1997): 227–69.

———. "Creating Mexican Consumer Culture in the Age of Porfirio Diaz, 1876–1911." PhD diss., Texas Christian University, 2006.

Burian, Edward. "A Contemporary Response to the Conditions of Place: El Colegio Cervantes de Torreón, Coahuila in Torreon, Coah." *Texas Architect* 54, no. 1 (January/February 2004): 22–23, 59.

———, ed. *Modernity and the Architecture of Mexico*. Austin: University of Texas Press, 1997.

Burrows, Edwin G., and Mike Wallace. *Gotham: A History of New York City to 1898*. New York: Oxford University Press, 1999.

Butler, Andrew M., Adam Roberts, and Sherryl Vint, eds. *The Routledge Companion to Science Fiction*. New York: Routledge, 2009.

Calderón, Francisco R. "Los ferrocarriles." In *Historia moderna de México*, vol. 7, part 1, edited by Daniel Cosío Villegas. Mexico City: Editorial Hermes, 1974.

Calvillo Velasco, Max, and Lourdes Ramírez Palacios. *Setenta años de historia del Instituto Politécnico Nacional, tomo 1*. Mexico City: Dirección de Publicaciones del Instituto Politécnico Nacional, 2006.

Calvino, Italo. *Invisible Cities*. Translated by William Weaver. London: Vintage, 1997.

Carey, Elaine. *Plaza of Sacrifices: Gender, Power, and Terror in 1968 Mexico*. Albuquerque: University of New Mexico Press, 2005.

Carr, Barry. *El movimiento obrero y la política en México*. Mexico City: Editorial Era, 1981.

Carreño, Manuel Antonio. *Manual de urbanidad y buenas maneras*. 1854. Reprint, Mexico City: Editora Nacional, 1979.

Carter, Ian. *Railways in Britain: The Epitome of Modernity*. Manchester: Manchester University Press, 2001.

Casasola, Gustavo. *Historia gráfica de la revolución mexicana, 1900–1960*. Mexico City: Trillas, 1960.

Cassinelli, Andrés, and Javier Fernández. *Mexico: It's Not Too Late to Catch Up*. Stanford: Stanford University, 2007.

Castañeda, Quetzil. *In the Museum of Mayan Culture: Touring Chichén Itzá*. Minneapolis: University of Minnesota, 1996.

Castellot, Gonzálo. *La televisión en México, 1950–2000*. Mexico City: Edamex, 1999.

Castellot de Ballin, Laura. *Historia de la televisión en México: Narrada por sus protagonistas*. Mexico City: Alpe, 1993.

Castells-Talens, Antoni. "Mexican Nostalgia, Maya Identity: The Reinvention of Iconographic Nationalism in Indigenous-Language Radio." *Journal of Global Mass Communication* 2 (2009): 5–23.

———. "When Our Media Belong to the State: Policy and Negotiations in Indigenous-Language Radio in Mexico." In *Making Our Media: Global Initiatives Toward a Democratic Public Sphere, Creating New Communication Spaces*, edited by Clemencia Rodriguez, Dorothy Kidd, and Laura Stein, 249–70. Cresskill, NJ: Hampton Press, 2009.

Catlin, Stanton. "Mural Census." In *Diego Rivera: A Retrospective*, edited by Cynthia Newman Helms. Exhibition catalog. Detroit: Detroit Institute of Arts, 1986.

Chase, Stuart. *Mexico: A Study of Two Americas*. New York: MacMillan, 1935.

Chirinos Arrieta, Eduardo. "Oliendo las flatulencias de carburo: Poética de la desconfianza en un poema de José Juan Tablada." In *Nueve miradas sin dueño: Ensayos sobre la modernidad y sus representaciones en la poesía hispanoamericana y española*, 35–52. Lima: Fondo Editorial PUCP, 2004.

Chorba, Carrie. *Mexico, from Mestizo to Multicultural: National Identity and Recent Representations of the Conquest*. Nashville: Vanderbilt University Press, 2007.

Churchill, Ward. *Acts of Rebellion: The Ward Churchill Reader*. New York: Routledge, 2003.

Coatsworth, John H. *Growth against Development: The Economic Impact of Railroads in Porfirian Mexico*. Dekalb: Northern Illinois University Press, 1976.

———. "Railroads, Landholding, and Agrarian Protest in the Early Porfiriato." *Hispanic American Historical Review* 54, no. 1 (1974): 48–71.

Cockcroft, James D. *Mexico's Hope: An Encounter with Politics and History*. New York: Monthly Review Press, 1998.

*Contrato de 1930 entre Ferrocarriles Nacionales y la Alianza Ferrocarrilera Mexicana*. Mexico City: Imprenta de E. Limón, 1930.

Cordero Reiman, Karen. "The Best Maugard Drawing Method: A Common Ground for Modern Mexicanist Aesthetics." *Journal of Decorative and Propaganda Arts* 26 (2010): 45–79.

Córdova, Carlos A. *Agustín Jiménez y la vanguardia fotográfica mexicana*. Mexico City: Editorial RM, 2005.

Cornejo Portugal, Inés. *Apuntes para una historia de la radio indigenista en México: Las voces del Mayab*. Mexico City: Fundación Manuel Buendía, 2002.

Coronado Malagón, Marcela. "Las viajeras zapotecas del Istmo de Tehuantepec y el ferrocarril." *Cuadernos del Sur* 14, no. 27 (2009): 60–67.

Cotter, Joseph. *Troubled Harvest: Agronomy and Revolution in Mexico, 1880–2002*. Westport, CT: Praeger, 2003.

Cowan, Ruth Schwartz. "The Consumption Junction: A Proposal for Research Strategies in the Sociology of Technology." In *The Social Construction of Technological Systems*, edited by Wiebe E. Bijker, Thomas P. Hughes, and Trevor Pinch, 1–23. Cambridge, MA: MIT Press, 1989.

———. *More Work for Mother: The Ironies of Household Technology from the Open Hearth to the Microwave*. New York: Basic, 1983.

Craib, Raymond B. *Cartographic Mexico: A History of State Fixations and Fugitive Landscapes*. Durham, NC: Duke University Press, 2004.

Curiel, Fernando, ed. *Medias Palabras: Correspondencia, 1913–1959*. Mexico City: UNAM, 1991.

Curiel, Fernando. *La querella de Martín Luis Guzmán*. Mexico City: Oasis, 1987.

Dávalos, Marcela. *Los letrados interpretan la ciudad: Los barrios de indios en el umbral de la independencia*. Mexico City: INAH, 2010.

Davis, Diane E. "The Social Construction of Mexico City: Political Conflict and Urban Development, 1950–1966." *Journal of Urban History* 24, no. 3 (1998): 364–415.

———. *Urban Leviathan: Mexico City in the Twentieth Century*. Philadelphia: Temple University Press, 1994.

Davis, Mike. *Buda's Wagon: A Brief History of the Car Bomb*. London: Verso, 2007.

De Anda Alanís, Enrique X. *Candela*. Cologne, Germany: Taschen, 2008.

———. *La arquitectura de la revolución mexicana: Corrientes y estilos en la década de los veinte*. Mexico City: UNAM, 1990.
De la Peña, Carolyn. "Slow and Low Progress: Why American Studies Should Do Technology." *American Quarterly* 58, no. 3 (2006): 915–41.
De la Peña, Carolyn Thomas, and Siva Vaidhyanathan. *Rewiring the "Nation": The Place of Technology in American Studies*. Baltimore: Johns Hopkins University Press, 2007.
De la Peña, Moisés T. *El servicio de autobuses en el Distrito Federal*. Mexico City: Departamento del Distrito Federal, 1943.
de los Reyes, Aurelio. *Cine y sociedad en México, 1896–1930*. Vol. 1: *Vivir de sueños (1896–1920)*. Mexico City: UNAM/Cineteca Nacional, 1981.
———. "El cine en el noroeste de México." In *Regionalización en el arte: Teoría y praxis*, edited by José Guadalupe Victoria, Elisa Vargas Lugo, and María Teresa Uriarte, 165–79. Mexico City: Gobierno del Estado de Sinaloa/Instituto de Investigaciones Estéticas–UNAM, 1992.
———. *Filmografía del cine mudo mexicano*. Vol. 1, *1896–1920*. Mexico City: Filmoteca UNAM, 1986.
———. *Filmografía del cine mudo mexicano*. Vol. 2, *1920–1924*. Mexico City: Dirección General de Actividades Cinematográficas–UNAM, 1994.
———. *Filmografía del cine mudo mexicano*. Vol. 3, *1924–1931*. Mexico City: Dirección General de Actividades Cinematográficas–UNAM, 2000.
———, ed. *Historia de la vida cotidiana en México*. Vol. 5. Mexico City: Colmex/FCE, 2006.
de Noriega, Luis Antonio, and Frances Leach. *Broadcasting in Mexico*. London: Routledge and Kegan Paul, 1979.
Debroise, Olivier. *Mexican Suite: A History of Photography in Mexico*. Translated by Stella de Sá Rego. Austin: University of Texas Press, 2001.
Del Río, Fanny, and Carlos Vargas. *El Autotransporte*. Mexico City: Secretaría de Comunicaciones y Transportes, 1988.
Delpar, Helen. *The Enormous Vogue of Things Mexican: Cultural Relations between the United States and Mexico, 1920–1935*. Tuscaloosa: University of Alabama Press, 1992.
Department of Commerce and Labor, Bureau of Manufactures. *Monthly Consular and Trade Reports, March, 1908, No. 330*. Washington, DC: Government Printing Office, 1908.
Derbez, Alain. *El jazz en México: Datos para una historia*. Mexico City: Fondo de Cultura Económica, 2001.
d'Estrabau, Gilberto. *El ferrocarril*. Mexico City: Secretaría de Comunicaciones, 1988.
Díaz, Clara. *La nueva Trova*. La Habana, Cuba: Letras Cubanas, 1994.
Dickerman, Leah. "Leftist Circuits." In *Diego Rivera: Murals for the Museum of Modern Art*, edited by Leah Dickerman and Anna Indych-López. New York: MoMA, 2011.
Dirección General de Estadística. *Séptimo censo general de población*. Mexico City: Secretaría de Economía–Dirección General de Estadística, 1950.
Doane, Mary Ann, *The Emergence of Cinematic Time: Modernity, Contingency, the Archive*. Cambridge, MA: Harvard University Press, 2002.
Domínguez y Pastor, Rafael. *Breves apuntes acerca de la higiene del enfermo*. Escuela Nacional de Medicina de México. Mexico City: Imprenta de Federico Gayosso, 1896.
Doyle, Kate. "Official Report Released on Mexico's 'Dirty War.'" National Security Archive Project, http://www.gwu.edu/~nsarchiv/NSAEBB/NSAEBB209/index.htm (accessed August 6, 2011).
Dulles, John W. F. *Yesterday in Mexico: A Chronicle of the Revolution, 1919–1936*. Austin: University of Texas Press, 1961.

Ebergenyi, Ingrid. *Primera aproximación al estudio del sindicalismo ferrocarrilero en México, 1917–1936*. Cuaderno de trabajo [working paper] no. 49. Mexico City: INAH, Dirección de Estudios Históricos, 1986.
Escorza Rodríguez, Daniel, and Heladio Vera. "Las cámaras Graflex en la campaña federal maderista contra Pascual Orozco, 1912." *20/10: Memoria de las Revoluciones* 10 (2010): 254–65.
Esquivel Obregón, Toribio. "La llegada del ferrocarril." In *Recordatorios públicos y privados*, 276–77. Mexico City: Universidad Iberoamericana, 1994.
*Estadísticas sociales del porfiriato, 1877–1910*. Mexico City: Talleres Gráficos de la Nación, 1956.
Estados Unidos Mexicanos. *Anuario estadístico: 1939*. Mexico City: Dirección General de Estadística, 1940.
EZLN (Ejército Zapatista de Liberación Nacional). *EZLN: Documentos y comunicados*. Mexico City: Era, 1995.
Felski, Rita. *The Gender of Modernity*. Cambridge: Harvard University Press, 1995.
*Fermín Revueltas: Constructor de espacios*. Exhibition catalog. Mexico City: INBA/Museo Mural Diego Rivera/Editorial RM, 2002.
Fernández, Claudia, and Andrew Paxman. *El Tigre: Emilio Azcárraga y su imperio Televisa*. Mexico City: Grijalbo, 2000.
Fernández Christlieb, Fatima. *La radio mexicana: Centro y regiones*. Mexico City: Juan Pablos, 1991.
———. *Los medios de difusión masiva en México*. Mexico City: Juan Pablos, 1982.
Ferrocarriles Nacionales de México. *Informes anuales from June 30, 1909, to June 30, 1925*. N.p.
———. *Octavo informe anual correspondiente al año social que terminó el 30 de junio de 1916*. Mexico City: American Book and Printing, 1916.
———. *Los salarios y la empresa*. Mexico City: Editorial Cultura, 1931.
Fink, James J. *The Automobile Age*. Cambridge: MIT Press, 1990.
Flores, Tatiana. "Clamoring for Attention in Mexico City: Manuel Maples Arce's Avant-Garde Manifesto Actual No. 1." *Review: Literature and Arts of the Americas* 69 (Fall 2004): 208–20.
Flores Olea, Víctor, and Rosa Elena Gaspar de Alba. *Internet y la revolución cibernética*. Mexico City: Océano, 1997.
Florescano, Enrique. "Mitos mesoamericanos: Hacia un enfoque histórico." *Vuelta* 207 (February 1995): 25–35.
Florescano, Enrique, Virginia García Acosta, and Magdalena A. Garcia Sánchez. *Mestizajes tecnólogicos y cambios culturales en México*. Mexico City: CIESAS, 2004.
Folgarait, Leonard. *Mural Painting and Social Revolution in Mexico, 1920–1940: Art of the New Order*. Cambridge, MA: Cambridge University Press, 1998.
Frampton, Kenneth. *Studies in Tectonic Culture*. Cambridge, MA: MIT Press, 2001.
Fry, Matthew. "Mexico's Concrete Block Landscape: A Modern Legacy in the Vernacular." *Journal of Latin American Geography* 7, no. 2 (2008): 35–58.
Fuentes, Carlos. *Christopher Unborn*. New York: Farrar Straus Giroux, 1989.
———. *Valiente mundo nuevo: Épica, utopía y mito en la novela hispanoamericana*. Mexico City: Fondo de Cultura Económica, 1990.
Fuentes, Gloria. *La radiodifusión: Historia de las comunicaciones y los transportes en México*. Mexico City: Secretaría de Comunicaciones y Transportes, 1987.
Fullerton, John, and Elaine King. "Local Views, Distant Scenes: Registering Affect in Surviving Mexican Actuality Films of the 1920s." *Film History* 17, no. 1 (2005): 66–87.

Gallo, Rubén. "John Dos Passos in Mexico." *Modernism/Modernity* 14, no. 2 (2007): 329–45.

———. *Mexican Modernity: The Avant-Garde and the Technological Revolution*. Cambridge, MA: MIT Press, 2006.

Gámez de León, Tania. "William Henry Jackson en México: Forjador de imágenes de una nación (1880–1907)." *Contratexto* 4, no. 5 (2007): 73–92, http://www.ulima.edu.pe/revistas/contratexto/pdf/04.pdf (accessed May 31, 2010).

Gamio, Manuel. *Introducción, síntesis y conclusiones de la obra La población del valle de Teotihuacan*. Mexico City: Dirección de Talleres Gráficos, 1922.

Ganados, Pável. *XEW 70 años en el aire*. Mexico City: Clío, 2000.

García, Gustavo, and Jose Felipe Coria. *Nuevo cine mexicano*. Mexico City: Clío, 1997.

García Cubas, Antonio. *El libro de mis recuerdos: Narraciones históricas, anecdóticas y de costumbres mexicanas anteriores al actual estado social*. 1904. Reprint, Mexico City: Editorial Porrúa, 1986.

García Riera, Emilio. *Historia documental del Cine Sonoro Mexicano*. Mexico City: Era, 1988.

Garduño Pulido, Blanca. "Mexico's Enduring Images." In *México: Una nación persistente*, 15. Photographs by Hugo Brehme. Mexico City: Conaculta-INBA, 1995.

Garrido, Juan S. *Historia de la música popular en México (1896–1973)*. Mexico City: Extemporáneos, 1974.

Garza Merodio, Gustavo G. "Technological Innovation and the Expansion of Mexico City, 1870–1920." *Journal of Latin American Geography* 5, no. 2 (2006): 109–26.

Gatica García, Rufino Alejandro. "Proyecto de un sistema de escuelas radiofónicas para el Instituto Nacional Indigenista." BS diss., Instituto Politécnico Nacional, Mexico City, 1996.

Gauss, Susan M. "Masculine Bonds and Modern Mothers: The Rationalization of Gender in the Textile Industry in Puebla, 1940–1952." *International Labor and Working-Class History* 63 (Spring 2003): 63–80.

Geijerstam, Claes af. *Popular Music in Mexico*. Albuquerque: University of New Mexico Press, 1976.

Gerdes, Dick. "Point of View in *Los de abajo*." *Hispania* 64, no. 4 (1981): 557–63.

Gibbon, Thomas Edward. *Mexico under Carranza: A Lawyer's Indictment of the Crowning Infamy of Four Hundred Years of Misrule*. New York: Doubleday, 1919.

Gillespie, Jeanne L. "The Case of the *China Poblana*." In *Imagination beyond Nation: Latin American Popular Culture*, edited by Eva Bueno and Terry Caesar, 19–37. Pittsburgh: University of Pittsburgh Press, 1998.

Gojman de Backal, Alicia. *Historia del correo en México*. Mexico City: M. A. Porrúa, 2000.

Gonzales, Michael J. "Imagining Mexico in 1910: Visions of the Patria in the Centennial Celebration in Mexico City." *Journal of Latin American Studies* 39, no. 3 (2007): 495–533.

González de Bustamante, Celeste. "1968 Olympic Dreams and Tlatelolco Nightmares: Imagining and Imaging Modernity on Television." *Mexican Studies/Estudios Mexicanos* 26, no. 1 (2010): 1–30.

———. "Club de señoritas: Productions of Mexican femininity in the 1950s." *Studies in Latin American Popular Culture* 28 (2010): 132–40.

———. "Dependency and Development: The Importance of TV News in the History of Mexican Television." *Revista Galaxia* (São Paulo) 18 (2009): 247–62.

———. *"Muy buenas noches": Mexico, Television, and the Cold War*. Lincoln: University of Nebraska Press, 2012.

González Gómez, Ovidio. "Construcción de carreteras y ordenamiento del territorio." *Revista Mexicana de Sociología* 52, no. 3 (July–September 1990): 49–67.

González Mateos, Adriana. "El fifí y su chofer: Control social, homosexualidad y clase en un periodico del México posrevolucionario." *Signos Literarios* 2 (July–December 2005): 103–25.

González Mello, Renato. *La máquina de pintar: Rivera, Orozco, y la invención de un lenguaje.* Mexico City: UNAM, 2008.

González Rivera, Manuel. "Health Education in Mexico." *American Journal of Public Health* 37 (1947): 849–56.

Goubert, Jean-Pierre. *The Conquest of Water: The Advent of Health in the Industrial Age.* Princeton: Princeton University Press, 1988.

Gray, Frank. "*The Kiss in the Tunnel* (1899), G. A. Smith, and the Emergence of the Edited Film in England." In *The Silent Cinema Reader*, edited by Lee Grieveson and Peter Krämer, 51–62. London: Routledge, 2004.

Greene, Alison. "Cablevision (Nation) in Rural Yucatán: Performing Modernity and *Mexicanidad* in the Early 1990s." In *Fragments of a Golden Age: The Politics of Culture in Mexico since 1940*, edited by Gilbert M. Joseph, Anne Rubenstein, and Eric Zolov, 415–51. Durham, NC: Duke University Press, 2001.

Griffiths, Gloria Fraser. "La postal mexicana." *Artes de México* 48 (1999): 8–15.

Groys, Boris. "U-Bahn als U-Topie." *Kursbuch* (Rowohlt, Berlin) 112 (June 1993): 1–9.

Grunstein, Arturo. "Perspectivas gerenciales sobre el problema laboral de los Ferrocarriles Nacionales de México en la posrevolución, 1920–1935." *TST: Revista de Historia de los Transportes, Servicios y Telecomunicaciones* 14 (2008): 42–89.

———. "Surgimiento de los Ferrocarriles Nacionales de México (1900–1913): ¿Era inevitable la consolidación monopólica?" In *Historia de las grandes empresas en México, 1850–1930*, edited by Carlos Marichal and Mario Cerutti, 65–106. Mexico City: Fondo de Cultura Económica, 1997.

Guajardo, Guillermo. "'A pesar de todo, se mueve': El aprendizaje tecnológico en México, ca. 1860–1930." *Iztapalapa: Revista de ciencias sociales y humanidades* 18, no. 43 (1998): 305–28.

———. "Aprendizajes de innovación y negocios en el petróleo y los ferrocarriles de México, 1952–1992." In *Innovación y Empresa: Estudios históricos de México, España y América Latina*, edited by Guillermo Guajardo, 203–24. Mexico City: CEIICH/UNAM–Fundación Gas Natural, 2008.

———. "La tecnología de los Estados Unidos y la 'Americanización' de los ferrocarriles estatales de México y Chile, ca. 1880–1950." *TST: Revista de Historia de los Transportes, Servicios y Telecomunicaciones* 9 (2005): 110–30.

———. "Tecnología y campesinos en la Revolución Mexicana." *Mexican Studies/Estudios Mexicanos* 15, no. 2 (1999): 291–322.

Gudiño, María Rosa. "Educación higiénica y consejos de salud para campesinos en El Sembrador y El Maestro Rural, 1929–1934." In *Curar, sanar y educar: Enfermedad y sociedad en México, siglos XIX y XX*, edited by Claudia Agostoni, 71–97. Mexico City: IIH–UNAM/BUAP, 2008.

Guerrero, Julio. *La génesis del crimen en México.* 1901. Reprint, Mexico City: Consejo Nacional para la Cultura y las Artes, 1996.

Guerrero, Manuel, and Victoria Isabela Corduneaunu. "Trust, Credibility, and Relevance in the Consumption of Information among Mexican Youth: Third Generation TV Audiences." In *Empowering Citizenship through Journalism, Information, and Entertainment in Iberoamerica*, edited by Manuel Guerrero and Manuel Chavez, 157–97. Mexico City: University of Miami, Michigan State University, and Universidad Iberoamericana, 2009.

Gunning, Tom. "The Cinema of Attractions: Early Film, Its Spectator, and the Avant-Garde." In *Early Cinema: Space, Frame, Narrative*, edited by Thomas Elsaesser with Adam Barker, 56–62. London: British Film Institute, 1990.

Gutiérrez Chong, Natividad. *Nationalist Myths and Ethnic Identities: Indigenous Intellectuals and the Mexican State*. Lincoln: University of Nebraska, 1999.

Haber, Stephen H. *Industry and Underdevelopment: The Industrialization of Mexico, 1890–1940*. Stanford: Stanford University Press, 1989.

Hale, Charles. *The Transformation of Liberalism in Late Nineteenth-Century Mexico*. Princeton: Princeton University Press, 1989.

Halkin, Alexandra. "Outside the Indigenous Lens: Zapatistas and Autonomous Videomaking." In *Global Indigenous Media: Cultures, Poetics, and Politics*, edited by Pamela Wilson and Michelle Stewart, 160–80. Durham, NC: Duke University Press, 2008.

Hayes, Joy Elizabeth. "National Imaginings on the Air: Radio in Mexico, 1920–1950." In *The Eagle and the Virgin: National and Cultural Revolution in Mexico, 1920–1940*, edited by Mary Kay Vaughan and Stephen E. Lewis, 246–58. Durham, NC: Duke University Press, 2006.

———. *Radio Nation: Communication, Popular Culture, and Nationalism in Mexico, 1920–1950*. Tucson: University of Arizona Press, 2000.

Hendy, David. *Radio in the Global Age*. Cambridge, UK: Polity Press, 2000.

Henrique Cardoso, Fernando. "Associated-Dependent Development: Theoretical and Practical Implications." In *Authoritarian Brazil*, edited by Alfred Stephan, 142–76. New Haven: Yale University Press, 1973.

Henríquez, Cristina, and Melba Pría. *Regiones indígenas tradicionales: Un enfoque geopolítico para la seguridad nacional*. Mexico City: Instituto Nacional Indigenista, 2000.

Hernández, Vicente Martín. *Arquitectura doméstica de la ciudad de México (1890–1925)*. Mexico City: UNAM, 1981.

Hernández Lomeli, Francisco. "Obstáculos para el establecimiento de la televisión comercial en México (1950–1955)." *Comunicación y Sociedad* 28 (September–December 1996): 147–71.

Hernández Navarro, Luis, and Ramón Vera Herrera, eds. *Acuerdos de San Andrés*. Mexico City: Era, 1998.

Herrera, Jorge Luis. "El amor es una reflexión, un volver atrás." *El Universo del Buhón* 60 (2005).

Herring, Hubert Clinton, and Herbert Weinstock, eds. *Renascent Mexico*. New York: Covici, Friede, 1935.

Hershfield, Joanne. *Imagining la Chica Moderna: Women, Nation, and Visual Culture in Mexico, 1917–1936*. Durham, NC: Duke University Press, 2008.

Hobsbawm, Eric. "Introduction: Inventing Traditions." In *The Invention of Tradition*, edited by Eric Hobsbawm and Terence Ranger, 1–14. Cambridge: Cambridge University Press, 1983.

Hoeg, Jerry. "*La ciudad de los prodigios* de Eduardo Mendoza frente a una visión latinoamericana de ciencia, cultura y tecnología." *Revista Iberoamericana* 73, no. 221 (2007): 861–70.

———. *Science, Technology, and Latin American Narrative in the Twentieth Century and Beyond*. Bethlehem, PA: Lehigh University Press, 2000.

Huesca Rebolledo, Sabás. "La noticia por televisión." In *Apuntes para una historia de la televisión mexicana*, vol. 2, edited by Miguel Ángel Sánchez de Armas, 67–112. Mexico City: Revista Mexicana/Televisa, 1998.

Hughes, Lloyd H. *Las Misiones Culturales Mexicanas y su programa*. Paris: UNESCO, 1951.

Hughes, Thomas Parke. *Human-Built World: How to Think about Technology and Culture.* Chicago: University of Chicago Press, 2004.
Ibarguengoitia, Jorge. *Instrucciones para vivir en México.* Mexico City: Joaquín Mortiz, 1990.
Industria Eléctrica Mexicana (IEM). *Guía y recetario del refrigerador IEM.* Mexico City: IEM subsidiary of Westinghouse, n.d.
Instituto Nacional de Estadística y Geografía (INEGI). *Anuarios estadísticos de las entidades federativas.* Mexico City: INEGI, 1998.
Instituto Nacional Indigenista. *Primeras jornadas de la radiodifusión cultural indigenista.* Mexico City: Instituto Nacional Indigenista, 1996.
Jameson, Fredric, and Ian Buchanan. *Jameson on Jameson: Conversations on Cultural Marxism.* Durham, NC: Duke University Press, 2007.
Jiménez, José Olivio, and Carlos J. Morales. *La prosa modernista hispanoamericana: Introducción crítica y antología.* Madrid: Alianza, 1998.
Johnson, Ann. "Revisiting Technology as Knowledge." *Perspectives on Science* 13, no. 4 (2005): 554–73.
Joseph, Gilbert M., Anne Rubenstein, and Eric Zolov, eds. *Fragments of a Golden Age: The Politics of Culture in Mexico since 1940.* Durham, NC: Duke University Press, 2001.
Juárez López, José Luis. "Innovaciones en la cocina." *Cuadernos de Nutrición* 2 (2001): 56.
Kamminga, Harmke, and Andrew Cunningham. *The Science and Culture of Nutrition, 1840–1940.* Amsterdam: Rodopi, 1995.
Kaplan, Amy. "Manifest Domesticity." *American Literature* 70, no. 3 (September 1998): 583–84.
Kevles, Bettyann Holtzmann. *Naked to the Bone: Medical Imaging in the Twentieth Century.* Camden, NJ: Rutgers University Press, 1996.
Kismaric, Susan. *Manuel Alvarez Bravo.* New York: Museum of Modern Art, 1997.
Klich, Lynda. "Estridentópolis: Achieving a Post-revolutionary Utopia in Jalapa." *Journal of Decorative and Propaganda Arts* 26 (2010): 102–27.
———. "Revolution and Utopia: *Estridentismo* and the Visual Arts (1921–1927)." PhD diss., Institute of Fine Arts, New York University, 2008.
Knight, Alan. *The Mexican Revolution.* Vol. 1, *Porfirians, Liberals, and Peasants.* Lincoln: University of Nebraska Press, 1990.
———. "Popular Culture and the Revolutionary State in Mexico, 1910–1940." *Hispanic American Historical Review* 74, no. 3 (August 1994): 393–444.
———. "Racism, Revolution, and Indigenismo: Mexico, 1910–1940." In *The Idea of Race in Latin America, 1870–1940*, edited by Richard Graham, 71–114. Austin: University of Texas Press, 1990.
———. "Revolutionary Project, Recalcitrant People: Mexico, 1910–40." In *The Revolutionary Process in Mexico: Essays on Political and Social Change, 1880–1940*, edited by Jaime E. Rodriguez O., 227–64. Los Angeles: University of California Press, 1990.
———. "State Power and Political Stability in Mexico." In *Dilemmas of Transition*, edited by Neil Harvey, 29–63. London: Institute of Latin American Studies, University of London and British Academic Press, 1993.
———. "The Weight of the State in Modern Mexico." In *Studies in the Formation of the Nation State in Latin America*, edited by James Dunkerley, 212–53. London: Institute of Latin American Studies, 2002.
Koolhaas, Rem, and Bruce Mau. *S, M, L, XL: Small, Medium, Large, Extra Large.* New York: Monacelli Press, 1995.
Kowal, Donna M. "Digitizing and Globalizing Indigenous Voices: The Zapatista Movement."

In *Critical Perspectives on the Internet*, edited by Greg Elmer, 105–26. Lanham, MD: Rowman & Littlefield Publishers, 2002.

Kracauer, Siegfried. "Travel and Dance." In *The Mass Ornament: Weimar Essays*, edited and translated by Thomas Y. Levin, 65–73. Cambridge, MA: Harvard University Press, 1995.

Kram Villarreal, Rachel. "Gladiolas for the Children of Sanchez: Ernesto P. Uruchurtu's Mexico City, 1950–1968." PhD diss., University of Arizona, 2008.

Kuntz, Sandra, and Paolo Riguzzi. "La gran empresa de cabeza: Ferrocarriles Nacionales de México, 1908–1937." In *Empresas y grupos empresariales en América Latina, España y Portugal*, edited by Mario Cerutti and Javier Vidal, 115–21. Monterrey: Universidad Autónoma de Nuevo León, 2003.

Kuntz Flicker, Sandra. "The Export Boom of the Mexican Revolution: Characteristics and Contributing Factors." *Journal of Latin American Studies* 36 (2004): 267–96.

Landow, George P. *Hypertext 3.0*. Baltimore: Johns Hopkins University Press, 2006.

Lanier, Jaron, *You Are Not a Gadget: A Manifesto*. New York: Knopf, Borzoi Books, 2010.

Largo Farías, René. *La Nueva Canción Chilena*. Mexico City: Cuadernos de la Casa de Chile No. 9, 1977.

Lawrence, D. H. *The Plumed Serpent*. London: Wordsworth Editions, 1995.

Lear, John. "Mexico City: Space and Class in the Porfirian Capital, 1884–1910." *Journal of Urban History* 22, no. 4 (May 1996): 454–92.

———. *Workers, Neighbors, and Citizens: The Revolution in Mexico City*. Lincoln: University of Nebraska Press, 2001.

Legrás, Horacio. "El ateneo y los orígenes del estado ético en México." *Latin American Research Review* 38, no. 2 (2003): 34–60.

Lerner, Jesse. "La exportación de lo mexicano: Hugo Brehme en casa y en el extranjero." *Alquimia* 16 (2002–2003): 30–38.

Levi, Heather. *The World of Lucha Libre: Secrets, Revelations, and Mexican National Identity*. Durham, NC: Duke University Press, 2010.

Levy, Steven. *Hackers: Heroes of the Computer Revolution*. New York: Penguin, 2001.

Lewis, Bart L. "Modernism." In *Mexican Literature: A History*, edited by David William Foster, 139–70. Austin: University of Texas Press, 1994.

Lewis, Oscar. *Five Families: Mexican Case Studies in the Culture of Poverty*. New York: Basic Books, 1959.

Limón Rojas, Miguel. "El indigenismo: Un imperativo nacional" [Indigenism: A national imperative]. In *INI 40 Años*, edited by L. Herrasti Maciá, 81–101. Mexico City: Instituto Nacional Indigenista, 1988.

Linsley, Robert. "Utopia Will Not Be Televised: Rivera at Rockefeller Center." *Oxford Art Journal* 17, no. 2 (1994): 48–62.

List Arzubide, Germán. *Esquina: Poemas de Germán List Arzubide*. Mexico City: Librería Cicerón, 1923.

Lomas, David. "Remedy or Poison: Rivera, Medicine, and Technology." *Oxford Art Journal* 30, no. 3 (October 2007): 454–83.

Lomnitz-Adler, Claudio. *Deep Mexico, Silent Mexico: An Anthropology of Nationalism*. Minneapolis: University of Minnesota Press. 2001.

———. "Times of Crisis: Historicity, Sacrifice, and the Spectacle of Debacle in Mexico City." *Public Culture* 15, no. 1 (2003): 127–47.

López, Rick A. "The Noche Mexicana and the Exhibition of Popular Arts: Two Ways of Exalting Indianness." In *The Eagle and the Virgin: Nation and Cultural Revolution in Mexico*,

*1920–1940*, edited by Mary Kay Vaughan and Stephen E. Lewis, 23–42. Durham, NC: Duke University Press, 2006.

López Ferman, Lilia Isabel. "Las cosas que nos transformaron: Usos, apropiaciones y significados de la tecnología doméstica en la ciudad de México (1940–1970)." MA thesis, Centro de Investigaciones y Estudios Superiores en Antropología (CIESAS), 2005.

López Lozano, Miguel. *Utopian Dreams, Apocalyptic Nightmares: Globalization in Recent Mexican and Chicano Narrative*. West Lafayette: Purdue University Press, 2008.

López Pardo, Gustavo. *La administración obrera de los Ferrocarriles Nacionales de México*. Mexico City: UNAM–El Caballito, 1997.

Lorey, David E. *The University System and Economic Development in Mexico since 1929*. Stanford: Stanford University Press, 1993.

Lozano Herrera, Rubén. *Las veras y las burlas de José Juan Tablada*. Mexico City: Universidad Iberoamericana, 1995.

Magallanes Blanco, Claudia. *The Use of Video for Political Consciousness-Raising in Mexico: An Analysis of Independent Videos about the Zapatistas*. Lampeter, UK: Mellen Press, 2008.

Mahieux, Viviane. "The Chronicler as Streetwalker: Salvador Novo and the Performance of Genre." *Hispanic Review* 76, no. 2 (2008): 155–77.

Marroquí, José María. *La ciudad de México*. Vol. 2. Mexico City: Ediciones La Europea, 1900.

Marroquín y Rivera, Manuel. *Memoria descriptiva de las obras de provisión de aguas potables para la ciudad de México*. Mexico City: Imprenta y Litografía de Müller Hnos., 1914.

Marshall, Barbara. "Critical Theory, Feminist Theory, and Technology Studies." In *Modernity and Technology*, edited by Thomas J. Misa, 108. Cambridge, MA: MIT Press, 2004.

Martínez Medellín, Francisco J. *Televisa: Siga la huella*. Mexico City: Claves Lationoamericanas Publishers, 1989.

Mata Temoltzin, Victor, and Antonio Casanueva Fernández. *La economía mexicana y los ferrocarriles (1910–1920)*. Puebla: Gobierno del Estado de Puebla, 1999.

Matute, Álvaro. "De la tecnología al orden doméstico en el México de la posguerra." In *Siglo XX: La imagen, ¿espejo de la vida?*, edited by Aurelio de los Reyes. Vol. 5 of *Historia de la vida cotidiana en México*. Mexico City: El Colegio de México–FCE, 2006.

———. *Las dificultades del nuevo estado: Historia de la revolución mexicana, 1917–1924*. Vol. 7. Mexico City: El Colegio de México, 1995.

McCaa, Robert. "Missing Millions: The Demographic Costs of the Mexican Revolution." *Mexican Studies/Estudios Mexicanos* 19, no. 2 (Summer 2003): 367–400.

McSherry, Corynne. "Todas las voces: Indigenous Language Radio, State Culturalism, and Everyday Forms of Public Formation." *JILAS–Journal of Iberian and Latin American Studies* 5, no. 2 (December 1999): 99–132.

Meikle, Graham. "We Are All Boat People: A Case Study in Internet Activism." *Media International Australia* 107 (May 2003): 9–18.

Mejía Barquera, Fernando. "Anexo: Cronologías." In *Apuntes para una historia de la televisión mexicana*, vol. 1, edited by Miguel Ángel Sánchez de Armas, 513–80. Mexico City: Revista Mexicana/Televisa, 1998.

———. "Del canal 4 a Televisa." In *Apuntes para una historia de la televisión mexicana*, vol. 1, edited by Miguel Ángel Sánchez de Armas, 19–98. Mexico City: Revista Mexicana de Comunicación, 1998.

———. *La industria de la radio y la política del Estado Mexicano*. Vol. 1, *1920–1960*. Mexico City: Fundación Manuel Buendía, 1989.

Mendoza Áviles, Mayra. "La colección Hugo Brehme." *Alquimia* 16 (2002–2003): 41–43.

Meyer, Jean. "La ciudad de México, ex de los palacios." In *La reconstrucción económica: Historia de la revolución mexicana, IV Periodo 1924–1928*, vol. 10, edited by Enrique Krauze, 273–79. Mexico City, El Colegio de México, 1977.

———. "Revolution and Reconstruction in the 1920s." In *Mexico since Independence*, edited by Leslie Bethell, 201–40. Cambridge: Cambridge University Press, 1991.

Meyer, Lorenzo. "La encrucijada." In *Historia General de México*, edited by Daniel Cosío Villegas, 1275–1355. Mexico City: COLMEX, 1998.

Meyer, Pedro. *Herejías*. 1995. Reprint, Barcelona: Fundación Pedro Meyer and Lundwerg, 2008.

———. *Truths and Fictions: A Journey from Documentary to Digital Photography*. New York: Aperture, 1995.

Middlebrook, Kevin. *The Paradox of Revolution: Labor, the State, and Authoritarianism in Mexico*. Baltimore: John Hopkins Press, 1995.

Middleton, P. Harvey. *Industrial Mexico: 1919 Facts and Figures*. New York: Dodd, Mead and Company, 1919.

Miller, George, and Dorothy Miller. *Picture Postcards in the United States, 1893–1918*. New York: Clarkson N. Potter, 1976.

Miller, Marjorie, and Juanita Darling. "Emilio Azcárraga and the Televisa Empire." In *A Culture of Collusion: An Inside Look at the Mexican Press*, edited by William A. Orme Jr., 59–70. Miami: North-South Center Press, 1997.

Miquel, Ángel. *Acercamientos al cine silente mexicano*. Cuernavaca: Universidad Autónoma del Estado de Morelos/Ediciones Sin Nombre, 2005.

———. *Disolvencias: Literatura Cine y Radio en México (1900–1950)*. Mexico City: Fondo de Cultura Económica, 2005.

———. "Documentales de la revolución maderista." Paper delivered at the conference Cine mudo en Iberoamérica: Naciones, narraciones, centenarios, Universidad Nacional Autónoma de México, Mexico City, April 2010.

———. *Salvador Toscano*. Mexico City: Universidad de Guadalajara/Gobierno del Estado de Puebla/Universidad Veracruzana/UNAM, 1997.

Misa, Thomas J. "Findings Follow Framings: Navigating the Empirical Turn." *Synthese* 168, no. 3 (2009): 357–75.

Monroy Nasr, Rebeca. *Historias para ver: Enrique Díaz, fotorreportero*. Mexico City: UNAM-INAH, 2003.

Monsiváis, Carlos. "Los Hermanos Mayo: . . . y en una reconquista feliz de otra inocencia." *La Cultura en México*, a supplement of *Siempre!*, August 12, 1981, 2–8.

———. *Mexican Postcards*. London: Verso, 1997.

Moreno, Julio. "J. Walter Thompson, the Good Neighbor Policy, and Lessons in Mexican Business Culture, 1920–1950." *Enterprise & Society* 5, no. 2 (2004): 254–80.

———. *Yankee Don't Go Home! Mexican Nationalism, American Business Culture, and the Shaping of Modern Mexico, 1920–1950*. Chapel Hill: University of North Carolina Press, 2003.

Moreno Rivas, Yolanda. *Historia de la música popular mexicana*. Mexico City: CONACULTA–Alianza Editorial Mexicana, 1979.

Mraz, John. "Close-up: An Interview with the Hermanos Mayo, Spanish-Mexican Photojournalists (1930s–present)." *Studies in Latin American Popular Culture* 11 (1992): 195–218.

———. *Looking for Mexico: Modern Visual Culture and National Identity*. Durham, NC: Duke University Press, 2009.

———. *Nacho López: Mexican Photographer*. Minneapolis: University of Minnesota Press, 2003.

———. *Photographing the Mexican Revolution: Commitments, Testimonies, Icons*. Austin: University of Texas Press, 2012.

Mraz, John, and Jaime Vélez Storey. *Uprooted: Braceros in the Hermanos Mayo Lens*. Houston: Arte Público Press, 1996.

Muñoz, Maurilio. *Mixteca nahua-tlapaneca*. Mexico City: INI, 1963.

Musacchio, Humberto. *Ciudad quebrada*. Mexico City: Océano Publishers, 1985.

Musser, Charles. "Moving towards Fictional Narratives: Story Films Become the Dominant Product, 1903–1904." In *The Silent Cinema Reader*, edited by Lee Grieveson and Peter Krämer, 87–102. London: Routledge, 2004.

Nervo, Amado. "El automóvil de la muerte." In *Obras completas: Tomo II*, 597. Madrid: Aguilar, 1951.

———. *Obras completas*. Vol 2. Madrid: Aguilar, 1967.

Niblo, Stephen R. *Mexico in the 1940s: Modernity, Politics, and Corruption*. Wilmington, DE: SR Books, 1999.

Nieto, Rafael. *Polémica laborista*. Rome: Failli, 1926.

Noriega Hope, Carlos. *Carlos Noriega Hope: 1896–1934*. Mexico City: Instituto Nacional de Bellas Artes, 1959.

Norris, James D. *Advertising and the Transformation of American Society, 1865–1920*. New York: Greenwood Press, 1990.

Novo, Salvador. "El Joven." In *Toda la prosa*, 535–53. Mexico City: Empresas Editoriales, 1964.

———. *New Mexican Grandeur*. 5th ed. Mexico City: Petróleos Mexicanos, 1967.

———. "Radioconferencia sobre el radio." In *Viajes y ensayos I*, 39–40. Mexico City: Fondo de Cultura Económica, 1996.

Nye, David. *Technology Matters: Questions to Live With*. Cambridge: MIT Press, 2006.

Oldenzeil, Ruth. *Making Technology Masculine: Men, Women, and Modern Machines in America, 1879–1945*. Amsterdam: Amsterdam University Press, 1999.

Oles, James. "Industrial Landscapes in Modern Mexican Art." *Journal of Decorative and Propaganda Arts* 26 (2010): 134.

Olmos, Alejandro. "Algunos protagonistas de la televisión." In *Apuntes para un historia de la televisión Mexicana II*, edited by Miguel Ángel Sánchez de Armas and Maria del Pilar Ramirez, 277–331. Mexico City: Revista Mexicana de Comunicación, Televisa, 1998.

Olsen, Patrice Elizabeth. *Artifacts of Revolution: Architecture, Society, and Politics in Mexico City, 1920–1940*. Lanham, MD: Rowman and Littlefield Publishers, 2008.

O'Rourke, Kevin H., and Jeffrey G. Williamson. *Globalization and History: The Evolution of a Nineteenth-Century Atlantic Economy*. Cambridge: MIT Press, 1999.

Orozco Gómez, Guillermo. "La televisión en México." In *Histórias de la televisión en América Latina: Argentina, Brasil, Colombia, Chile, México, Venezuela*, edited by Guillermo Orozco Gómez and Nora Maziotti. Barcelona: Gedisa, 2002.

Ortega, Julio. Introduction to *Antología del cuento latinoamericano del siglo XXI: Las horas y las hordas*. Mexico City: Siglo Veintiuno, 1997.

———. "Prólogo." In *Antología del cuento latinoamericano del siglo XXI: Las horas y las hordas*, ed. Julio Ortega, 11–16. Mexico City: Siglo XXI, 2001.

Ortiz Gaitán, Julieta. "Arte, publicidad y consumo en la prensa: Del porfirismo a la posrevolución." *Historia Mexicana* 48, no. 2 (1998): 411–35.

———. *Imágenes del deseo: Arte y publicidad en la prensa ilustrada mexicana (1894–1939)*. Mexico City: UNAM, 2003.
Ortiz Hernán, Sergio. "De estaciones, trenes y paisajes." In *De las estaciones*, 40. Mexico City: Secretaría de Comunicaciones y Transportes, Ferrocarriles Nacionales de México, y Museo Nacional de los Ferrocarriles Mexicanos, 1995.
Pacheco, José Emilio. *Battles in the Desert and Other Stories*. Translated by Katherine Silver. New York: New Directions, 1987.
Padilla, Ignacio. "McOndo y el Crack: Dos experiencias grupales." In *Palabra de América*, ed. Roberto Bolano, Guillermo Cabrera Infante, Jorge Franco Ramos, Rodrigo Fresan, and Pere Gimferrer Torrens, 147. Barcelona: Seix Barral, 2004.
Pani, Alberto J. *La higiene en México*. Mexico City: Impr. de J. Ballescá, 1916.
Parlee, Lorena. "The Impact of United States Railroad Unions on Organized Labor and Government Policy in Mexico (1880–1911)." *Hispanic American Historical Review* 64, no. 3 (1984): 461–75.
Peñafiel, Antonio. *Memoria sobre las aguas potables de la ciudad de México*. Mexico City: Oficina Tipográfica de la Secretaría de Fomento, 1884.
Pérez Montfort, Ricardo. *Expresiones populares y estereotipos culturales en México, Siglos XIX y XX: Diez ensayos*. Mexico City: CIESAS, 2007.
———. "Fragmento de historia de las drogas en México, 1870–1920." In *Hábitos, normas y escándalo: Prensa y criminalidad durante el Porfiriato tardío*, edited by Ricardo Pérez Montfort, 143–210. Mexico City: CIESAS, 1997.
———. "Indigenismo, hispanismo y panamericanismo en la cultura popular mexicana de 1920 a 1940." In *Cultura e identidad nacional*, edited by Roberto Blancarte, 343–83. Mexico City: UNAM, 1994.
Perló Cohen, Manuel. *El paradigma porfiriano: Historia del desagüe del Valle de México*. Mexico City: UNAM, 1999.
Piccato, Pablo. *City of Suspects: Crime in Mexico City, 1900–1931*. Durham, NC: Duke University Press, 2001.
———. "Urbanistas, Ambulantes, and Mendigos: The Dispute for Urban Space in Mexico City, 1890–1930." In *Reconstructing Criminality in Latin America*, edited by Carlos Aguirre and Robert Buffington, 113–48. Wilmington, DE: Scholarly Resources, 2000.
Pilcher, Jeffrey M. "Fajitas and the Failure of Refrigerated Meatpacking in Mexico: Consumer Culture and Porfirian Capitalism." *The Americas* 60, no. 3 (January 2004): 411–29.
———. "Industrial Tortillas and Folkloric Pepsi: The Nutritional Consequences of Hybrid Cuisines in Mexico." In *Food Nations: Selling Taste in Consumer Societies*, edited by Warren Belasco and Philip Scranton, 222–39. New York: Routledge, 2001.
———. "Josefina Velázquez de León: Apostle of the Enchilada." In *The Human Tradition in Mexico*, edited by Jeffrey M. Pilcher, 199–209. Wilmington, DE: SR Books, 2003.
———. *¡Que vivan los tamales! Food and the Making of Mexican Identity*. Albuquerque: University of New Mexico Press, 1998.
———. *The Sausage Rebellion: Public Health, Private Enterprise, and Meat in Mexico City, 1890–1917*. Albuquerque: University of New Mexico Press, 2006.
Poniatowska, Elena. *La noche de Tlatelolco: Testimonios de historia oral*. Mexico City: Ediciones Era, 1971.
———. *Todo México*. Vol. 5. Mexico City: Diana, 1999.
Porter, Michael. "Attitudes, Values, Beliefs, and the Microeconomics of Prosperity." In *Cul-

*ture Matters: How Values Shape Human Progress*, edited by Lawrence E. Harrison and Samuel P. Huntington, 14–28. New York: Basic Books, 2000.

Prantl, Adolfo, and José Groso. *La ciudad de México: Novísima guía universal de la capital de la República Mexicana; Directorio clasificado de vecinos y prontuario de la organización y funciones del gobierno federal y oficinas de su dependencia*. Mexico City: Juan Buxó y Compañía Editores–Librería Madrileña, 1901.

Presidencia de la República. *50 años de Revolución Mexicana en cifras*. Mexico City: Presidencia de la República–NAFINSA, 1963.

Pruneda, Alfonso. *Higiene de los trabajadores, por el Dr. Alfonso Pruneda*. Mexico City: Ediciones de la Universidad Nacional de México, 1937.

Ramírez, Mari Carmen. "The Ideology and Politics of the Mexican Mural Movement: 1920–1925." PhD diss., University of Chicago, 1989.

Ramos Rodríguez, José Manuel. "Indigenous Radio Stations in Mexico: A Catalyst for Social Cohesion and Cultural Strength." *Radio Journal* 3 (2005): 155–69.

———. "La Voz de los sin voz: Emergencia de la radio comunitaria en México." *Revista Iberoamericana de Comunicación* 10 (2006): 13–22.

Rashkin, Elissa J. *The Stridentism Movement in Mexico: The Avant-Garde and Cultural Change in the 1920s*. Lanham, MD: Lexington Books/Rowman and Littlefield Publishers, 2009.

Redfield, Robert. *Tepoztlan, a Mexican Village: A Study of Folk Life*. Chicago: University of Chicago Press, 1930.

Revueltas, José. *México 68: Juventud y revolución*. Mexico City: Ediciones Era, 1978.

Reyes, Alfonso. *Obras completas*. Vol. 11. Mexico City: Fondo de Cultura Económica, 1960.

Reyes Palma, Francisco. "Arte funcional y vanguardia (1921–1952)." In *Modernidad y modernización en el arte mexicano, 1920–1960*. Mexico City: INBA, MUNAL, 1991.

Richardson, William. "The Dilemmas of a Communist Artist: Diego Rivera in Moscow, 1927–1928." *Mexican Studies/Estudios Mexicanos* 3, no. 1 (Winter 1987): 49–69.

———. "Maiakovskii en México." *Historia Mexicana* 29, no. 4 (April–June 1980): 623–39.

Riggins, Stephen Harold. *Ethnic Minority Media: An International Perspective*. Newbury Park, CA: SAGE Publications, 1992.

Riguzzi, Paolo. "Los caminos del atraso: Tecnología, instituciones e inversión en los ferrocarriles mexicanos, 1850–1900." In *Ferrocarriles y vida económica en México (1850–1950): Del surgimiento tardío al decaimiento precoz*, edited by Sandra Kuntz and Paolo Riguzzi, 31–97. Mexico City: Colegio Mexiquense–Ferrocarriles Nacionales–UAM, 1996.

Rivera, Diego. "AKhRR i stil proletarskogo revoliutsionnogo iskusstva (ot-krytoe pismo v redaktsiiu)." *Revoliutsiia i kultura*, no. 6 (1928): 43.

———. *Diego Rivera: Illustrious Words, 1922–1957*. Vol. 2. Mexico City: Editorial RM, 2008.

Robinett, Jane. *This Rough Magic: Technology in Latin American Fiction*. New York: P. Lang, 1994.

Rodriguez, Clemencia. *Fissures in the Mediascape: An International Study of Citizens' Media*. Cresskill, NJ: Hampton Press, 2001.

Rodríguez, José Antonio. "La construcción de un imaginario." In *Hugo Brehme: Fotograf-Fotógrafo*, edited by Michael Nungesser, 28–43. Berlin: Verlag Willmuth Arenhövel, 2004.

Rodríguez, José María. "Informe que rinde el Jefe del Departamento de Salubridad de los trabajos efectuados por el Departamento a su cargo en 1917 al C. Presidente de la República." In *Memoria de los trabajos efectuados por el Departamento de Salubridad Pública en el año de 1917*, vi–vii. Mexico City: Imprenta Victoria, 1918.

Romero, Héctor Manuel. *Historia del transporte en la Ciudad de México: De la trajinera al metro*. Mexico City: Secretaría General de Desarrollo Social, 1987.
Ronfeldt, David, John Arquilla, Graham E. Fuller, and Melissa Fuller. *The Zapatista Social Netwar in Mexico*. Washington, DC: RAND Arroyo Center, 1998.
Rosenblum, Naomi. *A World History of Photography*. New York: Abbeville Press, 1989.
Rubenstein, Anne. *Bad Language, Naked Ladies, and Other Threats to the Nation: A Political History of Comic Books in Mexico*. Durham, NC: Duke University Press, 1998.
Ruffinelli, Jorge. "Trenes revolucionarios: La mitología del tren en el imaginario de la Revolución." *Revista Mexicana de Sociología* 51, no. 2 (1989): 285–303.
Ruiz, Luis E. *Tratado elemental de higiene*. Mexico City: Oficina Tipográfica de la Secretaría de Fomento, 1904.
Sánchez Estevez, José Luis. "Luis Márquez Romay y su obra: Apuntes sobre la búsqueda del nacionalismo cultural en México." Licenciatura thesis, Centro Universitario de Ciencias Humanas, 1990.
Sánchez Flores, Ramón. *Historia de la tecnología y la invención en México*. Mexico City: Fomento Cultural BANAMEX, 1980.
Sánchez Ruiz, Enrique E. "Los medios de difusión masiva y la centralización en México." *Mexican Studies/Estudios Mexicanos* 4, no. 1 (1988): 25–54.
Saragoza, Alex. "Azcárraga and the Origins of Mexican Television." In *The State and the Media in Mexico: The Origins of Televisa* (forthcoming).
———. "The Selling of Mexico: Tourism and the State, 1929–1952." In *Fragments of a Golden Age: The Politics of Culture in Mexico since 1940*, edited by Gilbert Michael Joseph, Anne Rubenstein, and Eric Zolov, 91–115. Durham, NC: Duke University Press, 2001.
Sarlo, Beatriz. *The Technical Imagination: Argentine Culture's Modern Dreams*. Translated by Xavier Callahan. Stanford, CA: Stanford University Press, 2008.
Schell, Patience A. "Gender, Class, and Anxiety at the Gabriel Mistral Vocational School, Revolutionary Mexico City." In *Sex in Revolution: Gender, Politics, and Power in Modern Mexico*, edited by Jocelyn Olcott, 112–26. Durham, NC: Duke University Press, 2006.
Scherer García, Julio, and Carlos Monsiváis. *Parte de Guerra, Tlatelolco 1968: Documentos del General Marcelino García Barragán, los hechos y la historia*. Mexico City: Siglo/Aguilar, 1999.
Schivelbusch, Wolfgang. "Railroad Space and Railroad Time." *New German Critique* 14 (Spring 1978): 31–40.
Schmidt, Arthur. "Making It Real Compared to What? Reconceptualizing Mexican History since 1940." In *Fragments of a Golden Age: The Politics of Culture in Mexico since 1940*, edited by Gilbert Joseph, Anne Rubenstein, and Eric Zolov, 23–68. Durham, NC: Duke University Press, 2001.
Schneider, Luis Mario. *El estridentismo o una literatura de la estrategia*. Mexico City: INBA, 1970.
Secretaría de Agricultura y Recursos Hidráulicos–Comisión del Río Balsas. "Anteproyecto para la instalación de una radiodifusora en Tlapa, Gro." Unpublished internal document. Mexico City, 1977.
Secretaría de Comunicaciones y Transportes. *Anuario estadístico*. Mexico City, 2000, 2008.
Secretaría de Comunicaciones y Transportes–Ferrocarriles Nacionales de México. *Caminos de hierro*. Mexico City: Secretaría de Comunicaciones y Transportes–Ferrocarriles Nacionales de México, 1996.

Segre, Erica. "Reframing the City: Images of Displacement in Mexican Urban Films of the 1940s and 1950s." *Journal of Latin American Cultural Studies* 10, no. 2 (2001): 205–22.

Serafini, Dom. "TV in Mexico: Maintaining the Status Quo is a Failure." *Video Age International*, June 1, 2004, http://www.allbusiness.com/legal/laws/168258-1.html (accessed 1 May 2010).

Silva, Máximo. *Higiene popular: Colección de conocimientos y consejos indispensables para evitar las enfermedades y prolongar la vida, arreglada para uso de las familias*. Mexico City: Departamento de Talleres Gráficos, 1917.

Silva Herzog, Jesús. *Breve historia de la revolución Mexicana*. Vol. 1, *Los antecedentes y la etapa maderista*. 1960. Reprint, Mexico City: Fondo de Cultura Económica, 2007.

Simonett, Helena. "Popular Music and the Politics of Identity: The Empowering Sound of Technobanda." *Popular Music and Society* 24, no. 2 (2000): 1–24.

Sinclair, John. "Neither West nor Third World: The Mexican Television Industry within the NWICO Debate." *Media, Culture and Society* 12, no. 3 (1990): 343–60.

Smith, Laurel C. "Mobilizing Indigenous Video: The Mexican Case." *Journal of Latin American Geography* 5, no. 1 (2006): 113–28.

Sorensen, Diana. *A Turbulent Decade Remembered: Scenes from the Latin American Sixties*. Palo Alto, CA: Stanford University Press, 2007.

Stokes, Martin. "Introduction: Ethnicity, Identity, and Music." In *Ethnicity, Identity, and Music: The Musical Construction of Place*, edited by Martin Stokes, 1–28. Oxford: Berge, 1994.

Taylor, Lawrence Douglas. "Dinamiteros en acción: La conjura contra el Ferrocarril de Chihuahua en 1912." *Nuestro Siglo* (April–June 2002): 87–93.

Taylor, Philip, Jr. "The Mexican Elections of 1958: Affirmation of Authoritarianism?" *Western Political Quarterly* 13, no. 3 (1960): 722–44.

Tello Peón, Berta. "Intención decorativa en los objetos de uso cotidiano de los interiores domésticos del porfiriato." In *El arte y la vida cotidiana: XVI Coloquio Internacional de Historia del Arte*, 139–54. Mexico City: Instituto de Investigaciones Estéticas, UNAM, 1995.

Tenorio-Trillo, Mauricio. "1910 Mexico City: Space and Nation in the City of the Centenario." *Journal of Latin American Studies* 28, no. 1 (February 1996): 75–104.

———. "The Cosmopolitan Mexican Summer, 1920–1949." *Latin American Research Review* 32, no. 3 (1997): 224–42.

Thirion, Jordy Micheli, and Rubén Oliver Espinoza. "Changing Patterns in Mexican Science and Technology Policy (1990–2003): Still Far from Economic Development." In *Changing Structure of Mexico: Political, Social, and Economic Prospects*, edited by Laura Randall, 197–212. Armonk, NY: M. E. Sharpe, 2006.

Toledo, Francisco, Enrique Florescano, and José Woldenberg. *Cultura mexicana: Revisión y retrospectiva*. Mexico City: Taurus, 2008.

Tomes, Nancy. *The Gospel of Germs: Men, Women, and the Microbe in American Life*. Cambridge: Cambridge University Press, 1988.

Tortolero, Alejandro. *De la coa a la máquina de vapor: Actividad agrícola e innovación tecnológica en las haciendas mexicanas, 1880–1914*. Mexico City: Siglo XXI, 1995.

Trejo Delabre, Raúl, ed. *Las redes de Televisa*. Mexico City: Claves Latinoamericanas, 1988.

Trujillo Muñoz, Gabriel. *Biografías del futuro: La ciencia ficción mexicana y sus autores*. Mexico City: UABC, 2000.

———. *El futuro en llamas: Cuentos clásicos de la ciencia ficción mexicana*. Mexico City: Grupo Editorial Vid, 1997.

Valle Arizpe, Artemio. *Calle vieja y calle nueva*. Mexico City: Editorial Jus, 1949.
Van Hoy, Teresa. *A Social History of Mexico's Railroads: Peons, Prisoners, and Priests*. New York: Rowman & Littlefield Publishers, 2008.
Vargas, Lucila. *Social Uses and Radio Practices: The Use of Participatory Radio by Ethnic Minorities in Mexico*. Boulder, CO: Westview, 1995.
Vasconcelos, José. "Cómo se forma un ferrocarrilero." *Ferronales* 37, no. 6 (1960): 16–17.
———. *Ulises Criollo*. Mexico City: Ediciones Botas, 1935.
Vaughan, Mary Kay. *Cultural Politics in Revolution: Teachers, Peasants, and Schools in Mexico, 1930–1940*. Tucson: University of Arizona, 1997.
Vaughan, Mary K., and Stephen E. Lewis. *The Eagle and the Virgin: Nation and Cultural Revolution in Mexico, 1920–1940*. Durham, NC: Duke University Press, 2006.
Vela, Arqueles. *Evolución histórica de la literatura universal*. Mexico City: Ediciones Fuente Cultural, 1941.
Velasco, Antonio. *Medicina doméstica o tratado elemental y práctico del arte de curar: Obra muy importante, útil y provechosa para las familias*. Mexico City: Oficina Tipográfica de la Secretaría de Fomento, 1886.
Velasco García, Jorge H. *El Canto de la Tribu*. Mexico City: CONACULTA, 2004.
Velázquez de León, Josefina. *Como cocinar en los aparatos modernos*. Vol. 1. Mexico City: Academia de Cocina Velázquez de León, 1950.
Verani, Hugo J. "The Vanguardia and Its Implications." In *The Cambridge History of Latin American Literature*, vol. 2, *The Twentieth Century*, ed. Roberto González Echevarría and Enrique Pupo-Walker. Cambridge: Cambridge University Press, 1996.
Vigarello, Georges. *Lo limpio y lo sucio: La higiene del cuerpo desde la Edad Media*. Madrid: Alianza Editorial, 1991.
Villanueva, René. *Cantares de la Memoria: Recuerdos de un folclorista*. Mexico City: Planeta, 1994.
Villaseñor, Víctor Manuel. *Memorias de un hombre de izquierda*. Vol. 2. Mexico City: Grijalbo, 1976.
Virilio, Paul. *L'horizont negatif*. Paris: Editions Galilée, 1984.
Wajcman, Judy. *Feminism Confronts Technology*. University Park: Pennsylvania State University Press, 1991.
———. *TechnoFeminism*. Cambridge, UK: Polity Press, 2004.
Waters, Wendy. "Remapping Identities: Road Construction and Nation Building in Postrevolutionary Mexico." In *Nation and Cultural Revolution in Mexico, 1920–1940*, edited by Mary Kay Vaughan and Stephen E. Lewis, 221–43. Durham, NC: Duke University Press, 2006.
———. "Re-mapping the Nation: Road Building as State Formation in Post-revolutionary Mexico, 1925–1940." PhD diss., University of Arizona, 1999.
Weston, Edward. *The Daybooks of Edward Weston*. Vol. 1. Millerton, NY: Aperture, 1973.
Wilkie, James. *La Revolución mexicana: Gasto federal y cambio social*. Mexico City: Fondo de Cultura Económica, 1987.
Williams, Rosalind. "Opening the Big Box." *Technology and Culture* 48 (2007): 104–16.
Winner, Langdon. *Autonomous Technology: Technics-out-of-Control as a Theme in Political Thought*. Cambridge: MIT Press, 1977.
Wionczek, Miguel S. "Electric Power: The Uneasy Partnership." In *Public Policy and Private Enterprise in Mexico*, edited by Raymond Vernon, 19–110. Boston: Harvard University Press, 1964.

Witherspoon, Kevin B. *Before the Eyes of the World: Mexico and the 1968 Olympics*. DeKalb: Northern Illinois University, 2008.
Wittkower, Rudolf. "The Problem of Harmonic Proportion in Architecture." In *Architectural Principles in the Age of Humanism*. New York: W. W. Norton & Company, 1971.
Wolfe, Bertram. *The Fabulous Life of Diego Rivera*. New York: Stein and Day, 1963.
Womack, John. "The Mexican Revolution, 1910–1920." In *Mexico since Independence*, edited by Leslie Bethell, 185. Cambridge: Cambridge University Press, 1991.
Wosk, Julie. *Women and the Machine: Representations from the Spinning Wheel to the Electronic Age*. Baltimore, MD: The Johns Hopkins University Press, 2002.
XEPET. *Consejo consultivo de XEPET "La voz de los mayas": Sexta reunión plenaria*. Peto, Yucatan, Mexico: Radio XEPET, 1993.
Yehya, Naief. *Historias de mujeres malas*. Mexico City: Plaza & Janés, 2001.
———. *Tecnocultura*. Mexico City: Tusquets, 2008.
Zamora, Francisco José. *Guía indispensable del forastero en la ciudad de México y calendario para 1905*. Mexico City: J. Ricardo Garrido y Hermanos Editores, 1905.
Zárate Miguel, Guadalupe. "El Ferrocarril y la modernidad en la cultura popular: El caso de los exvotos pintados del santuario de la Virgen de los Dolores de Soriano." In *Memoria del IV Encuentro de investigadores del ferrocarril*, 9–16. Mexico City: Conaculta, 2000.
Zolov, Eric. "Discovering a Land 'Mysterious and Obvious': The Renarrativizing of Postrevolutionary Mexico." In *Fragments of a Golden Age: The Politics of Culture in Mexico since 1940*, edited by Gilbert M. Joseph, Anne Rubenstein, and Eric Zolov, 234–72. Durham, NC: Duke University Press, 2001.
———. *Refried Elvis: The Rise of the Mexican Counterculture*. Berkeley: University of California Press, 1999.
———. "Showcasing the 'Mexico of Tomorrow': Mexico and the 1968 Olympics." *The Americas* 61, no. 2 (October 2004): 159–88.

# Contributors

**Claudia Agostoni** is a historian and full-time researcher at the Instituto de Investigaciones Históricas of the Universidad Nacional Autónoma de México (UNAM). She has published articles on the social history of medicine and public health and is the author of *Monuments of Progress: Modernization and Public Health in Mexico City, 1876–1910* and the editor of *Curar, sanar y educar: Enfermedad y sociedad en México, siglos XIX y XX*.

**Sandra Aguilar-Rodríguez** is an assistant professor in Latin American history at Moravian College. Her work has been published in various journals, both in the United States and Latin America, such as the *Radical History Review*, *The Americas*, and *Revista de Estudios Sociales*. She is currently working on a book manuscript that examines eating and consumption patterns in mid-twentieth-century Mexico.

**Edward R. Burian** is an architect and an associate professor at the University of Texas at San Antonio. He edited and wrote *Modernity and the Architecture of Mexico*, which was translated into Spanish as *Modernidad y arquitectura en México*. His forthcoming book, *The Architecture and Cities of Northern Mexico, from Independence to Present*, explores the undervalued architectural culture of the region.

**Antoni Castells-Talens** is a researcher at the Universidad Veracruzana's Centro de Estudios de la Cultura y la Comunicación. His work on indigenous media has been published in *Revista Mexicana de Sociología*; *Communication, Culture, and Critique*; *Journal of Global Mass Communication*; and *Comunicación y Sociedad*.

**J. Brian Freeman** is an assistant professor of history at Fairleigh Dickinson

University. His work has been published in various journals, including *Studies in Latin American Popular Culture* and the *Journal of Latino–Latin American Studies*. He is currently working on a book manuscript on the history of the automobile in twentieth-century Mexico.

**Celeste González de Bustamante** is an assistant professor in the University of Arizona School of Journalism and has a dual faculty appointment at the Center for Latin American Studies. She is the author of *"Muy buenas noches": Mexico, Television, and the Cold War* and the coeditor of *Arizona Firestorm: Global Immigration Realities, National Media, and Provincial Politics*.

**Guillermo Guajardo** is a research professor at the Centro de Investigaciones Interdisciplinarias en Ciencias y Humanidades, Universidad Nacional Autónoma de México. He is the author of *Trabajo y tecnología en los ferrocarriles de México: Una visión histórica, 1850–1950* and *Tecnología, Estado y ferrocarriles en Chile, 1850–1950*.

**Joanne Hershfield** is a professor and chairperson in the Department of Women's and Gender Studies at the University of North Carolina–Chapel Hill. She is the author of *Imagining la Chica Moderna: Women, Nation, and Visual Culture in Mexico, 1917–1936*; *The Invention of Dolores del Río*; and *Mexican Cinema/Mexican Woman, 1940–50*.

**Anna Indych-López** is an associate professor of art history at The Graduate Center and The City College of New York. She is the author of *Muralism without Walls: Rivera, Orozco, and Siqueiros in the United States, 1927–1940* and the coauthor, with Leah Dickerman, of *Diego Rivera: The Murals for the Museum of Modern Art*, among other publications on modern art in Mexico.

**Lynda Klich** is a distinguished lecturer in the Art Department and Macaulay Honors College at Hunter College of the City University of New York. She specializes in modern Latin American art history and is currently preparing a book manuscript on *estridentismo*, a vanguard art and literary movement active in Mexico during the 1920s. She was guest editor of the Mexico theme issue of the *Journal of Decorative and Propaganda Arts*.

**Viviane Mahieux** is an assistant professor of Spanish at the University of California, Irvine. She is the author of *Urban Chroniclers in Modern Latin America: The Shared Intimacy of Everyday Life* and the editor of a collection of chronicles by Cube Bonifant, entitled *Una pequeña marquesa de Sade: Crónicas selectas, 1921–1948*.

**Carlos Monsiváis** (1938–2010) was Mexico's foremost twentieth-century cultural critic. He wrote extensively about popular culture, nightlife, and pop music, as well as on literature and the history of Mexican film. His writings include *Amor perdido*, *Entrada libre*, *Los rituales del caos*, *Aires de familia*, and *Salvador Novo: Lo marginal en el centro*, among many others.

**John Mraz** is a research professor at the Instituto de Ciencias Sociales y Humanidades, Universidad Autónoma de Puebla (Mexico). Among his recent books are *Looking for Mexico: Modern Visual Culture and National Identity*; *Trasterrados: Braceros vistos por los Hermanos Mayo*; and *Nacho López, Mexican Photographer*. He directed the award-winning documentary videotapes *Innovating Nicaragua* and *Made on Rails: A History of the Mexican Railroad Workers*.

**Ricardo Pérez Montfort** is a research fellow at the Centro de Investigaciones y Estudios Superiores en Antropología Social (CIESAS) and a professor in the postgraduate division of Universidad Nacional Autónoma de México. His most recent books include *Expresiones populares y estereotipos culturales en México: Siglos XIX y XX, Diez Ensayos* and *Cotidianidades, Imaginarios y Contextos: Ensayos de Historia y Cultura en México, 1850–1950*. His documentary *Voces de la Chinantla* won the first prize at the Festival de la Memoria 2007 Latin American Film Festival in Tepoztlán, Morelos. He is currently head of the CIESAS Audiovisual Laboratory.

**José Manuel Ramos Rodríguez** is a professor and researcher at the Universidad Autónoma de Puebla (Mexico). His professional experience started thirty years ago as one of the founders of the Indigenist Radio System. He holds a PhD in social sciences and communication from Universidad Nacional Autónoma de México. His fields of research include indigenous language radio, community media, and indigenous appropriation of information and communication technologies.

**Paolo Riguzzi** is a researcher at the Colegio Mexiquense in Toluca, Mexico. He is author of *¿Reciprocidad imposible? La política del comercio entre México y Estados Unidos, 1857–1938* and coeditor of *Ferrocarriles y vida económica en México, 1850–1950*.

**Araceli Tinajero** is an associate professor of Spanish at the Graduate Center and City College of New York. She is the author of *Orientalismo en el modernismo hispanoamericano*, *El Lector: A History of the Cigar Factory Reader*, and *Kokoro, una mexicana en Japón*. Tinajero is the editor of *Cultura y letras cubanas en el siglo XXI*, *Exilio y cosmopolitismo en el arte y la literatura hispánica,* and *Orientalisms of the Hispanic and Luso-Brazilian World*. She is the cofounder of the Mexico Study Group at the Bildner Center for Western Hemisphere Studies.

**Erja Vettenranta** is a PhD candidate in the Department of Hispanic and Luso-Brazilian Literatures and Languages at The Graduate Center of the City University of New York. She is writing her dissertation on technology and culture in Hispanic contemporary literature. Her articles have been published in *Guaraguao: Revista de Cultura Latinoamericana*; *Ciberletras*; and *Transmodernity: Journal of Peripheral Cultural Production of the Luso-Hispanic World*.

**Juan Villoro** has been a professor of literature at Universidad Nacional Autónoma de México and a visiting professor at Yale, Princeton, and the Pompeu Fabra University at Barcelona. He has received the Herralde Prize for his novel *El testigo*, the King of Spain Prize for journalism, and the Antonin Artaud Prize for his short-story collection *Los culpables*, among others. His play *Filosofía de vida* is a long-running hit in Buenos Aires. His novel for juvenile readers, *El libro salvage*, sold more than two million copies in Spanish and has been translated into French, Italian, and Portuguese. He is a weekly columnist for the Mexican newspaper *Reforma*.

**David M. J. Wood** is an associate researcher at the Instituto de Investigaciones Estéticas, Universidad Nacional Autónoma de México, and a member of the editorial board of the *Journal of Latin American Cultural Studies*. His research on Mexican, Colombian, and Bolivian cinema has been published in numerous edited volumes and in journals such as *Quarterly Review of Film and Video*, *Cosmos and History*, and *Secuencia*. His current research project is on historical and experimental compilation film in Mexico.

**Naief Yehya** is an industrial engineer, journalist, writer, and cultural critic. He is a contributor to *La Jornada*, *Zócalo*, *Replicante*, *Deep*, *Luvina*, *Letras Libres*, and *Art Nexus*, among other publications. He has written three novels and a short-story collection entitled *Historias de mujeres malas*, as well as *El cuerpo transformado: Cyborgs y nuestra descendencia tecnológica en la realidad y la ciencia ficción*; *Guerra y propaganda*; *Pornografía, sexo mediatizado y panico moral*; and *Tecnocultura, el espacio íntimo transformado en tiempos de paz y de Guerra*.

# Index

agriculture, 4, 9, 44, 56, 81, 84, 85, 181, 284, 285, 291
Agustín, José, 306
Alemán, Miguel, 9, 10, 84, 89n30, 90, 92, 93, 95, 96, 113
Alemán Velasco, Miguel, 113, 161
Alva de la Canal, Ramón, 264, 270, 271, 281n27
Álvarez Bravo, Manuel, 80–81
Alzate, Antonio de, 252–53
Arreola, Juan José, 305
automobiles, 1, 5, 9, 14, 36, 65–66, 81, 97–98, 118, 151, 214–28, 228n1, 229n13, 229n17, 231n53, 243, 247n39, 265, 278
aviation, 2, 5–8, 78, 151, 206, 273–75, 278, 283, 290, 292, 300n8, 303–4
Ávila Camacho, Manuel, 9
Azcárraga Jean, Emilio, 105
Azcárraga Milmo, Emilio, 105, 108n41, 113, 116–17, 120, 123n1
Azcárraga Vidaurreta, Emilio, 9, 93, 95–96, 98, 105, 107n26, 113–14, 163
Azcárraga, Raúl, 144, 155–56
Azuela, Mariano, 303, 305

Barragán Schwarz, Rodolfo, 320–21, 329n8–9
Barthes, Roland, 84, 89n34

Beals, Carleton, 222–23
blenders, 10, 14, 44–45, 47–51, 53n26, 161
Bonifant, Cube, 146
Borges, Jorge Luis, 260
Bray, Francesca, 60, 66, 68n20
Brehme, Hugo, 79–80, 87–88nn15–16, 222
Brenner, Anita, 222
Brunet, Meade, 95

Calles, Plutarco Elías, 7–8, 150, 227, 267, 271
Calvino, Italo, 250
Campa, Valentín, 172
Candela, Félix, 318–19, 321, 330n11
Cantinflas, 83
Cárdenas, Lázaro, 8–9, 56–57, 67n4, 240, 304
Carranza, Venustiano, 219, 237, 251, 267
Carrillo, Rafael, 79, 84
Casasola, Agustín Víctor, 77
Caso, Antonio, 151
Castellot, Gonzálo, 91, 96
Catholicism, 7, 26, 49, 57, 97, 114, 120, 165, 186, 206, 258, 289–90, 318–19
Chaplin, Charlie, 266
Charlot, Jean, 264, 272–77
Chase, Stuart, 1, 15n2, 222–23
Chávez, Carlos, 92

Chávez, Oscar, 166–67, 171
cinema, 7–8, 10, 26, 33–34, 37–38, 73, 83, 95, 100, 102, 111–13, 115, 118, 124, 151, 160–62, 169, 171, 197–210, 211n10, 211n13, 212n20, 212n27, 212n35, 213n41, 255–56, 266, 325
Cowan, Ruth Schwartz, 62

de Icaza, Francisco, 149
de la Huerta, Adolfo, 147, 150–51, 159n17, 221, 238
de la Madrid, Miguel, 11
de la Mora y Palomar, Enrique, 318–19
del Río, Dolores, 83
Díaz Ordaz, Gustavo, 10–11, 97, 102–3
Díaz, Enrique, 81, 84, 88n26
Díaz, Porfirio, 3, 4, 6, 16n11, 25, 27, 74, 75, 76–77, 143, 197–200, 203, 214, 215, 234, 236, 266
Diego, Juan, 258
Domínguez y Pastor, Rafael, 29
Dos Passos, John, 222

earthquake (1985), 11–12, 96, 175, 306
Echeverría, Luis, 11, 172
Eisenstein, Sergei, 208
El Buen Tono, 5, 145, 148, 158n8
electricity, 2–3, 5–6, 8–10, 28, 43–56, 58–59, 61–64, 67n4, 81, 91, 96, 128–30, 132–33, 145, 164, 188, 199, 208, 254, 258, 265, 267–71, 278, 284, 286–87, 289, 291, 305, 307, 309, 318, 327, 330n15
engineers and engineering, 4, 9, 25, 27–29, 38, 68n16, 92, 106n2, 249, 323, 325, 327, 330n11
Esquivel Obregón, Toribio, 235, 244n7
Esquivel, Laura, 305
*estridentismo*, 14, 145, 151–53, 155–56, 263–67, 269–79, 279n1–4, 281n39, 303–4

Félix, María, 83, 93, 100
Figueroa, Rubén, 119
flappers, 8, 151, 153
Ford, Henry, 218
Fox, Vicente, 13
Freeman, Greydon, 125–26

Fuchs, Ira, 125–26
Fuentes, Carlos, 12, 305

Galindo, Alejandro, 10
Galindo, Marco Aurelio, 146
Gallardo, Salvador, 270–71, 281n27
Gamio, Manuel, 221
García, Héctor, 83, 88n26
Gates, Bill, 124, 136
Gilberto, Joao, 170
Gómez Bolaños, Roberto ("Chespirito"), 117
Gómez Fernández, Enrique, 144
González, Eduardo, 257
González, Roberto, 174
González Camarena, Guillermo, 92, 95–96, 104, 113
Gorostiza, José, 304
Gruening, Ernest, 1, 15n1
Gutiérrez Nájera, Manuel, 303
Guzmán, Martín Luis, 143, 146–48, 149, 150–54, 156–58, 159n17, 237, 239

Haley, Bill, 162
Henríquez Guzmán, Miguel, 97
Henríquez Ureña, Pedro, 151
Hermanos Mayo. *See* Mayo brothers
Hidalgo, Miguel, 253, 257
highways, 9–10, 85–86, 98–99, 162, 226–27, 247n39, 248n45, 267
home economics, 43–45
Humphrys, Mark, 125

Ibargüengoitia, Jorge, 233, 242, 247n42
Infante, Pedro, 83, 111, 114, 163
Instituto Nacional Indigenista (INI), 178, 190, 190n3
Internet, 13–14, 73, 121, 124–37, 138n8, 139n29, 309

Jara, Heriberto, 264
Jara, Víctor, 166, 170, 175
jazz, 8, 154, 164, 222, 266
Jenkins, William, 95
Jiménez, Agustín, 80–81, 88n21
Jiménez, José Alfredo, 162
Jobim, Antonio Carlos, 170
Jobs, Steve, 124

Kahlo, Frida, 75, 257, 296, 298, 317, 318
Kahlo, Guillermo, 75
Koolhaas, Rem, 250, 328n2

Lara, Agustín, 163
Latapí, Aurora Eugenia, 81
Laure, Mike, 163
Lawrence, D. H., 221, 223
Le Corbusier, 276, 317
León de la Barra, Francisco, 203
Lewis, Oscar, 102
Liceaga, Eduardo, 27
Lindbergh, Charles, 227
List Arzubide, Germán, 223–24, 265, 277, 281n26–27, 303–2
López, Nacho, 83, 87n12, 88n26
López Mateos, Adolfo, 10, 97–98, 100
López Obrador, Andrés Manuel, 129
López Portillo, José, 11
Lumière Brothers, 197

Madero, Francisco, 6, 76, 203–5, 255–56
Malda, Gabriel, 36
manufacturing, 4, 29, 33, 46, 61, 217–18
Manzanero, Armando, 163, 164
Maples Arce, Manuel, 145, 153–56, 223–24, 263–67, 270, 272–74, 275–278, 281n28, 282n47, 303–4
Marinetti, F. T., 218, 265, 304
Marcos, Subcomandante. *See* Subcomandante Marcos
Márquez, Luis, 79–80, 87n13
Mayakovsky, Vladimir, 222
Mayo brothers, 81–83, 88n26–27
metro, 14, 249–60. *See also* subway
Meyer, Pedro, 84–86, 89n35
Milanés, Pablo, 170
mining and metallurgy, 1, 4, 56, 285, 320, 330n16
Modotti, Tina, 80–81, 87n14, 222, 270, 281n26, 296
Monsiváis, Carlos, x, 12, 14, 83, 109n62, 123, 251
Monterde García Icazbalceta, Francisco, 146

Negrete, Jorge, 111, 163
Nervo, Amado, 217–18, 303

Nicola, Noel, 170
Nieto, Rafael, 238
Noriega Hope, Carlos, 143, 146, 151–58
North American Free Trade Agreement (NAFTA), 12, 128
Novo, Salvador, 10, 92, 144, 146, 152, 214, 219, 223–24, 228n1, 231n59, 304

Obregón, Álvaro, 145, 147, 149, 150, 206, 220, 221, 225, 266–67, 273
O'Farrill, Rómulo, Jr., 98
O'Farrill, Rómulo, Sr., 90, 93, 95–96, 98, 99, 107n26, 113
O'Gorman, Juan, 317
Oldenzeil, Ruth, 68n16
Olympic Games, 10–11, 15, 20n43, 99, 103–4, 109n62, 161, 168, 316, 319–22
O'Reilly, Tim, 127
Orozco, José Clemente, 283–84
Orozco, Pascual, 203
Orwell, George, 186, 255
Owen, Gilberto, 304

Pacheco, José Emilio, 12, 306
Pani, Alberto J., 27
Parra, Ángel, 166, 168, 170, 175
Parra, Porfirio, 39n9
Parra, Violeta, 166, 186
Paz, Octavio, 259
Pellicer, Carlos, 223, 304
Petróleos Mexicanos (PEMEX), 98–99, 227
photography, 3, 5, 14, 63–66, 73–86, 87n12, 87n14, 88n16, 88n26–27, 89n34–35, 151, 197, 206, 222, 226, 245, 270, 272, 275, 277, 281n26, 289, 307
Polo, Marco, 250, 251
Ponce, Manuel M., 149, 155
Portes Gil, Emilio, 271
Pruneda, Alfonso, 27, 34, 36

Quintanilla, Luis, 145, 223–24, 303, 304

Radio Corporation of America (RCA), 63–64, 95, 162
radio, 1, 7–8, 10, 13–14, 18, 26, 33, 37–38, 48–49, 55, 63–64, 91–97, 102, 113–14,

119, 123, 131, 143–58, 158n6, 158n9, 160–76, 178–90, 190n3, 192n31, 222, 264–65, 267, 270, 272, 283, 291–93, 304, 321, 330n15
railroads, 3–5, 7–9, 14, 75, 197–210, 210n2, 211n6–7, 233–44, 244n7, 245n9, 246n26, 246n31, 247n34, 247n39, 247n43, 248n45, 278, 289, 303, 317, 330n16
Redfield, Robert, 56
refrigeration, 5, 14, 43–48, 50–51, 53n15, 53n19–20, 55–56, 64, 97, 161
Revueltas, Fermín, 264, 267–70, 272, 278
Revueltas, José, 109n62, 305
Reyes, Alfonso, 147, 239, 254, 304
Reyes, Judith, 166, 167, 168
Reyes Espíndola, Rafael, 74
Riolobos, Daniel, 163
Rivera, Diego, 14–15, 19n34, 221, 235, 272, 283–299, 299n4–5, 300n7, 300n10, 300n15, 317–18
rock & roll, 100, 161–65, 169, 171, 183
Rodríguez, José Maria, 35
Rodríguez, Silvio, 170
Ruiz, Luis E., 27
Ruíz Cortines, Adolfo, 93–94, 97, 100, 102
Rulfo, Juan, 305, 308

Saldaña, Jorge, 103
Salinas de Gortari, Carlos, 12, 96, 325
Salinas Pliego, Ricardo, 105
satellites, 99–101, 104–5, 113, 129, 251–52
Secretaría de Educación Publica (SEP), 126, 171, 180, 264, 283–84, 286–94, 297–99
Serrat, Joan Manuel, 164
sewing machines, 4, 56–57, 59, 291
Simpson, Eyler, 8
Siqueiros, David Alfaro, 283–84
Slim, Carlos, 136, 137
soap operas. See *telenovelas*
Sosa, Mercedes, 170
Stridentism. See *estridentismo*
Subcomandante Marcos, 129, 137
subway, 10, 14, 249–60

Superior Council of Public Health, 27–28, 34–35

Tablada, José Juan, 218
Tannenbaum, Frank, 234
telegraph, 3, 7, 80–81, 208–9, 223, 234, 267, 270–71, 276, 281n28, 304
*telenovelas*, 91, 97–98, 100–101, 112–13, 115–18, 121, 161, 181
telephones, 5, 8, 80, 136, 188, 222, 249, 267, 270–72, 275–76, 296, 304
Televisa, 90, 95, 100–101, 103–5, 113, 175–76, 326
television, 1, 14, 37, 73, 85, 90, 90–105, 106n2–3, 106n5, 106n9, 107n25–26, 111–23, 158n11, 160–62, 169, 171, 175, 188, 310, 321–22, 326, 330n15
Tolsá, Manuel, 317, 318
Toor, Frances, 222
Torres, Juan, 163
Toscano, Salvador, 199, 201, 203, 204, 205–7, 211n10–11, 212n26–27
Trotsky, Leon, 293
Tubau, María, 149

Vallejo, Demetrio, 241
Vargas, Pedro, 111, 163
Vasconcelos, José, 147, 149–50, 179, 218, 221, 264
Vela, Arqueles, 152–53, 223, 279, 282n47, 303
Velasco, José María, 268
Velázquez de León, Josefina, 48–49
Veloso, Caetano, 170
Villa, Francisco ("Pancho"), 7, 76–77, 146, 203, 237, 303
Villa, Lucha, 163
Villaurrutia, Xavier, 146, 152, 304
Volpi, Jorge, 306

Wajcman, Judy, 59
Weston, Edward, 80, 81, 87n14, 222, 272, 275, 277
Wozniak, Steve, 124

X-rays, 8, 154, 259, 286, 288–89, 300n7

Yáñez, Agustín, 305
Yupanqui, Atahualpa, 166, 170

Zapata, Emiliano, 77, 219, 295

Zapatistas (Chiapas), 12, 96, 128–29, 173, 179, 183–87, 189–90
Zedillo, Ernesto, 12, 242
Zabludovsky, Jacobo, 96–97, 99